GEOTHERMAL ENERGY: AN ALTERNATIVE RESOURCE FOR THE 21ST CENTURY

GEOTHERMAL ENERGY: AN ALTERNATIVE RESOURCE FOR THE 21ST CENTURY

HARSH GUPTA

Raja Ramanna Fellow
National Geophysical Research Institute
Hyderabad, India

SUKANTA ROY

National Geophysical Research Institute
Hyderabad, India

Amsterdam • Boston • Heidelberg • London • New York • Oxford • Paris
San Diego • San Francisco • Singapore • Sydney • Tokyo

Elsevier
Radarweg 29, PO Box 211, 1000 AE Amsterdam, The Netherlands
The Boulevard, Langford Lane, Kidlington, Oxford OX5 1GB, UK

First edition 2007

Library of Congress Cataloging-in-Publication Data
A catalog record for this book is available from the Library of Congress

British Library Cataloguing in Publication Data
A catalogue record for this book is available from the British Library

ISBN-13: 978-0-444-52875-9
ISBN-10: 0-444-52875-X

Printed and bound in The Netherlands

07 08 09 10 11 10 9 8 7 6 5 4 3 2 1

Dedicated to Siddharth, Kabir and Soumik

PREFACE

Hot springs have been used for balneological purposes from times immemorial. However, the use of Earth's heat as a source of energy began in the early 20th century when electricity was produced for the first time from geothermal steam at Larderello, Italy in 1904. With the passage of time, the global population has increased from 2.59 billion in 1951 to 6.3 billion in 2003, and the global energy consumption has increased from 2,710 million metric tons of coal equivalent to 15,178 million metric tons of coal equivalent during the same time. Alternative sources of energy are gaining importance and geothermal energy is one of them. Being a much cleaner source of energy, it deserves a special attention. Today, more than 20 countries generate electricity from geothermal resources and about 60 countries make direct use of geothermal energy. For several small countries, such as Iceland, Philippines, etc., geothermal energy constitutes a significant percentage of total electricity production. As of now, globally, the installed capacity is over 8,000 MW_e, and indirect use of geothermal energy amounts to over 15,000 MW_t. Considering that during 1971 total global geothermal electricity production was only 700 MW_e, the growth has been spectacular. A 10-fold increase in geothermal energy use is foreseeable at the current technology level.

Geothermal Energy: An Alternative Resource for the 21st Century provides a readable and coherent account of all facets of geothermal energy development. This is an updated version of *Geothermal Resources: An Energy Alternative* published by the Elsevier in 1980 (Gupta, 1980). The first chapter deals with the ever-growing need of energy due to increase in global population and the increase in per capita consumption of energy. New methods of power generation are briefly commented upon, and the role that geothermal energy is playing and the future scenarios are presented. The second chapter deals with Earth's structure, temperature and heat flow, and the plate tectonics hypothesis. All geothermal fields and prospects are located in the vicinity of the plate boundaries. The basic concepts of heat transfer and quantitative relations for the heat flow for some geometrically simple bodies are provided in the third chapter. The fourth chapter deals with classification of different kinds of geothermal systems. A suite of exploration techniques, viz. geological, geochemical, geophysical, airborne surveys are dealt with in some detail in the fifth chapter. How to make use of the currently available exploration techniques in a prospect area, and how to develop an exploration approach are also commented upon in this chapter. In the sixth chapter, assessment and exploitation of geothermal resources including drilling, reservoir physics and engineering and production technology are discussed. The Cerro Prieto geothermal field in Mexico, with an installed capacity of 720 MW_e, is among the three largest producing geothermal fields in the world. A detailed case study of this field is presented in the seventh chapter. The eighth chapter provides the worldwide status of the geothermal resource utilization as of now. A very new and advanced topic of using enormous amount of thermal energy stored in the tropical oceans for generation of electricity and potable water is discussed in the ninth chapter. An up to-date bibliography is provided at the end.

Important concepts and current technology developments have been dealt in this book for graduate students, scientists, engineers or investors seeking current knowledge. Geothermal energy development draws experts from diverse traditional disciplines such as geology, geophysics, engineering, investors, etc. These experts are usually abreast in their own disciplines, but to be able to converse with those with different background, a basic knowledge of other disciplines is required. This book meets this urgent requirement.

The material presented in this book is based on extensive laboratory and field investigations carried out by the authors. Visits to producing geothermal fields and interactions with a large number of specialists have been very useful. They are too many to list. The first author was directly involved with the development of projects dealing with the extraction of heat energy stored in tropical waters of the Indian Ocean. A special acknowledgement is due to Drs. M. Ravindran, S. Kathiroli, Purnima Jalihal and Robert Singh of the National Institute of Ocean Technology, Chennai, India. A special mention must be made of Professor F. Rummel, Professor S.K. Singh and Professor D. Chandrasekharam who readily provided material requested for. In the preparation of this text, P. Radhakrishnan provided immense help. Others who helped include M/s M. Jayarama Rao, Sharad Kumar Tank, O. Prasad Rao, G. Ramachander, M. Kranthi Kumar, V. Rajasekhar, P. Sundar Rao, G.S. Vara Prasad and M. Uma Anuradha. Harsh Gupta would like to acknowledge Department of Atomic Energy, Government of India for providing the Raja Ramanna Fellowship. We thank Dr. V.P. Dimri, Director, National Geophysical Research Institute, Hyderabad for providing facilities.

We are thankful to Mr. Friso Veenstra of Elsevier for urging us to write this book and to Linda Versteeg, Joyce Happee and others at Elsevier and Macmillan India Production team for excellent coordination during the production of this book and the good quality of production.

We would like to thank our wives Manju Gupta and Rakhee Roy and our children for support. Sukanta Roy is extremely grateful to his parents, Sachi Kanta Roy and Dipali Roy, for their encouragement.

Hyderabad, India
May, 2006

Harsh Gupta
Sukanta Roy

CONTENTS

Chapter 1

THE ENERGY OUTLOOK

INTRODUCTION

With the increase in world population, industrialization and improvement in the standard of living, there has been a continuous increase in consumption of energy. In the absence of substantial historical data to estimate world population prior to the seventeenth century, circumstantial evidence is often used. It is generally believed that prior to 8000 B.C., agriculture was not known and our ancestors at that time made living by hunting and gathering. Speculating on the basis of the population densities of the hunting and gathering tribes of today, and that only about one-third of the world's total land area could be used for such living, the world's population of 8000 B.C. is estimated to have been 5 million people. An examination of archeological remains and census figures for agricultural societies suggest a world population of about 250 million people at the time of Christ, which doubled to a population of about 500 million by A.D. 1650. As a consequence of cultural, agricultural, industrial and medical revolutions, time for the population to double has reduced from about 1500 years (approximated for the period 8000 B.C.–A.D. 1650) to 45 years during the period 1930–1975. Table 1.1 shows the alarmingly decreasing trend in the doubling time. With the current rate of population growth, the United Nations forecast that the world population would exceed 6 billion before entering the twenty-first century has come true. However, it is unlikely that by 2010, world population will reach 8 billion as anticipated in 1975.

Table 1.1. World population and the doubling time (United Nations, Statistical Office, *Demographic Year Book*)

Year	Estimated world population (in millions)	Approximate time for population to double (in years)
8000 B.C.	5	1,500
A.D. 1650	500	200
A.D. 1850	1,000 (1 billion)	80
A.D. 1930	2,000 (2 billions)	45
A.D. 1975	4,000 (4 billions)	35
A.D. 2010	8,000 (8 billions)?	35?

In the early times, say 400,000 years ago in the "early man" era, the energy consumption of an individual was limited to what he ate (approximately 2000 calories) and this energy was obtained from the sun through vegetable and animal life. Gradually he tamed animals and used them to bear his burden of labor. Invention of the wheel improved the efficiency in the use of energy. Then man learned to harness the energy of wind and water, which became two prime movers. At about the time of the beginning of the Christian era, water and windmills were invented and were put to a variety of uses. In the seventeenth century, coal began to be used for heating and the mining industry developed in Europe. The invention and improvement of the steam engine in the eighteenth century made power available readily and ushered in the industrial revolution. By the turn of the nineteeth century, all coalmines in England were equipped with steam engines to haul coal and men, and to pump water. Use of these steam engines for motive power on mine railroads in 1828 was so successful that railroads spread all over the world in the next 4–5 decades.

Michael Faraday's discovery of electromagnetic induction in 1831 laid the foundation of the immense electrical industry of today. The invention of electric light by Edison, although it has a more profound effect on living, does not constitute a major factor in energy consumption. Then came the invention of the internal combustion engine, which has proved itself to be extremely robust and reliable, and with it there has been an ever-increasing demand for oil and discoveries of oil fields.

In the recent years, the increase in energy consumption has been at least as dramatic as the population explosion. Table 1.2 shows an estimate of the world's energy consumption and production for the years 1980, 1985, 1990, 1995, 2000 and 2003. These statistics have been obtained from the data available for 220 countries (*International Energy Annual, 2003* of Energy Information Administration). The population estimates in Table 1.2 have been obtained from the Population Division of the Department of Economic and Social Affairs of the United Nations Secretariat (*World Population Prospects: The 2004 Revision, 2005*). The global trend of a continuously increasing demand for energy is evident from this table. There has been an almost 50% jump in world energy consumption during the period 1980–2003, arguably because of (i) about 42% increase in world population and (ii) worldwide improvement in the standard of living, particularly in the developed and developing countries. Fig. 1.1 shows the past and projected future growth of the world population, based on the United Nations' "medium fertility projection" and of total and per capita energy consumption. The present world population of about 6.5 billion is expected to reach 7.9 billion in 2025 and 9.1 billion in 2050 despite the projected declining fertility rates in the next few decades. The projected population growth has been attributed largely to the least developed countries and to a lesser extent, the developing countries. These are the regions where energy consumption is expected to grow exponentially due to emerging economies and increasing standards of living. In the world scenario, energy consumption is projected to increase further, about 57% by 2025.

Table 1.2. World population, total primary energy consumption and production (Data sources: United Nations, Statistical Office, *World Energy Supply Series*; World Population Prospects: The 2004 Revision, 2005; International Energy Annual, 2003 of the Energy Information Administration)

Year	Population (billion)	Total production (10^6 metric tons of coal equivalent)	Total consumption (10^6 metric tons of coal equivalent)	Consumption per capita (kilograms of coal equivalent)
1951	2.59	2,822	2,710	1,075
1961	3.08	4,418	4,329	1,387
1966	3.41	5,621	5,506	1,653
1971	3.78	7,260	7,096	1,931
1973	3.94	8,485	7,885	2,074
1980	4.44	10,358	10,205	2,311
1985	4.84	11,057	11,114	2,310
1990	5.28	12,598	12,508	2,384
1995	5.69	13,121	13,165	2,315
2000	6.08	14,303	14,395	2,364
2003	6.30	15,038	15,178	2,402

It is interesting to note how closely per capita income and per capita energy consumption are related. Per capita income is often used as a measure of wealth of a population. The United States of America, the most affluent country in the world today, with ~4.6% of the world's total population, consumes energy amounting to about one-fourth (23.4%) of the world's total energy consumption (*Annual Energy Review, 2004* of the Energy Information Administration). When comparing countries with vastly different socio-economic scenarios, a more useful statistic relative to per capita income is the Human Development Index (HDI). The HDI is a composite index computed by the United Nations that measures the average achievements in a country in three basic dimensions of human development: a long and healthy life, as measured by life expectancy at birth; knowledge, as measured by the adult literacy rate and the combined gross enrollment ratio for primary, secondary and tertiary schools and a decent standard of living, as measured by GDP per capita in purchasing power parity (PPP) U.S. dollars. HDI and per capita energy consumption for a global representative sample of countries covering industrialized, developing and poorly developed regions for the year 2002 are plotted in Fig. 1.2. The global sample contains the 60 most populous countries, which represent nearly 90% of world population and 90% of world energy consumption (Pasternak, 2000). Among the 60 countries, four countries, Afghanistan, Taiwan, North Korea and Iraq for which HDI values are not reported, have been excluded in Fig. 1.2. The industrialized countries are way ahead of others both in terms of energy consumption as well

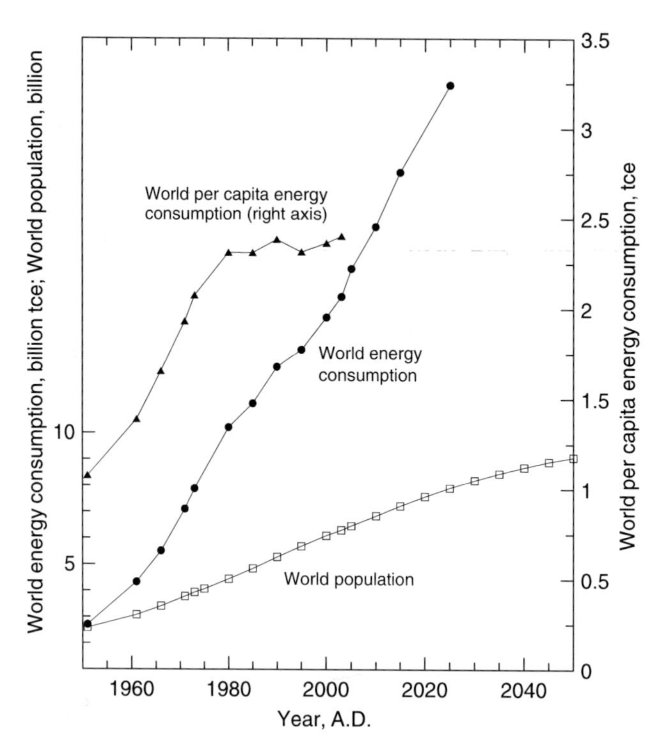

Fig. 1.1. World population, energy consumption and per capita energy consumption details and future scenarios. Energy is expressed in tons of coal equivalent (tce) to facilitate comparison with previous estimates.

as human development index. The differences between developed countries and least developed countries are startling. The correlation between HDI and per capita energy consumption suggests that in general, HDI reaches the highest values (0.90–0.95) when per capita energy consumption exceeds about 100 Btu per annum. The correlation also points to large, additional energy requirements in the next two to three decades as the developing economies strive to compete with the developed economies and the least developed nations move ahead towards "developing nations" status. Projections for three economic growth scenarios, low, medium and high, visualize ever-growing energy demands in the coming decades that would result from the improving standard of living in developing countries, even if the developed countries control their consumption pattern and make serious efforts to hold their energy requirements at the present-day level. Obviously, a very sincere effort needs to be made to cut down on population increase, limit our energy consumption and to find new sources of energy to maintain present standards of living in the developed countries and improve those of the developing countries.

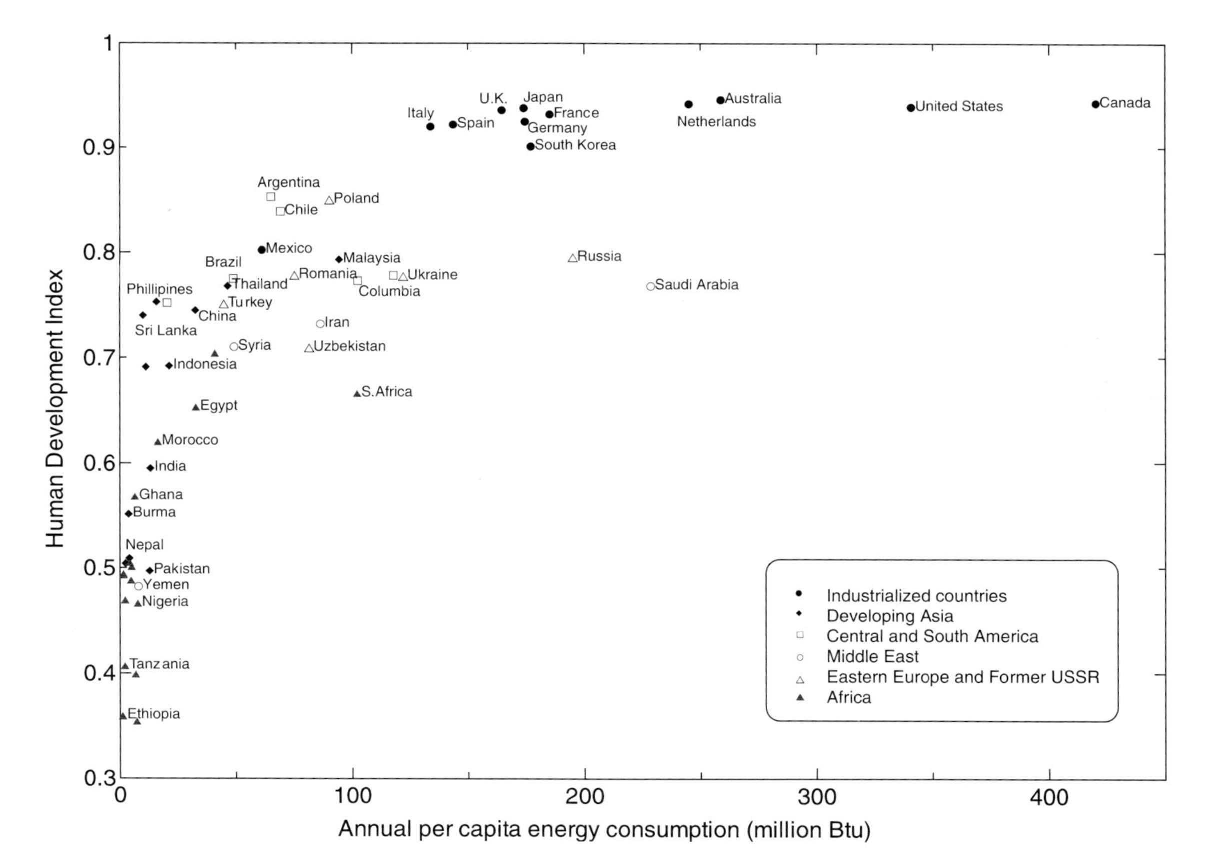

Fig. 1.2. The United Nations' human development index and annual per capita energy consumption for 60 most populous countries, 2002 (Data sources: Human Development Report, 2004, UNDP; International Energy Annual, 2003, Energy Information Administration).

The existing energy supply and demand situation is such that very soon all the energy that could be obtained from all available sources would need to be harnessed. Especially, conventional sources, such as oil and coal, would be depleted at an alarming rate. Growing concern about the environmental impact of the various energy resources places certain restrictions on their usage. The Kyoto Protocol, which became a legally binding treaty in 2005, requires participating countries to reduce their carbon emissions collectively to an annual average of about 5% below their 1990 level over the 2008–2012 period. To improve the energy situation, efforts need to be made in the following three directions, besides moderating the present-day increasing demands.

New Methods for Recovery, Power Generation and Distribution

The Earth is endowed with a large, but finite, quantity of fossil fuels. Today, an estimated 63% of the world's total energy consumption is provided by oil and natural gas. The world demand for oil is projected to grow from 78 million barrels per day in 2002 to 103 million barrels per day in 2015 and 119 million barrels per day in 2025, i.e., by about 1.9% annually. This is slightly lower than 2.3% growth projected for natural gas use (International Energy Outlook, 2005 of the Energy Information Administration). It is believed that most of the world's prolific petroliferous basins have been already identified and explored, and nearly all of the largest oil producing fields are under varying stages of production. It is, therefore, imperative to look for new methods of enhancing the recovery of "oil-in-place". The United States offers one good example where new methods for oil recovery hold great potential for meeting the country's energy requirements for the next several decades. The total recoverable oil using conventional methods is about 30 billion barrels (equivalent to about 2.5% of the world's total proven oil reserves), whereas an estimated 377 billion barrels of oil in place represents the "stranded" resource that could be the target for future enhanced oil recovery (EOR) efforts. Several EOR strategies such as thermal recovery, water and gas injection have been developed and tested to increase the oil recovery from current levels (10–30%) to about 60% of the established resource. The recent successes of CO_2 injection in depleting oil reservoirs to efficiently enhance the oil recovery have made it one of the most potential EOR tools. Reports of the U.S. Department of Energy indicate a possible CO_2-injected recovery of as much as 43 billion barrels of additional oil from what is left underground following the use of conventional techniques. Injection of CO_2 in the underground reservoirs provides additional benefits in reducing atmospheric CO_2 levels and restricting greenhouse-gas induced global warming.

Along with development of more efficient methods for energy production, it is equally important to practice energy conservation techniques. As the demand for energy continues to grow, new energy-efficient technologies in power generation, delivery and conversion can play an increasingly important role in moderating the growth in future energy demands and reducing environmental impacts. As much as 20% of the total energy consumed and 57% of the total electric power generated in the United States is used in power conversion processes using electrical motor systems. On the other hand, direct combustion processes increase efficiency while reducing emission of pollutants. There exists a huge additional potential for reduction in energy consumption in both industrial and domestic sectors by using energy-efficient designs and improvements in energy storage and management practices. Appropriate government legislations and incentives in both the developed as well as developing nations can bring about a big change towards more energy-efficient and environment-friendly technologies. Such measures will help extend the "fossil-fuel age" quite significantly and provide for a smoother transition to alternative energy resources.

Increased Utilization of the Renewable Energy Sources

A continuous flow of energy is received by the Earth from the Sun as direct heat and light as well as other forms of energy, such as the wind, ocean waves and tides and the temperature gradients in the ocean waters. Every year the Earth's surface receives about 10 times as much energy from sunlight as is contained in all the known reserves of coal, oil, natural gas and uranium combined. This energy is equivalent to 15,000 times the world's annual energy consumption by humans. Solar energy is clean, abundant and renewable. Although its vast energy potential was realized long ago, only a minute fraction of this free energy source is being utilized today accounting for less than 2% of the world primary energy production. Nuclear energy provides about 6% of the world primary energy. Other renewable energy resources indirectly derived from solar energy such as traditional biomass (wood and dry crop wastes) contribute about 10% and hydropower stations contribute about 7% to world primary energy production (Fridleifsson, 2003; Annual Energy Review, 2004 of the Energy Information Administration). On the contrary, mankind has exploited the finite energy resources of the Earth, which have been generated over periods much longer than that it is taking for man to deplete them.

The diffused nature of solar energy complicates its capture, concentration, storage and conversion processes into commercial energy forms. This makes it less cost-effective compared to fossil fuels. However, driven by impending oil crisis over the past two decades, solar energy has drawn the attention of a large number of scientists and engineers who are making efforts to convert solar radiation into forms such as electricity, heat and chemical fuels. The primary solar energy technologies include photovoltaics, concentrating solar power and solar heating and cooling systems. Presently, the United States of America and Japan are the largest users of solar energy. The largest solar electric generation plant in the world is located in California and supplies up to 354 MWe to greater Los Angeles area. As of 2004, Japan has 1200 MWe total installed generating capacity from solar photovoltaic plants. Gulf countries, richly endowed with oil and natural gas, have started to recognize the need to store their reserves for strategic purposes and look beyond oil for future energy requirements. Solar energy appears to be an attractive complementary energy source in most countries. Several developing countries, particularly in the regions endowed with rich sunlight resources round the year, have declared incentives to users of solar energy for their small energy needs. Such measures have resulted in increased utilization of solar energy. For example, Cyprus has the largest per capita use of solar energy, with 90% of the houses and a majority of apartments and hotels fitted with solar water heaters.

A great amount of heat energy is stored in the oceans, particularly in tropical regions. This energy can be converted into electricity through a process called as Ocean Thermal Energy Conversion (OTEC). The sun's heat warms up the surface water more relative to the deep ocean water resulting in a temperature difference between warm surface water and cold water at depths of about a kilometer or so, which creates a thermal energy resource available round the year. The temperature difference typically ranges from ~10 °C to ~25 °C, with the highest values occurring in equatorial ocean waters. In tropical regions (i.e., in the latitude band 20° S to 20° N), this difference is usually of the order of ~20 °C or more, making such regions more favorable for OTEC than the oceans at higher latitudes. There are mainly two versions of OTEC. In the open-cycle OTEC system, warm surface-ocean water is turned into vapor by depressurizing it, the pressure of the expanding vapor is then used to run a turbine and produce electricity. The vapor is condensed back into liquid by using deep cold ocean water, producing fresh, desalinated water as an important byproduct. In closed-cycle OTEC system, a working fluid with low-boiling point such as freon or ammonia is evaporated in a heat exchanger using warm seawater from the ocean surface. The expanding vapor is used to run a turbine in a similar way as in open-cycle OTEC system. The vapor is condensed by deep cold water, pumped back to the evaporator, and recycled. In addition, hybrid OTEC designs using combinations of open- and closed-cycle OTEC systems are possible. OTEC can be particularly useful in providing clean energy as well as fresh water to small island communities located in tropical oceans. Small-sized experimental plants producing up to a few hundred kilowatts of electric power have been successfully tested, as for example, at Keahole Point in Hawaii (Vega, 1995). A pilot-scale desalination plant of 1,00,000 l/day capacity using OTEC technique has been operating successfully in Kavaratti island in India. A 1 MW capacity, floating water OTEC plant is under testing in the eastern Indian offshore, off Tuticorin coast (Ravindran, 2005). Commercial exploitation of OTEC will depend largely on data obtained from design and testing of suitable scaled version pilot plants, cost–benefit ratio of present-day technology and an accurate assessment of environmental impacts caused primarily due to transportation of massive amounts of ocean water.

Besides producing thermal energy, oceans are a potential source of mechanical energy from the tides and waves. Tides are driven by the gravitational attraction of the moon, while waves are powered by the winds. Coastal areas experience two high and two low tides daily. The difference in water level between a high and a low tide, which is typically a few meters, is utilized to generate electricity. The only successful, industrial-scale tidal power station is located near Saint Malo in Britanny, France. It has been operating since 1966 and produces ~240 MW of power.

Geothermal energy is another abundant resource of energy, which has been successfully catering to both industrial and domestic energy requirements in several parts of the world over the past few decades. Geothermal energy uses the Earth's internal heat. Being a relatively clean and renewable resource, it has been a preferred choice for an alternative energy resource. It is produced when underground heat is transferred by water that is heated as it passes through hot rocks or shallow magma bodies located at depths of a few hundred metres to a few kilometers. The water is brought to the surface as hot water or steam through borewells drilled for the purpose. The water is often naturally occurring groundwater that seeps down along faults/fractures. In some cases, the water is artificially introduced by pumping down from the surface through borewells. Even though the extraction and transportation processes of geothermal energy are capital intensive, commercial exploitation of geothermal energy in recent times has become economically viable because of rapidly increasing prices of crude oil products. This form of energy will be described in the later sections.

Development and Utilization of Other Energy Sources and Techniques

Today there are alternatives to reduce the dependence on oil. Coal, the most abundant fuel, is getting harder to mine. The way it is practiced today, coal mining is hazardous and many advanced countries are plagued by labor shortages. Automatic machinery to make deep mining safer is not available. In the absence of proper reclamation, strip mining is also not very acceptable. Burning of coal is environmentally hazardous because of sulfur contamination and production of CO_2 and other greenhouse gases. The technology to remove pollutants from coal and convert it into a cleaner fuel is lacking. Therefore, although enough proven coal deposits exist to meet the world's energy demands for hundreds of years, technological barriers do not permit their fullest utilization in an environmentally acceptable way.

Natural gas is one of the least expensive, least polluting and most versatile fossil fuels. Its relative efficiency in comparison with other energy sources makes it even more desirable for electricity generation in many parts of the world. It has been projected to be the fastest-growing component of world energy consumption in the next two decades, by 2.3% annually on an average, compared with projected annual growth rates of 1.9% for oil consumption and 2.0% for coal consumption (International Energy Outlook, 2005 of the Energy Information Administration). Its share of total energy consumption is projected to grow from 23% to 25% during this time. About two-thirds of the increase in gas demand is in the industrial and power generation sectors. Global gas reserves are abundant, but of an uneven distribution. In the past, its low price relative to oil inhibited oil companies to drill deeper for it rather than to drill for oil. The global market for gas is much smaller than for oil because gas transport is costly and difficult due to a relatively low-energy content per unit volume. The rising demand of natural gas over the last few years has resulted in increased exploration efforts leading to discovery of large reserves and their production. The demand for natural gas has also revitalized efforts to produce synthetic natural gas from naphthol as well as coal.

Over the last few decades, nuclear energy has emerged as a significant addition to the traditional energy sources. It is about the only new energy technology developed commercially during the second half of the twentieth century. Extraordinary compactness is one of the unique properties of the nuclear fuels. Operating reactors use uranium fuels generating about 20,000 times as much heat as would an equivalent amount of coal. This ratio increases to 15 million times in breeder reactors. The U.S. Atomic Energy Commission has supported research and development amounting to several billion dollars for the next generation of breeder reactors. Nonetheless, there is a growing concern about the consequences of large-scale use of nuclear fission as a source of power as well as about the chances of a serious accident and subsequent radiation-caused damage. Long-term storage as well as disposal of radioactive wastes also causes serious problems. Natural disasters, such as earthquakes, technological failures and deliberate sabotage, as well as human carelessness, make nuclear reactors quite vulnerable. The possibilities of large-scale release of radioactive substances make nuclear-fission-generated power the most hazardous of all sources of energy. The radioactive fallout from the Chernobyl nuclear plant accident of April 26, 1986 in Ukraine may have been as high as 4.8 GBq/m^2 in the 10 km zone surrounding the plant and lies scattered over an estimated area of at least 150,000 km^2 in Belarus, Russia and Ukraine. According to reports of International Atomic Energy Agency, the explosion completely destroyed the reactor resulting in release of over 100 radioactive elements in the atmosphere, including the most damaging elements for the human body such as plutonium, iodine, strontium and cesium. Iodine has a half-life of 8 days, but Strontium-90 and Cesium-137 with half-lives of 29 years and 30 years, respectively, are present in the area even today. An estimated 56 people lost their lives due to acute radiation exposure in the immediate aftermath of the accident, about 200,000 people were evacuated and relocated. Several thousand people continue to suffer from diseases such as cancer resulting from long-term exposure. The long-term damage to the environment including flora and fauna are yet to be assessed. Therefore, although nuclear energy has enormous potential to contribute to global energy needs, it is essential to work on the safety standards of nuclear reactors and containment of the nuclear fallout in case of accidents as well as disposal of nuclear waste.

Experimental work has demonstrated that solid organic waste could successfully be converted into synthetic fuel. This is helpful to society in two ways: (1) it provides a reusable energy source, and at the same time, (2) it solves the problem of waste disposal. For an affluent and highly wasteful society like the U.S., it is estimated that waste-converted synthetic fuel could provide 3% of the crude oil requirement or 6% of the natural gas consumption of the entire United States. However, the conversion technologies are restricted due to a scarcity of solid waste.

In the recent years, a new resource of energy, gas-hydrates, is drawing worldwide attention. Gas-hydrates are crystalline compounds of water and methane molecules, and are formed at high pressure and low temperature. They occur naturally in the outer continental margins and the permafrost regions of the world. Various geological and geophysical parameters such as bathymetry, seabed temperature, sedimentation and total organic carbon content are useful in assessing the gas-hydrate stability zones in offshore regions. As the energy contained in gas-hydrates is estimated to be double the amount of total fossil fuel reserves, and one volume of hydrates produces ~164 volumes of gas at standard temperature and pressure, gas-hydrates are expected to be a major future energy resource. Besides, gas-hydrates act as cap rock that can trap "free-gas" underneath. Gas-hydrates have several other important implications on climate change, submarine geo-hazard and geothermal modeling. Therefore, the detection and quantitative assessment of gas hydrates are essential to evaluate its resource potential.

GEOTHERMAL ENERGY—AN ALTERNATIVE

One other potential source of energy is the Earth's heat. Unlike fossil fuels, it is considered to be a relatively clean and renewable energy resource. Although hot springs have been in use for centuries for balneological purposes, the use of the Earth's heat as an energy source only began early in the twentieth century when electricity was generated for the first time from geothermal steam at Larderello, Italy in 1904. By 1913, a 12.5 MW electric plant was in continuous operation there. The spread of the technology to other parts of the world had been rather slow during the first half of the twentieth century, being mostly confined to Italy. Later, interest developed in other parts of the world with intensive pioneering exploration being carried out in New Zealand, the United States of America and Japan, where electric power plants were commissioned in 1958, 1960 and 1961, respectively. Although geothermal water began to be used for large-scale municipal district heating service in Iceland in 1930, electricity production from steam started only in 1969. The utilization has increased rapidly during the last three decades mainly from variable capacity additions by Phillipines, United States, Italy, New Zealand, Iceland, Costa Rica, El Salvador, Guatemala and Russia. Development of geothermal energy registered the maximum growth rate of ~22.5% per 5 years between 1980 and 1990 and a slightly smaller rate of ~16.7% between 1990 and 2000 (Huttrer, 2001). Much progress in utilizing this very potential source of energy has been made during the recent years, and this will be discussed in the later sections.

Geothermal resources vary widely from one location to another, depending on the temperature and depth of the resource, the rock chemistry and the abundance of ground water. Geothermal resources are predominantly of two types: high temperature (>200 °C) such as found in volcanic regions and island chains, and moderate-to-low temperature (50–200 °C) that are usually found extensively in most continental areas. The type of geothermal resource determines the method of its utilization. High-temperature resources (dry steam/hot fluids) can be gainfully utilized to generate electric power, whereas the moderate-to-low-temperature resources (warm-to hot water) are best suited for direct uses. However, aided by modern technology, even the moderate temperature resources (~100 °C) are being utilized for generation of electric power using the binary-cycle method. The most extensive *direct use* of low-temperature geothermal resources (50–100 °C) is in space heating of individual buildings or entire districts in cold countries. Geothermal water is pumped through a heat exchanger, where it transfers its heat to city water supply systems. A second heat exchanger transfers the heat to the building's heating system. Another common *direct use* is in heating or cooling buildings using geothermal heat pumps, which utilize the relatively stable temperature at a depth of a few meters in the ground. These pumps circulate water or other liquids through pipes buried in a continuous loop. In winter, the difference between warm underground temperature and the cold atmosphere is transferred through the buried pipes into the circulating liquid and then transferred again into the building. In summer, circulating fluid in the pipes collects heat from the building, thus cooling it, and transfers it into the Earth. In yet another use, inexpensive low-temperature geothermal waters are being piped under roads and sidewalks in Klamath Falls, Oregon, USA to keep them from freezing in winter. In several developing nations, devoid of adequate conventional fossil fuels, there is a high potential of geothermal resources. For example in Tibet, with no readily available fossil fuels, the Nagqu geothermal field provides a useful energy source for the local population with the help of a 1 MWe binary plant built in 1993. In big countries such as the United States of America, geothermal energy will not replace fossil fuels as a major energy resource, but would contribute significantly to the nation's energy requirements.

Although, geothermal energy has been used to generate electricity for about nine decades and technology for its commercial exploitation has improved over the past two decades, the easy availability of fossil fuels such as oil, gas and coal at relatively low prices is not conducive for rapid development of the geothermal industry. The situation has changed dramatically over the past few years. International oil prices have almost doubled, resulting in a better market for geothermal energy. Further, the world has been alerted of increased atmospheric concentrations of greenhouse gases such as carbon dioxide, methane and nitrogen oxides in present-day global warming scenarios and their potential impacts to the society at large. There has been a growing recognition that use of geothermal energy contributes only a fraction of atmospheric pollution when compared with fossil fuels such as coal and oil. The best example comes from Iceland, where geothermal energy accounts for about 50% of total primary energy use and 86% of all space heating, leading to a clean environment and improved quality of life (Fridleifsson, 2001). Today, besides being used in at least 21 countries to generate electricity totaling to about 8000 MWe (Huttrer, 2001), geothermal energy is being used in 58 countries for direct uses (space heating and cooling, health spas, fish farming, agricultural and industrial purposes) totaling over 15,000 MWt (Lund and Freeston, 2001). Phillipines, which had the second largest installed geothermal generating capacity (~1900 MWe) after the United States (~2200 MWe) in the year A.D. 2000, meets about 22–27% of its present-day electricity requirements from geothermal steam. In United States, which is the world's largest energy consumer, geothermal energy amounts to about 0.4% of its overall energy production. It is estimated that worldwide electric power generation from geothermal resources could increase by about tenfolds at the present technology levels. Several other countries are actively exploring and assessing their geothermal resources to meet their energy requirements and contribute to world's energy needs. Obviously, the future use of geothermal energy would very much depend on overcoming technical barriers both in production and utilization, and its economic viability compared to the other energy sources. Political will of administrators in encouraging an environmentally acceptable alternative energy resource, will also play a very important role.

Chapter 2

BASIC CONCEPTS

INTRODUCTION

In the previous chapter the world's growing demands for energy have been discussed and it has been suggested that in the future, with proper development, geothermal resources could make an important contribution to this requirement. We shall now briefly discuss the structure of the Earth, its thermal history and heat flow and the processes going on within it, which are responsible for the generation and transport of heat.

EARTH'S STRUCTURE

The Earth's structure can be approximated by a series of concentric spherical shells. The large-scale features of the Earth's internal structure are shown in Fig. 2.1. The core, constituted by the two innermost regions, has the greatest average density, exceeding $10^4\,\mathrm{kg\,m^{-3}}$. In spite of differing views on the details of the composition of the core, it is now fairly well accepted that iron–nickel alloy is the most probable constituent. However, the observed characteristics of the core do not match with its being purely iron-nickel—too dense—and the presence of some lighter material is postulated. Silicon has been proposed to be an alloying element in the core. For this to be true, it is necessary to assume the presence of suitable conditions, in the early history of the Earth, making it possible for large quantities of silicon to reduce—stripping away oxygen atoms and adding electrons. Sulfur is another light element that has been suggested as being present in the core. This would require a different set of conditions to exist in the early history of the Earth. The scope of this book does not permit any discussion on the origin of the core. Irrespective of its origin, certain aspects of the present structure of the core are well established from seismological evidence. The outer part is molten since it does not transmit shear waves. The study of compressional waves, which travel through the inner part of the core, shows higher velocities, leading to the suggestion that the inner core is solid. Results of the study of free oscillation of the Earth, as well as the detection of seismic waves that have traveled through the inner core as shear waves, confirm the above suggestion.

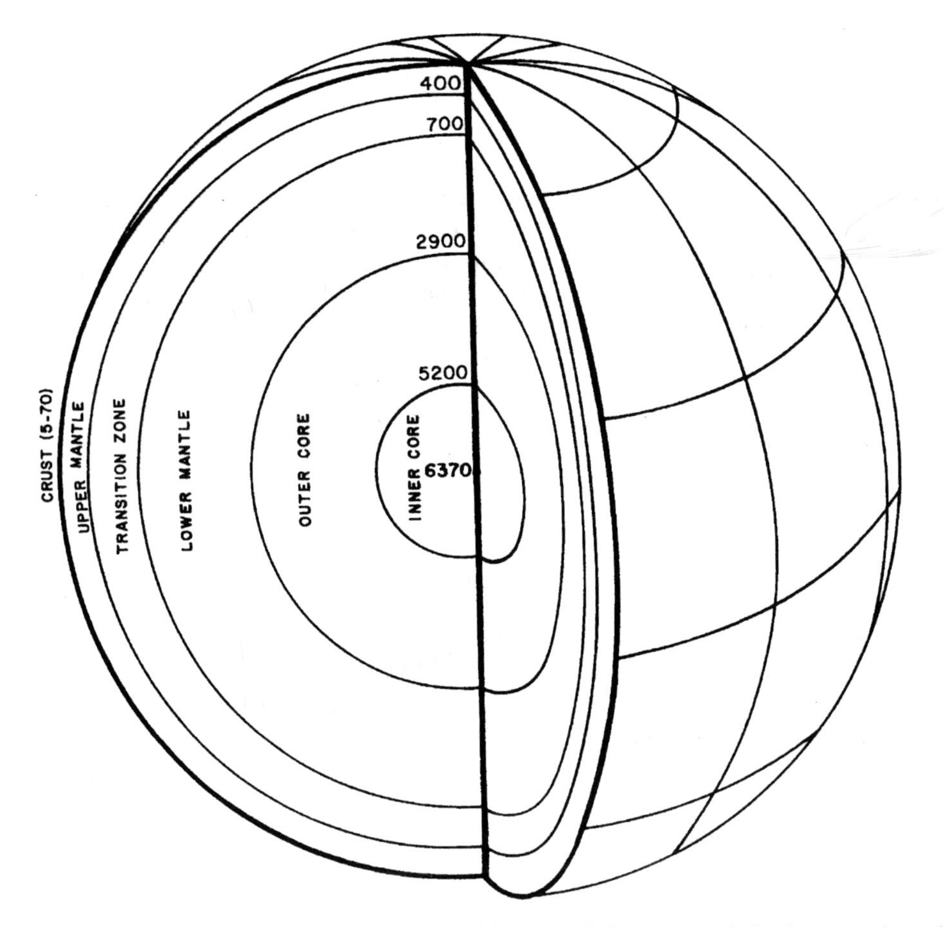

Fig. 2.1. Internal structure of the Earth. Thickness of the crust and depths to various discontinuities from the Earth's surface are given in kilometers.

The mantle overlies the core. It has an average density of $4.5 \times 10^3\,\mathrm{kg\,m^{-3}}$, indicating that its constituents are rocky rather than metallic. The composition of the mantle is not completely determined. However, based on its density, seismic wave velocities, and study of rocks that are believed to have come from the mantle, oxygen and silicon are believed to predominate, with magnesium and iron being the most abundant metallic ions. On the basis of seismic wave properties, the mantle could also be divided into a number of concentric shells. The lower mantle extends from a depth of about 700 km within the Earth to the top of the core at 2,900 km. As a result of the increase in pressure, the seismic velocity and density increase with depth in the lower mantle. The amount of iron in the silicate minerals also increases with depth, a factor that also contributes to the increase in density and velocity.

The sudden changes in seismic velocity in the transition zone extending from about 400 km depth to the top of the lower mantle are more likely related to alterations in the crystal structure than to changes in the composition. For example, at a pressure of approximately 14 GPa, corresponding to a depth of about 400 km, the atoms in the crystal structure of mineral olivine are rearranged into a more compact form resembling the mineral spinel. With each phase change in the transition zone, density and seismic velocity increase. It is interesting to note that the deepest known earthquakes, at about 700 km, coincide with the seismic velocity increase, probably associated with a phase change marking the boundary between the transition zone and the lower mantle.

The upper mantle extends from the base of the crust to a depth of about 400 km and is mainly composed of olivine, pyroxene, and garnet. These minerals would have the same stable crystalline structures at pressures reached within the upper mantle as they do on the surface of the Earth. The existence of an approximately 100-km-thick low-velocity zone for both shear and compressional waves, with its top at depths varying from 80 to 120 km, is one of the remarkable features of the upper mantle. This low-velocity zone is particularly well defined for the shear waves and is more prominent in the upper mantle below the oceanic basins. The low-velocity zone is probably caused by small pockets of molten material scattered throughout the mass of solid minerals. With the increase in pressure, the temperature at which parts of the mantle minerals begin to melt increases as shown in Fig. 2.2. The temperature in the mantle increases steeply with depth and can rise into the melting zone and fall again below the melting temperature at greater depths. Besides causing lower seismic velocities, the semi-molten material reduces the mechanical strength considerably in this zone.

The Mohorovićić discontinuity, named after its discoverer in the early twentieth century, and also known as the Moho for short, by definition separates the crust from the upper mantle. Seismic P-wave velocities suddenly increase from 6.5 km s^{-1} above the discontinuity to 8 km s^{-1} below it. The Moho at places may be due to phase change or a compositional boundary or both. The composition of the crust beneath the oceans differs considerably from that beneath the continents. The oceanic crust accounts for about 65% of the total Earth's surface and is covered with an average of 4-km-thick layer of water. Below the water layer, there exists about 0.5 km of sediments overlying an approximately 1.5-km-thick layer of basaltic volcanic rocks. The next 6 km, down to the Moho, mostly consist of metamorphic equivalents of the basaltic volcanic rocks and other iron- and magnesium-rich igneous rocks.

In contrast to the relatively uniform oceanic crust, the continental crust varies in thickness from less than 25 km under certain shield areas to more than 60 km below certain high mountains, with an average thickness of 35 km, and it has a complicated structure. At the surface, a great variety of rocks are exposed. These include sediments, such as clays, sandstones or limestones, and ancient shields, mostly composed of granites and volcanic lavas. The underlying basement is composed of granites

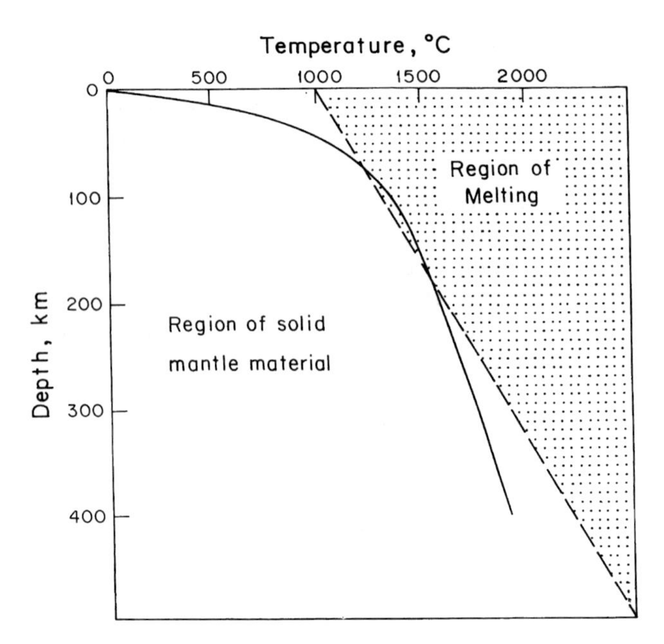

Fig. 2.2. Increase in melting temperature with depth within the Earth (modified from Sawkins et al., 1974). According to this model, the temperature passes into the zone of partial melting over a restricted depth.

formed either through freezing of the molten material or metamorphism of the sediments. Being a result of complicated thermal, mechanical, and chemical processes, the granites vary considerably in their mineralogical composition, and are generally characterized by a low melting point and a large portion of free silica in the form of crystalline quartz. With the increase of depth, the proportion of basalt increases, and at many places a second-order discontinuity, known as the Conrad discontinuity, which separates the granitic layer from the basaltic layer, has been inferred from the seismic wave travel-time studies. The depth of the Conrad discontinuity varies from one region to another, estimated to be one-third to two-thirds of the continental crustal thickness. Below this discontinuity, and extending to the Moho, a basaltic layer has been inferred.

In addition to recognizing the basic differences in the continental and the oceanic crust, such as the elevation, thickness, structure, and overall composition, understanding the transition between the oceanic and the continental crust is also very important. The continents extend as shallow continental shelves beyond the shoreline. The total area covered by the shelves amounts to about 6% of the total global surface. The edge of the continental shelves is followed by the continental slope leading into deeper water. Deposition of sediments at the base of continental slopes gives rise to the continental rise on the edge of the deep ocean basins.

THERMAL STRUCTURE OF THE EARTH

Along with the Earth's structure and composition, unraveling the temperature distribution within the Earth has been one of the fundamental research problems that impact our understanding of the evolution of the Earth. The present-day temperature distribution inside the Earth depends on (i) the original temperature distribution shortly after formation, (ii) the distribution and intensity of heat sources, both of which are time-dependent, and (iii) the mechanism of internal heat transfer—conduction, convection or both (Verhoogen, 1980). Even after several decades of theoretical and experimental research, the thermal structure of the Earth continues to be poorly understood and to an extent is speculative. The uncertainty is mostly because the temperature distribution is inseparable from the hypothesis of the Earth's origin and its thermal history of the last 4.5 billion years, both being debated (Lubimova, 1969; Verhoogen, 1980; Jeanloz and Morris, 1986; Stacey, 1992), and is also attributable to inadequate understanding of the physical properties of Earth materials at high pressures and temperatures that exist in the lower mantle and core. The composition and nature of the mineral phases present in the lower mantle and the composition of the core are also debated.

A probable model of temperature distribution with depth within the Earth is shown in Fig. 2.3. The constraints for this distribution have been derived from geophysical and petrological observations, and their interpretation using experimental laboratory studies on behavior of mantle- and core-constituent minerals at high pressures and temperatures. As discussed earlier, the core is made of iron

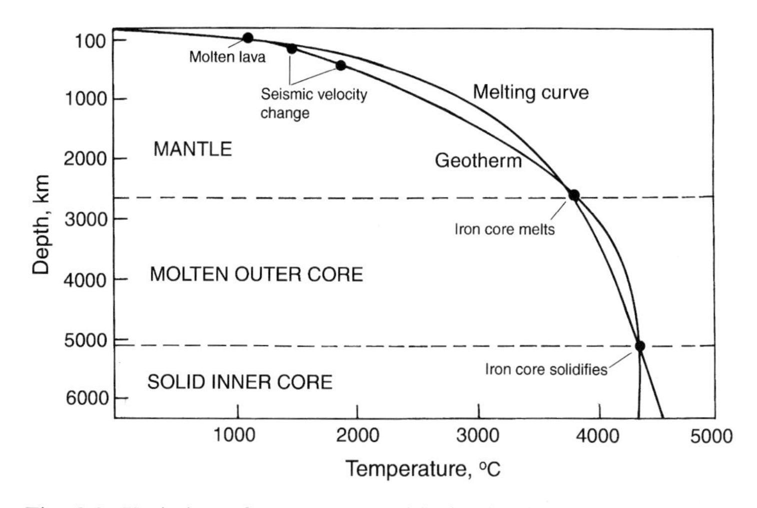

Fig. 2.3. Variation of temperature with depth within the Earth (modified from Press and Seiver, 1998). In deeper portions, overall uncertainties in temperatures could be as much as 1,000–1,500 °C.

alloyed with small quantities of light elements such as sulfur, oxygen and hydrogen, and the lower mantle comprises silicate perovskite as the dominant phase along with certain volatiles. Because the molten outer core is sandwiched between the mantle and the solid inner core, the temperature at the top of the outer core must exceed the melting point of the iron alloy comprising the core, while being less than the temperature required to melt the silicates and oxides of the lower mantle. As shown in Fig. 2.3, temperatures in the lower mantle and inner core are below the melting-point curve, while those in the outer core are above the melting-point curve.

In the last two decades, several high-pressure experimental studies on melting point and phase transformation of iron have provided additional constraints on the temperature inside the Earth's core (see, for example, comprehensive summaries by Jeanloz and Morris, 1986; Duffy and Hemley, 1995). During the same time, studies on melting temperatures of minerals in the Mg–Fe–Si–O system, for example, (Mg,Fe)SiO_3-perovskite, have resulted in important constraints on the temperatures in the lower mantle. Pressure and temperature data on mantle xenoliths constrain the temperatures in the depth range of 100–250 km. Best estimates of temperature at these depths range between 700 and 1,500 °C. The phase transition from olivine to spinel at the 400-km discontinuity constrains temperature estimates at the top of this transition zone to range from 1,400 °C to 2,000 °C. Temperatures at the inner core–outer core boundary (~330 GPa) are estimated to be in the range of 5,000–5,800 °C, while the temperature at the top of the outer core is estimated to be 3,500–4,700 °C. The melting curve of perovskite represents an upper bound to the temperature of the lower mantle. Experimental data constrain the maximum temperature in the upper half of the lower mantle to ~2,900 °C. Temperature estimates resulting from these studies show large differences, perhaps due to differences in experimental conditions, uncertainties in the amount of alloying components in the core and mantle, and uncertain effects of these components at high pressures. The overall uncertainties in estimated temperatures become larger as we go deeper in the Earth. On an average, uncertainties of the order of 1,000–1,500 °C are not improbable, considering our lack of knowledge of the Earth's composition.

HEAT FLOW AND TEMPERATURE DISTRIBUTION WITHIN THE LITHOSPHERE

It has been known for a long time that temperatures in mines exceed those at the surface. The rate of increase in temperature with depth in the Earth is called geothermal gradient. This results in an upward flow of heat from the Earth's interior to the surface, which is referred as heat flow. Heat flow in an area is determined by combining geothermal gradient with thermal conductivity of the rock formations over which the gradient has been computed, and is expressed in terms of heat flowing out over unit area per unit time, $mW\,m^{-2}$. When integrated over the Earth's surface, the total heat flow turns out to be $\sim 4.4 \times 10^{13}$ J every second

($\sim 1.4 \times 10^{21}$ J yr^{-1}), which is orders of magnitude greater than the energy associated with tidal dissipation, seismic strain release, volcanoes, magnetic field generation, and other geologic processes that result in energy loss from the Earth's interior. The determination of heat flow is one of the few measurable quantities that provide information on thermal state within the Earth.

At the present time, direct measurements of temperatures in the Earth's interior are limited to a depth of 12.261 km as obtained in the Kola super-deep borehole SG-3, located in the northern part of Baltic shield near the town of Zapolyarniy (in the northwest of Russia), where a temperature of ~180 °C has been measured. Another reliable measurement of temperature at a great depth has been made in the 9.101-km deep KTB borehole in Oberpfalz, Germany, ~265 °C at the bottom of the hole. Higher temperature at a shallower depth in the KTB borehole is a result of generally higher geothermal gradient and heat flow in the Oberpfalz area. Most measurements are commonly made to depths of a few hundred meters only. In deeper parts of the crust and the sub-crustal lithosphere, temperatures are computed on the basis of heat-flow data as the boundary condition at the surface, along with models for distribution of heat production and thermal conductivity with depth. Chapman (1986) computed a family of geotherms for different heat-flow conditions for the continental lithosphere, which provide realistic estimates of temperature distribution with depth on the basis of existing information on heat production in the crust and experimental data on thermal properties of rocks at temperatures representative of the deeper levels of the Earth (Fig. 2.4). Uncertainties associated with these estimates can be as large as several tens of degrees at depths > 50 km. There is scope for

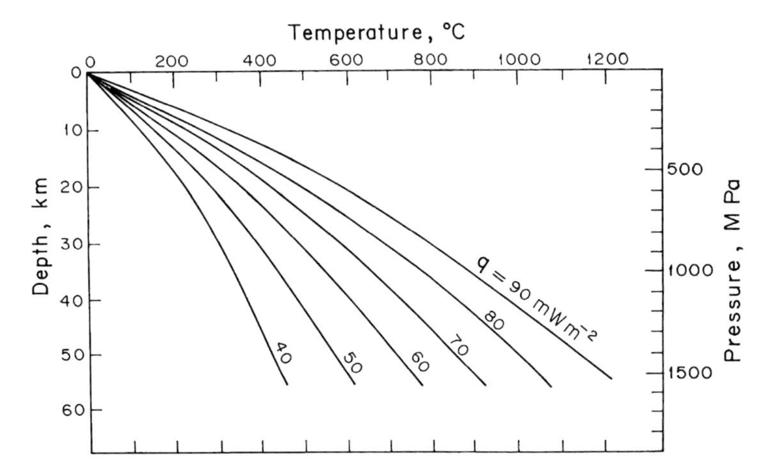

Fig. 2.4. Geotherms for the continental lithosphere computed on the basis of heat-flow data and realistic models for distribution of thermal properties with depth. Geotherms have been computed for a range of heat-flow values, 40–90 mW m^{-2}, representative of the continental lithosphere. (Modified after Chapman, 1986)

improving the estimates with better-constrained models for crustal structure, heat production, and thermal conductivity in the crust and upper mantle. The processes that determine the temperature profiles in the oceanic lithosphere are quite different from those in the continental lithosphere. The oceanic lithosphere contains only small quantities of radiogenic elements, and the temperature profiles are mainly determined by underlying convective heat transfer. The heat flux is the largest in the mid-ocean ridges and decreases with distance from the ridges.

The number of sites where heat-flow data have been acquired from temperature measurements in boreholes and thermal conductivity of rock formations has increased from a few hundred in the early 1960s to more than 20,000 in the early 1990s (Pollack et al., 1993). Several new heat-flow determinations have been made during the past decade both in continents and oceans. According to the latest global compilation of heat-flow data, the average heat flow is $65\,\mathrm{mW\,m^{-2}}$ for continental areas and $101\,\mathrm{mW\,m^{-2}}$ for oceanic areas, resulting in an areally weighted, global average of $87\,\mathrm{mW\,m^{-2}}$ (Pollack et al., 1993). It is instructive to examine the diverse nature of datasets that have contributed to computation of the average heat-flow values. A large number of measurements in oceans have been reported from mid-ocean ridges and their immediate vicinity where heat-flow values are severalfold higher relative to old oceans (> 50–60 Ma). In fact, more than 50% of the heat-flow sites are located in Oligocene to Quaternary crust, and more than 50% of the Earth's total heat loss comes from the Cenozoic oceanic lithosphere. Furthermore, most oceanic heat-flow values have been corrected for hydrothermal circulation, which is a subject of recent debate (Hofmeister and Criss, 2005). In the continents, the highest heat flow values have been reported from tectonically active areas, including volcanic areas, and lowest values from shields of Archaean age. The quality of the heat-flow dataset is not uniform. In certain datasets corrections have been made but not in all. In many cases, heat-flow values have been determined from temperature measurements in boreholes shallower than 100 m in depth. Geothermal gradients estimated from temperature–depth profiles measured in such boreholes are often perturbed by various transient effects such as fracture-controlled groundwater flow within the borehole column and diffusion of surface air-temperature variations that have taken place over the past few decades to few centuries.

Because of the extremely uneven distribution of heat-flow sites, both in continents and oceans, empirical heat flow estimators have been employed to estimate heat flow in regions devoid of measurements prior to computation of global averages. Mean heat flow computed for a geological province is assigned to geologic provinces of similar age where no heat flow observations exist. Heat flow vs. age plots are shown for oceanic and continental areas in Fig. 2.5. While a strong basis for correlation between heat flow and oceanic age exists, the age of continental crust is not a reliable heat-flow indicator because of the variability of heat flow in provinces of similar age. Uncertainties are also introduced by different dating methods used. Some methods yield the age of formation, while others indicate the age of last metamorphism. Continental rocks, unlike the oceanic crust, bear the impress of a great variety of

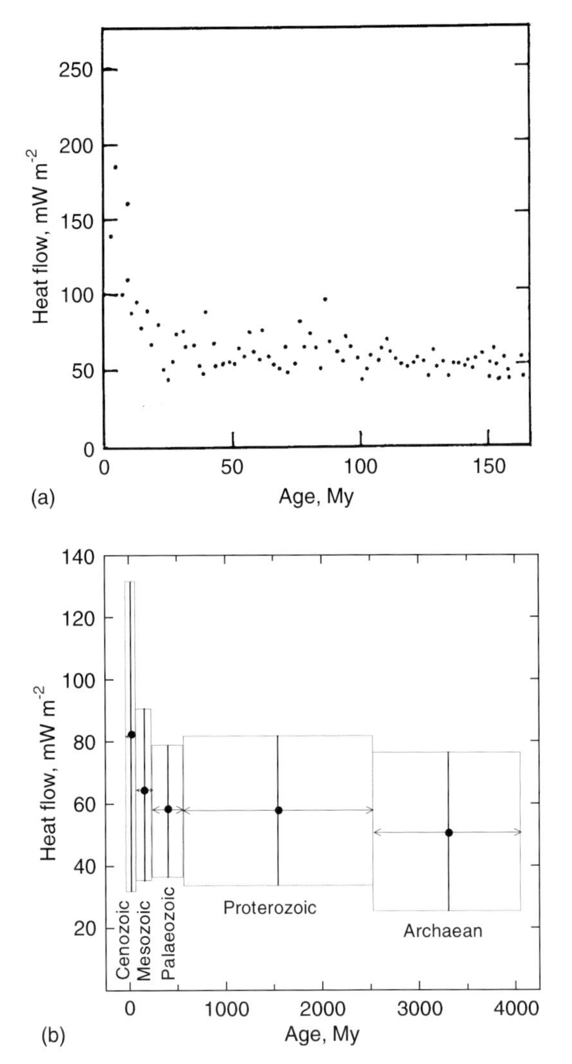

Fig. 2.5. Observed variations in (a) oceanic and (b) continental heat-flow values and their dependence on age (modified after Stein and Stein, 1992; Clauser, 2006). Source of data: Compilations by Stein and Stein (1992) for oceanic heat flow and by Pollack et al. (1993) for continental heat flow.

processes such as plutonism, volcanism, uplift, subsidence, folding, etc., all of which have variable surface thermal effects even for the same age (Rao et al., 1982). These fundamental premises defy a single, universal, heat flow–age relation, unlike in the case of oceans, where age of the oceanic crust can be uniquely defined as the time since its formation at a mid-oceanic ridge. Nevertheless, the average values provide an overall perspective useful in computing the total heat loss through the Earth's surface.

Localized zones of anomalously high heat flow, such as young orogenic zones and regions where cooling magma bodies occur at shallow depths, are the potential targets for exploration and exploitation of geothermal energy. These will be discussed in greater detail in the next section.

Sources for Heat Flow

The release of heat due to the cooling of the Earth and the heat produced by radioactivity appear to be primarily responsible for the heat flow observed at the surface of the Earth and the temperature distribution within it. The amount of radioactive elements present in the rocks releases enough heat to account for a major portion (typically, ~60% for continental crust) of the total heat flow observed on the Earth's surface. Of all the radioactive elements present, significant heat contributions are made by the decay of long-lived isotopes of uranium (^{238}U and ^{235}U), thorium (^{232}Th), and potassium (^{40}K). The two uranium and one thorium isotopes decay to stable isotopes of lead, releasing energy in the process.

$$^{238}U = {}^{206}Pb + 8\,{}^{4}He + 51.6\ MeV$$
$$^{235}U = {}^{207}Pb + 7\,{}^{4}He + 46.2\ MeV$$
$$^{232}Th = {}^{208}Pb + 6\,{}^{4}He + 42.6\ MeV$$

Potassium has three isotopes: ^{39}K, ^{40}K, and ^{41}K. Of these, only ^{40}K is radioactive, but it is a very rare isotope with an abundance of 0.0119%. ^{40}K decays to release energy by two processes, electron capture to ^{40}Ar and beta-decay to ^{40}Ca. Pertinent details of half-life and heat production by the most common radioactive elements in rocks are given in Table 2.1.

Heat production of a rock is the total heat produced by the radioactive isotopes of U, Th, and K. It is defined as the quantity of heat produced by radioactivity in unit volume of the rock per unit time, and is expressed in $\mu W\,m^{-3}$. It is estimated using

Table 2.1. Major radioactive isotopes found in rocks, their half-lives, and heat production (Rybach, 1976)

Isotope	Half-life (yr)	Heat production ($J\,kg^{-1}\,yr^{-1}$)
^{238}U	4.50×10^{9}	2.97×10^{3}
^{235}U	0.71×10^{9}	18.01×10^{3}
^{232}Th	13.9×10^{9}	0.83×10^{3}
^{40}K	1.30×10^{9}	0.92×10^{3}

the following relation (Rybach, 1976):

$$\text{Heat production } (\mu\text{W m}^{-3}) = 10^{-5}\,\rho(3.48C_{\text{K}} + 9.52C_{\text{U}} + 2.56C_{\text{Th}}),$$

where C_{U} and C_{Th} are uranium and thorium contents expressed in ppm, C_{K} the potassium content expressed in weight percent, and ρ the density of the rock expressed in kg m^{-3}. For example, a rock formation with density 2670 kg m^{-3} and containing the average concentrations (2 ppm uranium, 8 ppm thorium, and 2% potassium) would produce 1.24 μW m^{-3} of heat. Further, if these average elemental concentrations were to be representative of the continental crust with a thickness of 35 km, radioactivity would contribute ~43 mW m^{-2} to the heat flow measured at the surface. The distribution of heat production with depth in the crust has been a subject of extensive research. However, it is now well recognized that the upper part of the continental crust is rich in radioactive elements, whereas the lower crust and upper mantle are considerably depleted. In the case of oceanic crust, which is depleted in radioactive elements, heat flow is primarily attributable to cooling of the lithosphere with time.

VOLCANOES, EARTHQUAKES AND PLATE TECTONICS

As discussed earlier, the temperatures within the crust are typically much below the melting point of rocks and hence, normally no molten material should exist close to the surface of the Earth. Volcanoes are special features, which mostly occur around the island arcs and the mid-oceanic ridges. The recent conceptual development of the theory of plate tectonics, and its successful application in explaining most observed geological processes, provide an acceptable explanation for the restricted geographical distribution of the volcanoes. In the following, we shall briefly review the plate tectonics hypothesis to understand the occurrence of volcanoes and other regions of high heat flow.

The two basic assumptions of plate tectonics are very simple. The first is to assume that the outer portion of the Earth consists of a few large rigid plates, very much like caps on a sphere, which do not undergo any significant internal deformation. The second is to assume that each plate is in relative motion with respect to the other plates and that significant deformation takes place only at the plate boundaries. Plates are areas within which horizontally directed tectonic processes of any significance are not presently occurring. There could be small vertical movements within the plates. However, they would be of much smaller magnitude compared to large horizontal movements at the plate boundaries. The plate boundaries are commonly marked by narrow belts of seismic activity. The geographical distribution of the major plates is shown in Fig. 2.6.

The outer portion of the Earth, including the crust and extending to the low-velocity zone in the upper mantle, which constitutes the rigid plates, is known as the

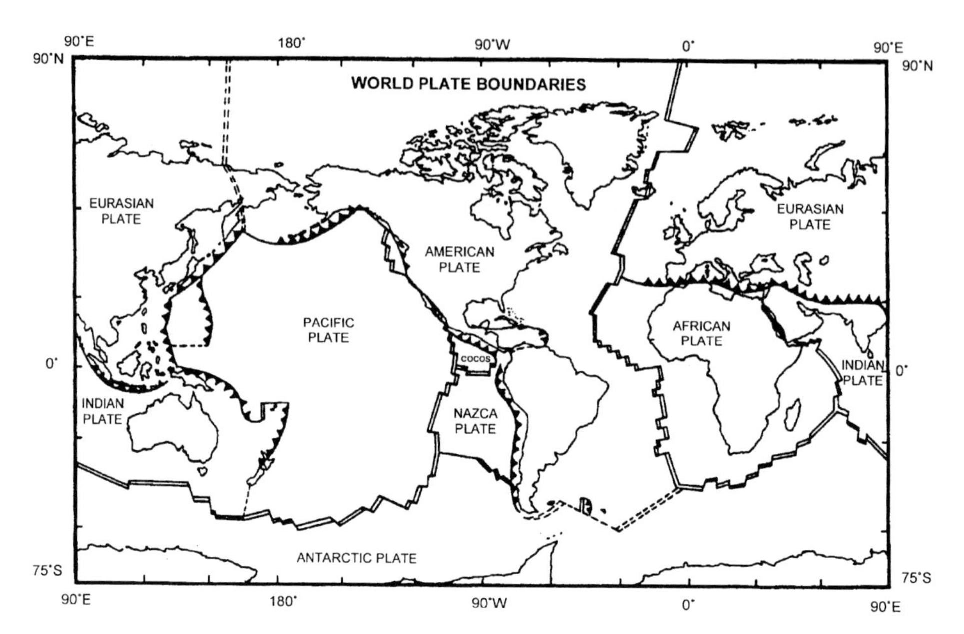

Fig. 2.6. Major plates of the world. Mid-oceanic ridges are shown by double parallel lines, transform faults are shown by single lines, and trenches and other subduction zones are shown by lines with teeth on one side. The direction of the teeth indicates the direction in which a plate descends. Dotted lines show uncertain boundaries. The seismic belts and the volcanoes are located in the vicinity of the plate boundaries.

lithosphere. The portion lying below the lithosphere, and probably extending to 700-km depth, is called the asthenosphere. In contrast to the rigid lithosphere, material in the asthenosphere is hot enough to deform and flow internally.

The processes within the Earth's interior can make the plates move apart leaving a gap between them, or move plates towards one another, shortening and/or buckling them, as well as making them slide past each other at their contact. With the receding of plates from one another, the hot and partially molten material from the asthenosphere rises to fill up the gap, resulting in sea-floor spreading. Gradually this semi-molten material cools down, becoming a part of the rigid plates on the two sides of the spreading zone. It appears that equal amounts of the new material are added on the two sides, and hence, the spreading is usually symmetrical. The newly added material takes time in cooling and shrinking, and the new lithosphere gradually moves away from the zone of spreading. In the meantime more of the lithosphere is added, causing the center of oozing to be elevated above the surrounding older sea floor. The elongated elevated zones thus created in the center of the sea-floor spreading are called mid-oceanic ridges, since they occupy central positions in the Atlantic and Indian Oceans. Besides their elevation, the other physical manifestations of the mid-oceanic ridges are the associated linear belt of earthquakes characterized by a normal faulting-type focal mechanism, high heat flow and a typical symmetric pattern of the observed magnetic anomalies.

The spreading rates of the diverging plates at the mid-oceanic ridges are estimated to vary from 2 to 20 cm yr^{-1}, the average being 6 cm yr^{-1}.

The processes that create and modify the continental crust are much more complicated compared to the generation of the oceanic crust. Because the surface area of the Earth seems to remain constant, it is implied that the creation of the lithosphere at divergent plate boundaries in the mid-oceanic ridges must be balanced by the destruction of the lithosphere at convergent plate boundaries. This shortening takes place through either the deflection of one of the plates into the asthenosphere, a process known as subduction, or through the crumpling and buckling of the two colliding plates. As mentioned earlier, the oceanic lithosphere is derived directly from underlying mantle material, and hence, its overall composition must be similar. Cooling makes it denser and therefore easier for it to sink back and be re-absorbed in the mantle. When two oceanic plates collide, one of them sinks and goes back to the mantle. At the collision of an oceanic plate with a continental plate, the oceanic plate is consumed in the subduction zone. In contrast with the oceanic lithosphere, the continental lithosphere is not easily forced down and consumed in the subduction zone since it includes the continental crust, which has a different composition from the mantle material and is much less dense. Therefore, when two continental plates collide, neither of them is consumed, and buckling and crumpling take place, resulting in great mountain ranges, such as the Himalaya. Fig. 2.7 shows typical subduction-zone models.

Oceanic trenches are distinct topographic features associated with the subduction of the oceanic lithosphere. The oceanic plate bends down over a distance of the order of 100 km as it enters the subduction zone, straightens out again, and penetrates the asthenosphere at a steep angle. The contact between the overlying and the sinking plate is down at the bend causing the trench (Fig. 2.7). The deepest known trenches in the western Pacific are more than 10-km deep. The deepest point in the Mariana trench, located near Japan, is 11,033-m deep and is the deepest known location of the Earth.

It has been estimated that at places the down-going lithospheric plate in the subduction zone can retain its identity up to 700 km of depth, when it finally gets heated up and mixes with the asthenospheric material. This process involves a duration of the order of 10 million years. The descending slab of oceanic lithosphere could cause additional melting in the mantle, or the upper oceanic crust part of the descending lithosphere itself melts in the 150- to 300-km depth range. The molten material rises through the overlying plate and breaks through the Earth's surface to form volcanoes. This explains the occurrence of most volcanoes over the subduction zones. When the overlying lithosphere, through which the molten material rises, is oceanic, the volcanoes build island arcs. When the overlying lithosphere is continental, volcanic mountain chains are created.

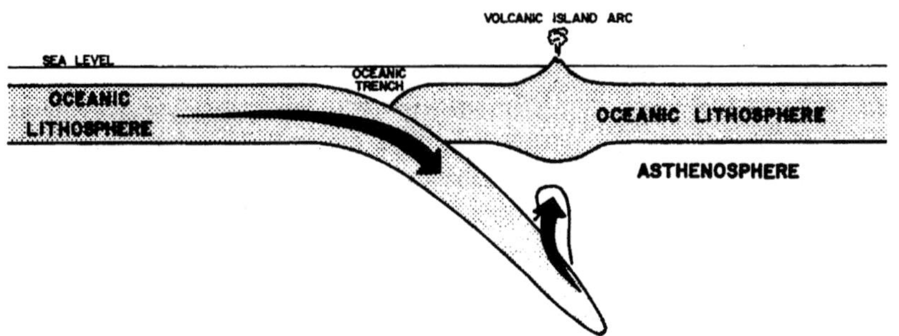

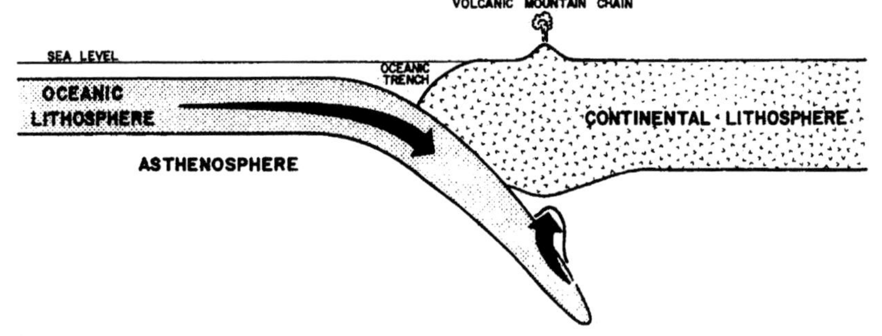

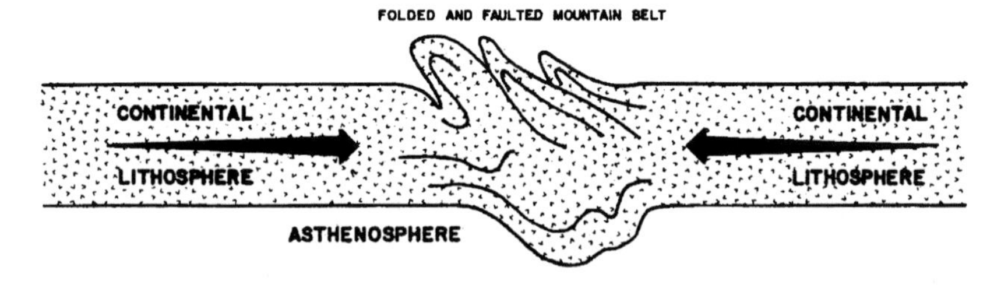

Fig. 2.7. Converging plates and the resultant physiographic features.

The two major earthquake belts, namely the circum-Pacific belt and the Alpide belt, which account for most of the earthquakes, lie in the zones of subduction and continent–continent collision, respectively. Earthquakes associated with the subduction. zones can distinctly be divided into four zones. In the outer zone, the earthquakes originate within the oceanic crust or the upper lithosphere of the sinking plate near the outer wall of the trench. The shallow zone includes earthquakes occurring at the contact of the two plates extending up to a depth of 100 km. These

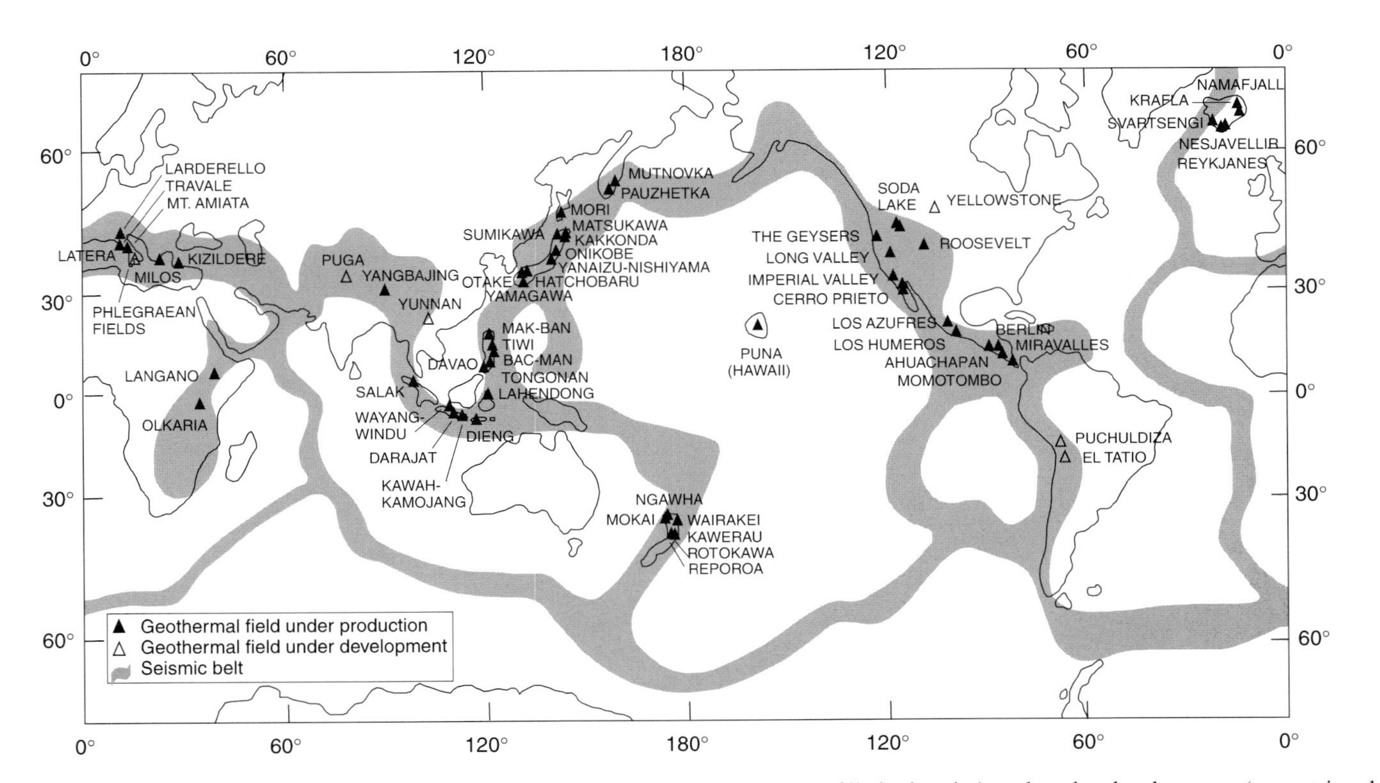

Fig. 2.8. Worldwide distribution of the geothermal fields under production (filled triangles) and under development (open triangles). This figure should be seen together with Fig. 2.6 to realize that all the geothermal fields are located near to the plate boundaries.

earthquakes are followed by the intermediate-depth earthquakes in the range 100–300 km, and they occur in the down-going slab rather than at its contact with the asthenosphere. Deep-focus earthquakes occurring in the 300–700 km depth range are not found in all subduction zones. All these earthquakes are related to the stiffness of the sinking lithosphere and hence indicate its penetration through the asthenosphere.

Earthquakes associated with the continent–continent plate collision are much more diffused and obviously do not show any pattern analogous to those observed at the oceanic plate–continental plate collision.

As explained in Figs. 2.6 and 2.8, plate boundaries, earthquakes, volcanoes, and regions of high heat flow, where geothermal fields are located, are all related.

Chapter 3

HEAT TRANSFER

INTRODUCTION

Heat can be transferred by three processes: conduction, convection and radiation. Conduction governs the thermal conditions in almost entire solid portions of the Earth and plays a very important role in the lithosphere. Convection dominates the thermal conditions in the zones where large quantities of fluids (mostly molten rocks) exist, and thus it governs the heat transport in the fluid outer core and the mantle. On a geological time scale, the mantle behaves like a viscous fluid due to the existence of high temperatures. Convection, which involves transfer of heat by the movement of mass, is a more efficient means of heat transport in the Earth compared to pure conduction. However, in several processes of the Earth's interior, both conductive as well as convective heat transfer play important roles. Radiation is the least important mode of heat transport in the Earth. The process of heat exchange between the Sun and the Earth, through radiation, control the temperatures at the Earth's surface. Inside the Earth, radiation is significant in the hottest parts of the core and the lower mantle only.

The loss of Earth's internal heat through the continental and oceanic lithosphere takes place primarily by conduction, except near the mid-oceanic ridges where convection due to hydrothermal circulation becomes significant. Cooling of magmatic intrusive bodies inside the crust and the upper mantle takes place both by conduction as well as convection. In anomalous geothermal locales such as near geysers, hot water springs and fumaroles, convective heat transfer through circulating hot waters in the shallow subsurface levels far exceeds the background heat conduction. Conduction governs the temperature distribution in the continental lithosphere as well as effects of sedimentation, burial, uplift and erosion on subsurface temperature distribution. Convection dominates heat transfer in the Earth's deep mantle and core.

TEMPERATURE, HEAT AND ITS STORAGE

Temperature of an object can be described as the property, which determines the sensation of hotness or coldness felt from contact with it. More unambiguously, using the Zeroth law of thermodynamics, temperature of a system is defined as the property that determines whether or not that system is in thermal equilibrium with any other system with which it is put in thermal contact (Finn, 1993). When two or more systems are in thermal equilibrium, they are said to have the same temperature. Temperature is most commonly measured in the Celsius (°C), Fahrenheit (°F) and Kelvin (K) scales. The first two scales are based on the melting point of ice and the boiling point of water. In the Kelvin scale, the limiting low temperature, called the absolute zero, is taken as the zero of the scale, and the triple point of water—where the ice, water and water vapor phases can co-exist in equilibrium, is equal to 273.16 K. The three scales are related as follows:

$$(\mathrm{K}) = (^\circ\mathrm{C}) + 273.16$$

$$(^\circ\mathrm{C}) = 5/9\,[(^\circ\mathrm{F}) - 32]$$

$$(^\circ\mathrm{F}) = 9/5(^\circ\mathrm{C}) + 32$$

Heat is defined as the energy transfer between two systems at different temperatures. Heat energy originates from other kinds of energy according to the first law of thermodynamics. It is important to distinguish between temperature and heat. Temperature is a property of matter, while heat is the energy that is flowing because of a temperature difference. According to the second law of thermodynamics, transmission of heat takes place from a body at a higher temperature to another body at a lower temperature. Under suitable conditions, heat can be transformed into yet other forms of energy. When subjected to heating, a body consumes the heat energy (ΔQ) through the increase of its internal energy (ΔU) and work done in doing so (ΔW), i.e., $\Delta Q = \Delta U + \Delta W$. Thermodynamic processes that occur without the gain or loss of heat to or from the system are called as adiabatic processes.

The calorie was originally considered as the basic unit of heat energy, being equal to the heat required to raise the temperature of 1 g of pure water from 14.5 to 15.5 °C at normal atmospheric pressure. This unit is used even today in several countries. After the widespread adoption of the SI system of units, heat is usually expressed in units of energy, i.e., joule (J), with the relation: 1 cal = 4.184 J.

Storage of heat energy is the change of enthalpy or heat content of a medium in the path of heat transmission. In accordance with the first law of thermodynamics, change of enthalpy (ΔQ) occurs as a result of a change in the temperature (ΔT) of the medium with time. The amount of heat that a body is able to store as a result of change in temperature, called free heat, depends upon its specific heat capacity (c). It is defined as the amount of heat required to raise the temperature of unit mass of a material by 1 °C, and it differs from one material to another. In the case of gases, the distinction between specific heat at constant pressure as well as at constant volume becomes important and it needs to be considered. Specific heat capacities of some common materials are listed in Table 3.1. The relationship between change in heat content and change in temperature of a body of mass m is generally expressed as

$$\Delta Q = mc\Delta T \tag{3.1}$$

Table 3.1. Specific heat capacities of some commonly used materials at 20 °C (Source: Tipler, 1999)

Substance	Specific heat capacity $\mathrm{J\,kg^{-1}\,K^{-1}}$
Aluminum	900
Bismuth	123
Copper	386
Brass	380
Gold	126
Lead	128
Silver	233
Tungsten	134
Zinc	387
Mercury	140
Alcohol (ethyl)	2,400
Water	4,186
Ice (−10 °C)	2,050
Granite	790
Glass	840

Several geothermal problems involve fluid-filled porous rocks, with water as a common fluid. The specific heat capacity of a rock with a_1 kilograms of dry weight and a_2 kilograms of water content is given by (after Kappelmeyer and Haenel, 1974)

$$c_{\text{wet}} = \frac{a_1 c_{\text{dry}} + a_2 c_{\text{water}}}{a_1 + a_2} \tag{3.2}$$

or simply by

$$c_{\text{wet}} = a_1 c_{\text{dry}} + (1 - a_1) c_{\text{water}} \tag{3.3}$$

when $(a_1+a_2) = 1$, i.e., 1 kg of wet porous rock containing a_2 kilograms of water. A typical value of specific heat capacity for dry rocks and soils is $\sim$840 $\mathrm{J\,kg^{-1}\,K^{-1}}$ while that for water is $\sim$4184 $\mathrm{J\,kg^{-1}\,K^{-1}}$. Therefore, with the increase of water content, the specific heat of the porous rocks increases.

It should be noted that heat added or removed during phase changes do not change the temperature. Therefore Eq. (3.1) is not valid when a phase change takes place, for example, from ice to water or from water to vapor. Large amounts of energy are required for such phase change. The amount of energy required to change a unit mass of a substance from the solid to the liquid state without changing its temperature is called the latent heat of fusion. A change from solid to liquid phase involves absorption of energy whereas a change from liquid to solid phase would involve release of energy by the same amount. The energy required to change a unit

Table 3.2. Thermal properties of a few commonly used materials (Source: Young, 1992)

Substance	Melting point (K)	Boiling point (K)	Latent heat of melting ($kJ\,kg^{-1}$)	Latent heat of vaporization ($kJ\,kg^{-1}$)
Helium	3.5	4.216	5.23	20.9
Hydrogen	13.84	20.26	58.6	452
Nitrogen	63.18	77.34	25.5	201
Oxygen	54.36	90.18	13.8	213
Ethyl alcohol	159	351	104.2	854
Mercury	234	630	11.8	272
Water	273.15	373.15	334	2,256
Sulfur	392	717.75	38.1	326
Lead	600.5	2,023	24.5	871
Antimony	903.65	1,713	165	561
Silver	1,233.95	2,466	88.3	2,336
Gold	1,336.15	2,933	64.5	1,578
Copper	1,356	2,840	134	5,069

mass of a liquid into the gaseous state at the boiling point is called the latent heat of vaporization. For example, about 334 $kJ\,kg^{-1}$ of heat is required for melting solid ice and about 2,260 $kJ\,kg^{-1}$ of heat is required for converting water to vapor at 1 atm. Typical values of melting point, boiling point and latent heats of some commonly used materials are listed in Table 3.2. The expression for change in heat content (ΔQ) in case of phase transformation is given by

$$\Delta Q = L\Delta m \tag{3.4}$$

where L is the latent heat of vaporization, expressed in $kJ\,kg^{-1}$, and Δm the mass of substance (in kg) transformed.

However, several geothermal problems involve change of enthalpy or heat content due to both changes of temperature and phase. In such cases, ΔQ can be expressed as

$$\Delta Q = \sum mc\Delta T + \sum L\Delta m \tag{3.5}$$

where the summation is carried out for a system comprising materials of different specific heat capacities and latent heats.

HEAT CONDUCTION

Thermal conduction takes place by the transfer of kinetic energy of molecules or atoms of a warmer body to those of a colder body. The transfer of kinetic energy takes place through movement of the valence electrons (also called conduction electrons) in an atom, a process analogous to electrical conduction. This type of conduction can take place in both solids and fluids. Inside the Earth, however, conduction of heat takes place mainly through poorly conducting solid rocks constituting the crust and the mantle, which are comprised of minerals having a very few conduction electrons. Another type of conduction, called lattice or phonon conduction, caused by lattice vibrations in the rocks, is primarily responsible for heat transfer in such cases. Detailed treatment of heat conduction is provided in several textbooks (e.g., Carslaw and Jaeger, 1959; Jacob, 1964); applications of heat conduction to problems in geothermics have been dealt by Kappelmeyer and Haenel (1974), Lachenbruch and Sass (1977), Haenel et al. (1988) and others. In this section we shall discuss some basic concepts, which are useful in understanding the heat flow and temperature distribution inside the Earth.

Fourier's Equation of Heat Conduction

When a temperature gradient exists within a body, heat energy will flow from the region of high temperature to the region of low temperature. This phenomenon is known as conductive heat transfer, and is described by Fourier's equation,

$$\vec{q} = -k\,\vec{\nabla}T \tag{3.6}$$

where $\vec{q}$ is the flow of heat per unit area per unit time (called as heat flow), k the thermal conductivity of the body (assumed isotropic) and $\vec{\nabla}T$ is the temperature gradient. The negative sign appears because heat flows in the direction of decreasing temperature. When applied to the heat flow of the Earth, we usually consider the heat flow toward the Earth's surface, i.e., $k(\partial T/\partial z)$, where the z-axis is taken vertically downward. $(\partial T/\partial z)$ is called the geothermal gradient, and is expressed in $^{\circ}\mathrm{C\,km^{-1}}$ or $\mathrm{mK\,m^{-1}}$. Thermal conductivity depends on the nature of the material through which the heat is flowing and is affected by physical conditions such as the temperature. It is expressed in $\mathrm{W\,m^{-1}\,K^{-1}}$. From Eq. (3.6) it follows that within an isotropic body, heat flow is a vector quantity with its direction normal to the surface of constant temperature. Heat flow is expressed in $\mathrm{mW\,m^{-2}}$.

In geothermics, one often has to deal with media having anisotropic thermal properties. This is particularly common while dealing with the problems of heat flow in sedimentary rocks where the properties in the bedding plane (plane of sedimentation) tend to differ from properties perpendicular to it. Similarly, one often has to deal with crystals and rock-forming minerals belonging to different systems, such as the monoclinic (feldspar, mica), orthorhombic (pyroxene, olivine), hexagonal (quartz, ilmenite) and other systems.

In an anisotropic media, the thermal conductivity could be represented by a symmetrical tensor of the second order of the type

$$\begin{bmatrix} k_{11} & k_{12} & k_{13} \\ k_{21} & k_{22} & k_{23} \\ k_{31} & k_{32} & k_{33} \end{bmatrix} \tag{3.7}$$

Assuming each component of the heat-flow vector to be linearly dependent on all components of the temperature gradient at that point, we could write the heat flow in the three mutually perpendicular directions for an anisotropic media as (Kappelmeyer and Haenel, 1974):

$$\begin{aligned} q_x &= -\left(k_{11}\frac{\partial T}{\partial x} + k_{12}\frac{\partial T}{\partial y} + k_{13}\frac{\partial T}{\partial z}\right) \\ q_y &= -\left(k_{21}\frac{\partial T}{\partial x} + k_{22}\frac{\partial T}{\partial y} + k_{23}\frac{\partial T}{\partial z}\right) \\ q_z &= -\left(k_{31}\frac{\partial T}{\partial x} + k_{32}\frac{\partial T}{\partial y} + k_{33}\frac{\partial T}{\partial z}\right) \end{aligned} \tag{3.8}$$

In the case of anisotropic layers, conductivity being different in the three mutually perpendicular directions coinciding with the x-, y- and z-axes, Eq. (3.8) simplifies to

$$\begin{aligned} q_x &= -k_1\frac{\partial T}{\partial x} \\ q_y &= -k_2\frac{\partial T}{\partial y} \\ q_z &= -k_3\frac{\partial T}{\partial z} \end{aligned} \tag{3.9}$$

In Eq. (3.9), k_1, k_2 and k_3 are conductivities in the x, y and z directions, respectively. In case the conductivity is the same in the xy plane ($k_1 = k_2$) and different in the z direction, a case seen very often in sedimentary rocks, the heat flow could be expressed as

$$\begin{aligned} q_x = q_y &= -k_1\frac{\partial T}{\partial x} = k_1\frac{\partial T}{\partial y} \quad \left(\text{since } \frac{\partial T}{\partial x} = \frac{\partial T}{\partial y}\right) \\ q_z &= -k_3\frac{\partial T}{\partial z} \end{aligned} \tag{3.10}$$

Differential Equation of Heat Conduction

Fourier's equation (3.6), with the energy conservation law, can be used to derive a differential equation describing the temperature field in a medium. In other words, the equation is the mathematical expression of the fact that the rate of increase of heat content of a small volume should be equal to the sum of the rate of heat generation in it and the rate of flow of heat into it across its surface. For the Earth, the rate of heat generation per unit volume could represent the effects of radioactive decay, phase change, frictional heating or chemical reaction. For a material of constant conductivity (isotropic) with a constant rate of heat generation A per unit time per unit volume, the differential equation can be written as (Carslaw and Jaeger, 1959):

$$\nabla(-k\nabla T) = A - \rho c \frac{\partial T}{\partial t} \tag{3.11}$$

where ρ is density, c the specific heat capacity at constant pressure, the product (ρc) is called the volumetric specific heat capacity, t the time. Eq. (3.11) can be simplified to

$$\nabla^2 T - \frac{1}{\alpha}\frac{\partial T}{\partial t} = -\frac{A}{k} \tag{3.12}$$

where $\alpha = k/\rho c$ is called thermal diffusivity; it is expressed in $\mathrm{m^2\,s^{-1}}$. In the one-dimensional case, for heat flow in the z direction, Eq. (3.12) reduces to

$$\frac{\partial^2 T}{\partial z^2} - \frac{1}{\alpha}\frac{\partial T}{\partial t} = -\frac{A}{k} \tag{3.13}$$

If there are no heat sources within the small volume, i.e., $A = 0$, Eq. (3.13) reduces to the Poisson's equation

$$\frac{\partial^2 T}{\partial z^2} = \frac{1}{\alpha}\frac{\partial T}{\partial t} \tag{3.14}$$

Further, at steady state, $\frac{\partial T}{\partial t} = 0$, Eq. (3.14) reduces to the Laplace's equation

$$\frac{\partial^2 T}{\partial z^2} = 0 \tag{3.15}$$

The differential equation of heat conduction (3.12) can be modified in different cases such as variable conductivity, or water flow with a finite velocity, and can be solved for understanding heat transfer processes in a wide variety of situations inside the Earth.

HEAT CONVECTION

Within fluids, the heat transfer takes place through a combination of molecular conduction and energy transportation created by the motion of fluid particles. This mode of heat transfer is known as *convection.* The heat exchange rate in fluids by convection is much higher than the heat exchange rate in solids through conduction. This difference becomes more prominent in geothermics because rocks have very low-thermal conductivities compared to metals and other solids.

Convection processes inside the Earth can be of two broad types: free and forced. *Free* or *natural convection* refers to the free motion of a fluid and is solely due to differences in the densities of the heated and cold particles of a fluid. The origin and intensity of free convection are solely determined by the thermal conditions of the process and depend on the kind of fluid, temperature, potential and volume of the space in which the process takes place. *Forced convection* occurs under the influence of some external force. Flow of water in hot springs and heat transport due to volcanic eruptions are examples of forced convection (advection). Forced convection depends on the physical properties of the fluid, its temperature, flow velocity, shape and size of the passage in which forced convection of fluid occurs. Generally speaking, forced convection may be accompanied by free convection, and the relative influence of the latter increases with the difference in the temperatures of individual particles of the fluid and decreases with the velocity of the forced flow. The influence of natural convection is negligible at high-flow velocity.

In problems dealing with the transmission of heat through the process of convection, the fluid under consideration is usually bounded on one or more sides by a solid. Let at any given time, T_s be the temperature of the solid at its boundary with the fluid and T_∞ the fluid temperature at a far-off yet unspecified point. In accordance with Newton's law of cooling, the amount of heat flowing would be proportional to the temperature difference and could be expressed as

$$q = h(T_s - T_\infty) \qquad (3.16)$$

where h is the heat transfer coefficient. The heat is transferred by convection and consequently the heat transfer coefficient depends, in general, upon the thermal boundary condition at the solid–fluid boundary. However, under many situations, h can be estimated satisfactorily when the fluid dynamics of the flow system is known.

When fluid flows past a stationary solid, the fluid viscosity causes the molecules adjacent to the solid to adhere to the solid surface and consequently velocity increases from zero at the solid surface to U_∞ at a far-removed point at the free stream. The velocity U reaches U_∞ asymptotically in accordance with the scheme shown in Fig. 3.1A. For practical applications, it is customary to define a layer thickness, S, where U attains an arbitrary value (say, 0.9 U_∞). When the viscosity is low, S is usually small compared to the dimensions of the *external flow* systems (e.g., a system in which the fluid medium extends to much larger dimensions compared to

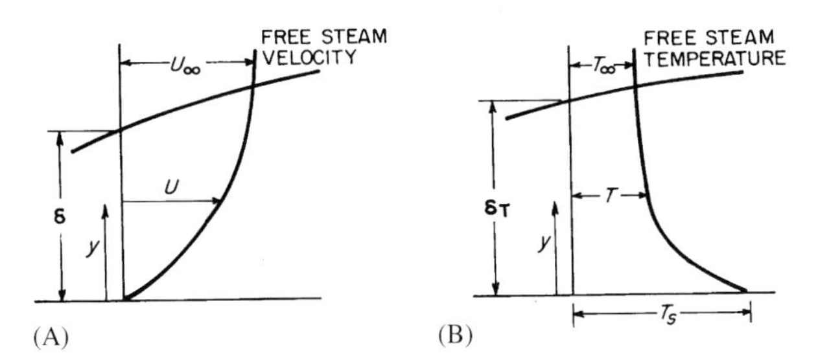

Fig. 3.1. Flow past a stationary wall. (A) Velocity boundary layer. (B) Temperature boundary layer.

the solid boundary area under consideration such as the flow over turbine blades, aircraft wings, etc.). In *internal flow* systems (e.g., flow in pipes, and ducts) the boundary layers are thin compared with the inlet aperture for a limited distance only and as the distance from the entry point increases, the boundary layers merge, leaving no free stream region.

The temperature variations for a fluid flowing past a solid wall behave very similarly to the velocity variation discussed above. Fig. 3.1B shows schematically the temperature variations from the wall to the free stream temperature. Because of the fluid particles adhering to the wall, the heat from the wall could only be transmitted through molecular conduction and could be expressed as

$$q = -k\left(\frac{\partial T}{\partial n}\right)_{\mathrm{s}} \tag{3.17}$$

where $(\partial T/\partial n)_{\mathrm{s}}$ is the temperature gradient of the fluid, normal to the solid surface. When the temperature field is known, the heat transfer coefficient could be estimated from

$$h = -k\left(\frac{(\partial T/\partial n)_{\mathrm{s}}}{T_{\mathrm{s}} - T_{\infty}}\right) \tag{3.18}$$

For external flow systems, T_∞ is the uniform free stream temperature and for the internal flow systems it represents the average of the bulk temperature of the fluid. Direct determinations of h from Eq. (3.18) are rare and it is mostly estimated from empirical or semi-empirical solutions of the fluid behavior.

In the general three-dimensional case of flow involving interaction between temperature field and flow, it is necessary to know the velocity in the three directions, temperature, pressure, density, viscosity and thermal conductivity of the fluid for an adequate description of the fluid behavior. To reduce complexities, convection problems are often solved assuming a steady state flow where the dependent variables at a place do not change with time. This simplification renders derivation of convection equations less difficult and is a fairly accurate description of a majority of thermal convection problems.

Estimates of heat transfer coefficients under certain simple physical conditions, as given by several authors, have been compiled by Kappelmeyer and Haenel (1974). Some of the commonly used cases are listed below.

At a surface of a square copper sheet of length of 0.5 m with air flowing parallel to the surface with a velocity U between 0 and $25\,\mathrm{m\,s^{-1}}$, the heat transfer coefficient h ($\mathrm{W\,m^{-2}\,K^{-1}}$) is estimated to be (Nusselt and Jurges, 1922)

$$h = 4.18\,(1.71\,U^{0.78} + 1.28\,U^{-0.6U}) \tag{3.19}$$

In case of a very rough surface and flow velocity of air exceeding $5\,\mathrm{m\,s^{-1}}$, h would be (Hiramatsu and Kokado, 1958)

$$h_{\mathrm{rough}} = 4.42\,(1.71\,U^{0.78} + 1.28\,U^{-0.6U}) \tag{3.20}$$

In geothermics, one often needs to calculate convective heat transfer in mine shafts and galleries. Stoces and Cernik (1931) have estimated the coefficient of surface heat transfer between rocks and the air in galleries as

$$h = 4.18\,(0.556 + 0.278\ U.h') \tag{3.21}$$

where h' varies from 5 to 10. Hiramatsu and Kokado (1958) have done some more sophisticated calculations taking into account the air resistivity at the wall of the galleries and the coefficient of friction in the air and found

$$h = 14.52\frac{U_{\mathrm{m}}\mu K_{\mathrm{a}}}{v} \tag{3.22}$$

where U_{m} is air velocity averaged over the section of the gallery, μ the coefficient of friction in the air, K_{a} the thermal conductivity of the air, and v the kinematic viscosity of air.

Experimental measurements carried out in mines have shown that the above relations yield reasonable values for the coefficients of surface heat transfer in several cases.

HEAT RADIATION

In the previous sections, we have discussed the transfer of heat through conduction and convection, the two processes requiring presence of a medium. The means by which energy is transmitted between bodies without contact and in the absence of intervening medium is known as *radiation.* Transmission of energy through radio waves, visible light, X-rays, cosmic rays, etc., all belong to this category, having different frequencies in the spectrum of *electromagnetic radiation.* Here we are concerned with the type of radiation which is principally dependent on the temperature of the body, known as *thermal radiation* and belonging mostly to the infrared and to a small extent to the visible portion of the electromagnetic radiation spectrum. The heat transferred into or out of an object by thermal radiation is a function of several components. These include its surface reflectivity, emissivity, surface area, temperature and geometric orientation with respect to other thermally participating objects. In turn, an object's surface reflectivity and emissivity is a function of its surface conditions (roughness, finish, etc.) and composition.

To account for a body's outgoing radiation (or its *emissive power*, defined as the heat flux per unit time), one makes a comparison to a perfect body, which absorbs the entire amount of heat radiation falling on its surface as well as emits the maximum possible thermal radiation at any given temperature. Such an object is known as a black body. The concept of *black body* is important in understanding the radiation of heat. According to Stefan–Boltzmann's law, heat emitted by a black body at any given temperature, q_b ($W\,m^{-2}$), is expressed as follows for a unit area in a unit time:

$$q_b = \sigma T^4 \tag{3.23}$$

where q_b is the heat flow through radiation from the surface of a black body, T the temperature, and σ a constant known as the Stefan–Boltzmann constant, with a theoretical value of $5.67 \times 10^{-8}\,W\,m^{-2}\,K^{-4}$. Because no material ideally fulfills the properties of absorption and emission of the theoretically defined black body, for practical purposes a new constant of emissivity, ε, is defined for real surfaces as

$$\varepsilon = \frac{q}{q_b} \tag{3.24}$$

q being the radiant heat from a real surface. The term ε generally implies total emissivity; measurements made normal to the radiating surface yield normal emissivity (ε_n). In Table 3.3, the normal emissivity of a few solids is listed. From Eqs. (3.23) and (3.24) the radiant heat flow from the surface of body with emissivity ε is given by

$$q = \varepsilon\sigma T^4 \tag{3.25}$$

The heat radiated in a unit time from a unit area of an isothermal surface A to another isothermal surface B with temperatures θ_A and θ_B and emissivities ε_A and ε_B would be (Kappelmeyer and Haenel, 1974)

$$q_{AB} = b_{AB}\left(T_A^4 - T_B^4\right) \tag{3.26}$$

The constant b_{AB} depends upon the geometry of isothermal surfaces A and B and their configuration with respect to one another. When the two isothermal surfaces

Table 3.3. Normal emissivity, ε_n, of a few solids

Material	θ (°C)	ε_n
Gold, polished	130	0.018
Silver	20	0.020
Copper, polished	20	0.030
Copper, black oxidized	20	0.78
Aluminum paint	100	0.20–0.40
Iron, bright abrased	20	0.24
Iron, red rusted	20	0.61
Iron, heavily rusted	20	0.85
Zinc, gray oxidized	20	0.23–0.28
Lead, gray oxidized	20	0.28
Clay, fired	70	0.91
Porcelain	20	0.92–0.94
Glass	90	0.94
Paper	95	0.92
Wood	70	0.935
Tar paper	20	0.93

are parallel to each other and are separated by a small distance compared to their dimensions, b_{AB} becomes

$$b_{AB} = \frac{\sigma}{\frac{1}{\varepsilon_1} + \frac{1}{\varepsilon_2} - 1} \tag{3.27}$$

When surface B encloses A, b_{AB} becomes

$$b_{AB} = \frac{\sigma}{\frac{1}{\varepsilon_1} + \frac{A}{B}\left(\frac{1}{\varepsilon_2} - 1\right)} \tag{3.28}$$

In case, a body with emissivity ε and temperature T is surrounded by a black body at temperature T_0, the approximate heat loss per unit area per unit time is given by

$$q = 4T_0^3\sigma\varepsilon(T - T_0) \tag{3.29}$$

Relation (3.29) is used quite often in geothermics. In accordance with this relation, with an average terrestrial heat flow of $80\,\mathrm{mW\,m^{-2}}$, and assuming that this entire heat is completely transferred to space by radiation, the temperature rise at the Earth's surface caused by the flow would be (with $T_0 = 288\,\mathrm{K}$, $\varepsilon = 0.9$, $\sigma = 5.67 \times 10^{-8}\,\mathrm{W\,m^{-2}\,K^{-4}}$)

$$T - T_0 = \frac{q}{4T_0^3\sigma\varepsilon} = 0.0164\,\mathrm{K} \tag{3.30}$$

Since the heat loss at the Earth's surface is not only through radiation, but also by convection, the real temperature rise observed at the Earth's surface due to terrestrial heat flow is normally found to be less than 0.016 K.

TEMPERATURE ESTIMATES IN SOME SIMPLE GEOLOGICAL SITUATIONS

Mathematical representation and analysis of precise thermal behavior within the Earth are extremely difficult. Modeling is helpful in estimating gross properties and geothermal behavior under varying geological conditions. In the following we estimate the temperatures at shallow crustal depths under some very simplified assumptions.

Let us first consider a geothermally undisturbed section of crust, i.e., a section not disturbed by recent volcanic activity or any other tectonic or mechanical activity capable of disturbing the temperature field. Assuming that the thermal conductivity k and heat production A_o due to radioactivity, etc., are both dependent on depth, z, the temperature variation with depth could be expressed as

$$\frac{\mathrm{d}}{\mathrm{d}z}\left[k(z)\frac{\mathrm{d}T}{\mathrm{d}z}\right] = A_0(z) \tag{3.31}$$

In the absence of a source of heat ($A_0 = 0$), the temperature T_z at depth z could be expressed as

$$T(z) = T_0 + q\int_0^z \mathrm{d}z/k(z) \tag{3.32}$$

where q is the terrestrial heat flow and T_0 the surface temperature.

One of the geological settings often encountered is that of a layered structure (e.g., a sedimentary rock country) in the crust. If the layer boundaries were to be horizontal and thermal conductivity in each of these layers were to be constant, the temperature at the bottom of any layer m would be

$$T(z_m) = T\left(\sum_{i=1}^{m} d_i\right) + T_0 + q\sum_{i=1}^{m}\frac{d_i}{k_i} \tag{3.33}$$

where d_i is the thickness and k_i the thermal conductivity of the ith layer. Suppose there were two layers of thicknesses d_1 and d_2 and conductivities k_1 and k_2 with horizontal interfaces and overlying a half space of conductivity k_3, then, in accordance with Eq. (3.33), the temperature at depth z would be

$$T(z) = T_0 + q\frac{z}{k_1} \quad \text{for } 0 \leqslant z \leqslant d_1$$

$$T(z) = T_0 + q\left(\frac{d_1}{k_1} + \frac{z - d_1}{k_2}\right) \quad \text{for } d_1 \leqslant z \leqslant (d_1 + d_2) \tag{3.34}$$

$$T(z) = T_0 + q\left(\frac{d_1}{k_1} + \frac{d_2}{k_2} + \frac{z - d_2}{k_3}\right) \quad \text{for } z \geqslant (d_1 + d_2)$$

In case the anomalous heat flow observed at the Earth's surface could entirely be attributed to the heat production contrast of an anomalous body underground, the relation derived by Simmons (1967) could be applied. The relation, analogous to the one between gravity and magnetics, is simply

$$\Delta q = \frac{gA}{2\pi G\rho} \tag{3.35}$$

where q is the heat flow produced by the anomalous body, g the acceleration due to gravity, A the contrast in heat production, G the gravitational constant, and ρ the density.

Relations obtained for some geometrically simple forms from Eq. (3.35) are as follows:

(A) *Sphere*

$$\Delta q = \frac{2}{3}\frac{Ar^3}{z^2}\left[1 + \left(\frac{x}{z}\right)^2\right]^{-3/2} \tag{3.36}$$

(B) *Vertical thin cylinder*

$$\Delta q = \frac{Ar^2}{2z}\left[1 + \left(\frac{x}{z}\right)^2\right]^{-1/2} \tag{3.37}$$

(C) *Vertical sheet*

$$\Delta q = \frac{Ah}{2\pi}\log\left[\left(1 + \left(\frac{x}{z_2}\right)^{-2}\right) \Big/ \left(1 + \left(\frac{x}{z_1}\right)^{-2}\right)\right] \tag{3.38}$$

The various notations used in Eqs. (3.36)–(3.38) are expressed in Fig. 3.2. Many more relations for simple two-dimensional as well as three-dimensional geothermal figures are given by Simmons (1967).

We shall now consider two examples of more realistic geological situations. In investigating geothermal fields over extended regions, the assumption of lateral homogeneity of thermal conductivity is often not valid. Changes in lateral conductivity affect the temperature field. When these changes could be approximated by simple geometrical boundaries, the corresponding temperature field could be estimated by application of mathematical approximation methods and use of modern computers.

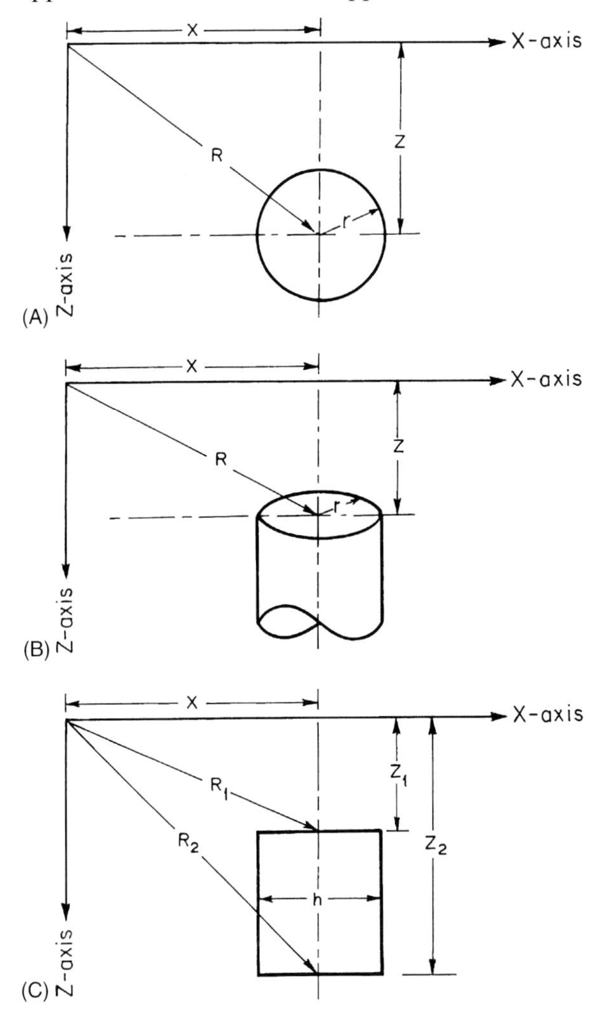

Fig. 3.2. Geometrical parameters used in relations (Eqs. 3.36 to 3.38) for (A) sphere, (B) vertical cylinder and (C) vertical sheet.

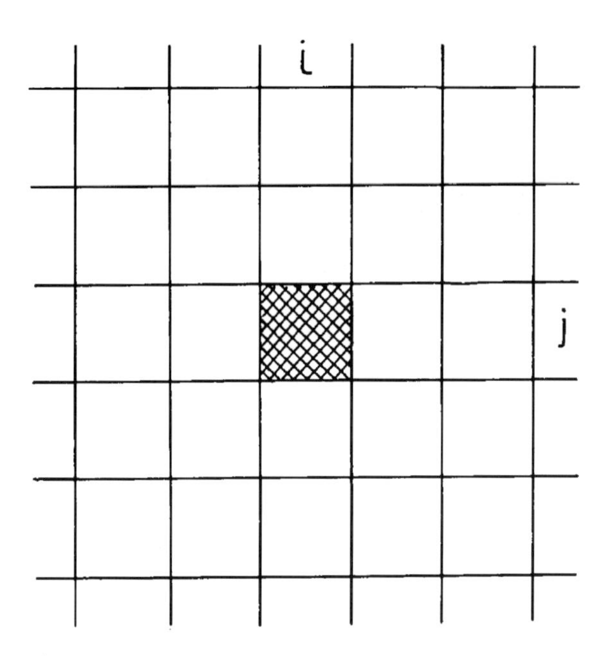

Fig. 3.3. Square grid pattern for calculation of temperature field using finite difference method.

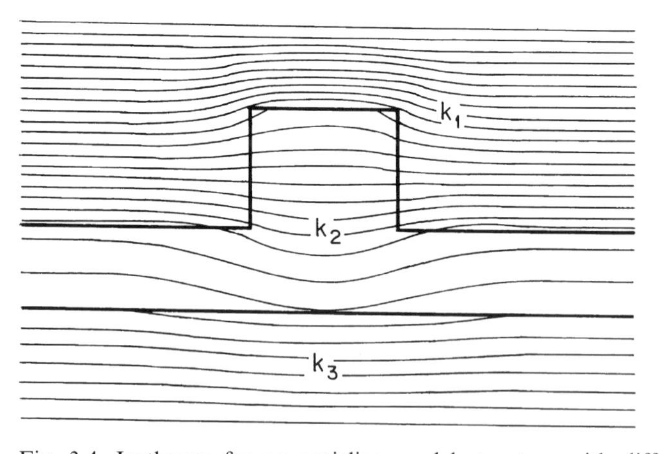

Fig. 3.4. Isotherms for an anticline model structure with different thermal conductivities (after Mundry, 1966).

In the finite difference method, the model is divided into a grid pattern. Let us consider the case when the model is divided into a square grid pattern with the length of a unit being Δd (Fig. 3.3). Using the Taylor series expansion, the derivatives occurring in the heat-diffusion equation are replaced by their finite differences. Let

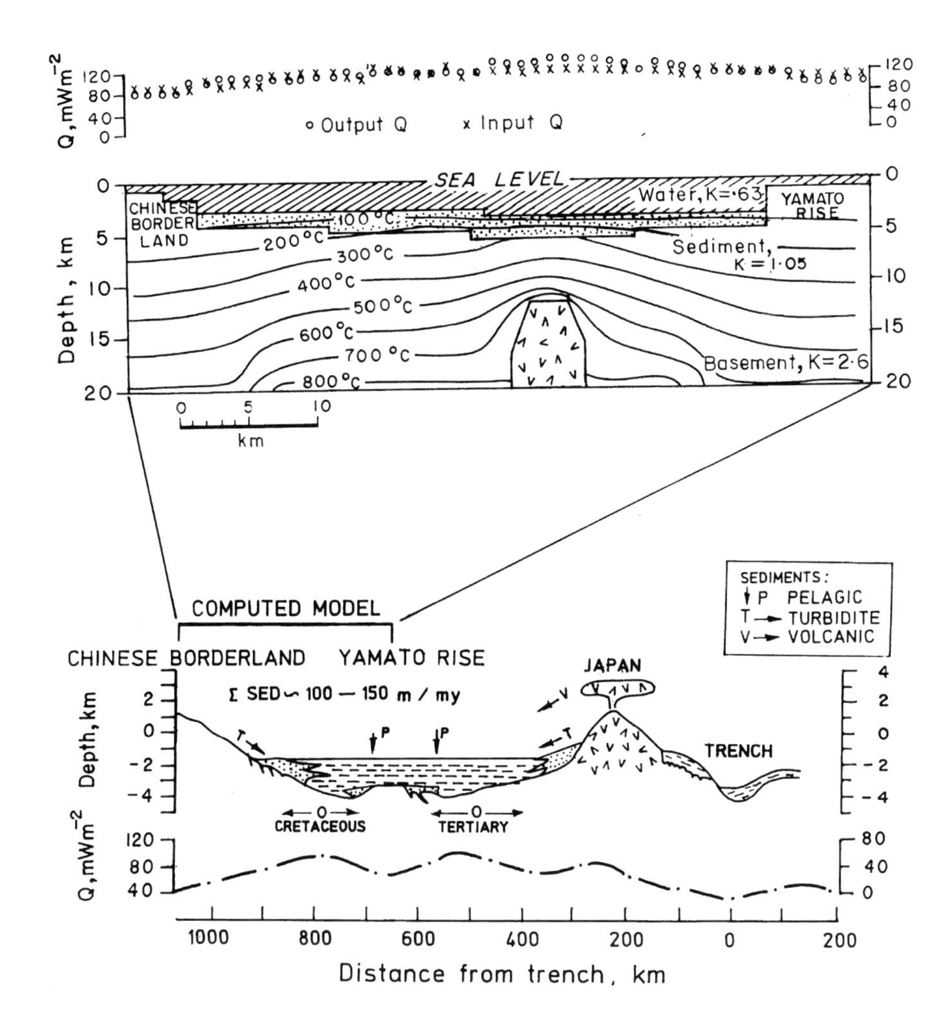

Fig. 3.5. Below: Generalized cross-section from Chinese border land through the Japan trench showing relevant geological features. Above: Computer-modeled, hand contoured plot of subsurface isothermal distribution beneath the Japan abyssal plain (modified from Schlanger and Combs, 1975).

$T_{i,j,n}$ denote the temperature of the grid square (i,j) at a time $n\Delta t$ ($n = 0, 1, 2, \ldots$, and Δt is the time interval), and $\Delta T_{i,j,n+1}$ the increase in $T_{i,j,n}$ during next Δt time. $\Delta T_{i,j,n+1}$ can be calculated using the following relation (Hagiwara, 1966)

$$\Delta T_{i,j,n+1} = \frac{k\Delta t}{\rho c(\Delta d)^2}\left[T_{i-1,j,n} + T_{i+1,j,n} + T_{i,j+1,n} + T_{i,j-1,n} - 4T_{i,j,n}\right] \tag{3.39}$$

where k is thermal conductivity, ρ the density, c the specific heat, and Δd the length of the grid square.

The isotherms for an anticlinal strucure (adapted from Mundry, 1966) are shown in Fig. 3.4. The middle deformed layer has the highest conductivity, causing the crowding of isothermal surfaces in the top and bottom layers of lower conductivity. The geometrical shape and thermal conductivity contrast adapted in Fig. 3.4 is akin to a salt dome type of geological situation. It may be noted that the temperature distribution within the good conductor depends upon the conductivity contrast and the thickness of the overburden. Another example involving complex geology is presented in Fig. 3.5 (modified from Schlanger and Combs, 1975). To investigate the temperature structure within the sedimentary prism in the Japan Basin, Schlanger and Combs (1975) used a finite difference approximation of the heat conduction equations and solved the resulting set of simultaneous equations by the method of successive overrelaxation. The thermal conductivity structure between the Chinese borderland and Yamato Rise was estimated from the isopach data and the geometry of the sedimentary prism. Finally, the subsurface isotherms were computed assuming a temperature of 4 °C at the sediment–water interface and 800 °C for the intrusive body so as to conform to the observed heat-flow profile over the Earth's surface.

Chapter 4

GEOTHERMAL SYSTEMS AND RESOURCES

INTRODUCTION

A geothermal resource can be simply defined as a reservoir inside the Earth from which heat can be extracted economically (cost wise less expensive than or comparable with other conventional sources of energy such as hydroelectric power or fossil fuels) and utilized for generating electric power or any other suitable industrial, agricultural or domestic application in the near future. A geothermal reservoir can contain heat both in the solid rock as well as in the fluids that fill the fractures and pore spaces within the rock. Estimates of geothermal resources are made on the basis of geological and geophysical data such as (i) depth, thickness and extent of geothermal aquifers, (ii) properties of rock formations, (iii) salinity and geochemistry of fluids likely present in the aquifers, and (iv) temperature, porosity and permeability of rock formations (Rummel and Kappelmeyer, 1993). A geothermal resource is distinct from a geothermal reserve, which refers to the part of a resource that can be extracted economically at the present price level. Reserves are confirmed on the basis of detailed reservoir datasets obtained invariably by deep drilling into potential resource areas. Therefore, the main factors in estimating reserves are the cost of drilling and the quality of available data on subsurface rock formations.

Radioactive decay of long-lived isotopes, particularly those of potassium, uranium and thorium, continuously generates heat within the Earth. The heat content of the Earth is estimated to be 1.3×10^{31} J. Heat is lost from the Earth's surface at an average rate of $\sim$80 mW m^{-2}. In most areas, this heat reaches the Earth's surface in a diffuse state, making it uneconomical to exploit this vast heat resource. It is believed that heat transfer below the lithosphere is mostly by convection and in the lithosphere by conduction. Rocks are relatively poor conductors of heat. Below nine-tenths of the Earth's surface, the thermal gradients vary from about 10 °C km^{-1} to about 60 °C km^{-1}. Consequently, the temperatures encountered at depths of a few kilometers are in excess of the Earth's surface temperatures by only an order of 100 °C and power generators working at such small temperature differences have very low efficiency. Recent advances in binary-cycle geothermal plant technologies have made it possible to utilize such moderate temperature resources at shallow depths for low-to-moderate scale power generation. With the improvement of drilling technology, it may become economical to drill deeper, say in excess of 5 km, and reach the required high temperatures of the order of 200–300 °C. Nevertheless, it is not sufficient only to reach the hot rock. Additionally, heat needs to be extracted by circulating fluids and for this process to be effective, there should exist an abundance of pore space and fissures for the fluids to permeate and circulate. At greater depths, the weight of the overlying rocks tends to close the pores and fissures, reducing the permeability considerably. Therefore, although hot rocks at depth exist almost everywhere, the current technological barriers, as well as unsuitable geological conditions make it uneconomical to extract geothermal energy at most places.

Under some geological situations, however, such as the plate boundaries and sometimes well within the plates (as e.g., in active or geologically young volcanoes associated with mantle hotspots such as in the Hawaii island), heat may be locally transferred within a few kilometers of the Earth's surface through the process of convection by magma or molten rocks. The magma has temperatures in the neighborhood of 1000 °C and interacts with the near-surface rocks, causing surficial manifestations of geothermal activity such as the hot springs, geysers and fumaroles. Under certain suitable geological conditions, the heat becomes trapped, forming heat reservoirs. In such areas, after drilling a few hundred meters, temperatures of the order of 200–300 °C are found and the regions could be suitable for harnessing the geothermal energy. The major producing geothermal fields of the world exploit such situations.

TYPES OF GEOTHERMAL SYSTEMS

The essential requirements for a geothermal system to exist are (1) a large source of heat, (2) a reservoir to accumulate heat, and (3) a barrier to hold the accumulated heat. There is a suite of geological conditions that could result in a variety of geothermal systems. Consequently, all geothermal fields differ from one another. However, depending upon certain common characteristics, these can broadly be classified into the following categories: (1) vapor-dominated, (2) hot water, (3) geopressured, (4) hot dry rock (HDR), and (5) magma. The above-mentioned categories are briefly discussed here.

Vapor-Dominated Geothermal Fields

Most of the presently exploited geothermal fields contain water at high pressures, and temperatures in excess of 100 °C. When this water is brought to the Earth's surface, the pressure is considerably reduced, generating large quantities of steam, and a mixture of saturated steam and water is produced. The ratio of steam to water varies from one site to another. Some of the best-known geothermal fields, such as Cerro Prieto (Mexico), Wairakei (New Zealand), Reykjavik (Iceland), Salton Sea (U.S.A.) and Otake (Japan), belong to this category. Since steam associated with water is produced in these fields, they are known as wet steam fields. There are a few other important geothermal fields such as Larderello (Italy) and The Geysers

(U.S.A.) which produce superheated steam with no associated fluids. Fields of the latter type are known as dry steam fields.

The basic requirements of the vapor-dominated geothermal fields, whether of dry steam or of wet steam type, include adequate supplies of water in addition to the three prerequisites mentioned earlier. Fig. 4.1 is a schematic model representation of a vapor-dominated field.

Heat source

The fact that vapor-dominated geothermal fields are situated in regions of recent (Miocene–Quaternary) volcanism, some of them being located on or close to volcanoes, has verified that magma is their source. Young, high-temperature (500–1000 °C) magma intrusions to within depths of a few to several kilometers from the Earth's surface allow the necessary heat to be accumulated in economical quantities. In hard compact rocks, faulting may provide a channel for the magma to reach the surface. Soft or plastic rocks, when present, can flow and block the fault space, causing the magma to spread at the contact between the soft and the hard rocks. Active volcanoes, fumaroles, hot springs and geysers are obvious surface manifestations of recent volcanic activity. In addition, certain geological environments, such as regions of Quaternary uplift and regions of Late Tertiary and Quaternary subsidence, are indicative of shallow magmatic intrusions.

Reservoir and water supply

In order to form a heat reservoir, the anomalous magmatic intrusion should encounter porous and permeable, water-filled rock strata. Within the reservoir, convection currents of hot water and/or steam are set up, providing a good heat exchange, and the temperature difference between the top and bottom of the reservoir is not very significant. A variety of rocks have been found to constitute good reservoirs. At Larderello (Italy), it is fractured limestone and dolomite; at The Geysers (U.S.A.), it is fissured graywacke; at Wairakei (New Zealand), it is pumiceous breccia and tuff; and at Cerro Prieto (Mexico), it is deltaic sands. Good reservoirs could also be formed at geological unconformities and formation boundaries, provided that they are permeable and have good hydraulic continuity and water supply.

The origin of geothermal fluids has been debated in the past. In addition to a meteoric origin, magmatic and juvenile origins for geothermal fluids have been suggested. However, recently conducted isotopic studies in geothermal fields have shown that at least 90% of the geothermal water has a meteoric origin. The permeable aquifers forming the reservoir must therefore have hydraulic continuity with large recharge areas for the rainwater to be available in continuous supply. The freshly supplied water is heated conductively at the impermeable base of the reservoir. Withdrawing of the heated reservoir fluid through boreholes, or its upward movement through vents and fissures, disturbs the hydrological balance. This is

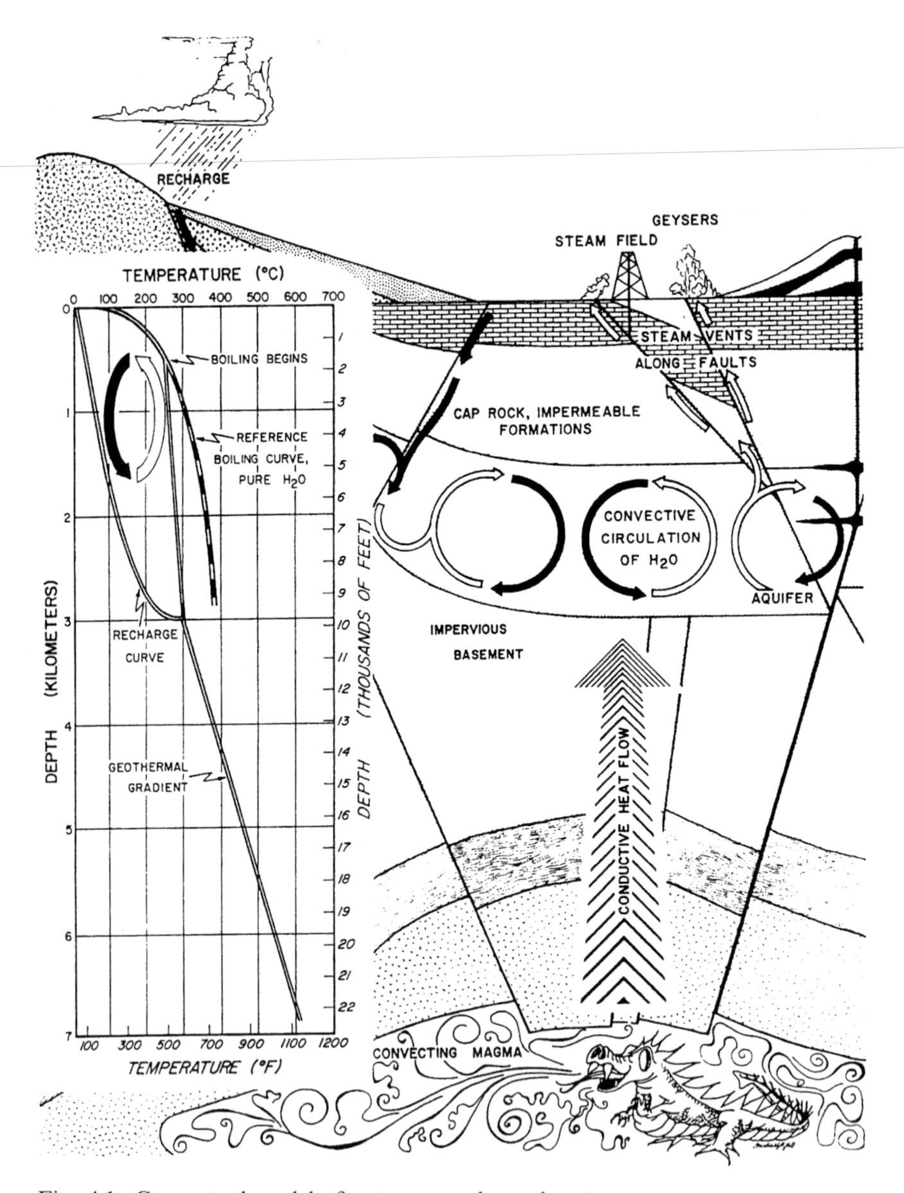

Fig. 4.1. Conceptual model of a steam geothermal system.

restored, fully or partially, by the inflow of new water. An idea of the amount of the inflow can be had from the fact that a natural steam field operating a 100 MW power plant lets out between 1,000 and 2,000 tons of water every hour. Some of the geothermal fields, such as the Larderello in Italy, have easily identifiable recharge areas.

At Larderello, the permeable reservoir terrain, consisting of Mesozoic limestones and dolomites, outcrops thereby providing an easy access to surficial water.

Cap rock—the barrier

An impermeable cap rock, or a cap rock with low permeability, overlying the reservoir, is necessary to prevent the escape of hot reservoir fluids through convection. The heat loss through conduction is not prevented by the cap rock. However, the amount of heat conducted is much smaller than that which could be lost through possible convection. Since volcanism is associated with tectonic movements causing fissures, ideal unfissured impermeable cap rock is nowhere to be found. The geochemical processes associated with geothermal fields, i.e., hydrothermal alteration of rocks and mineral deposition, are helpful in sealing off the fissures. Typical examples of cap rocks rendered impermeable through chemical action and deposition are seen at The Geysers and Otake geothermal fields. At The Geysers, calcite- and silica-filled fractures, up to 1 in. wide, are commonly seen. Evidence of hydrothermal alteration is presented by the bleaching of graywacke as well as by the absence of vegetation in patches. The geochemical and hydrothermal processes are complicated and vary from place to place.

At many other steam-producing fields, original impervious rocks constitute the cap rock. Examples are the lacustrine Huka Formation at Wairakei (New Zealand), the deltaic clay at Cerro Prieto (Mexico) and Salton Sea (California) and the Flysch Formation at Larderello (Italy).

Hot Water Geothermal Systems

In hot water geothermal fields, water-convection currents carry the heat from the deep source to the shallow reservoir. The bottom of the convective cell may be heated through conduction from hot rocks. The geology of hot water geothermal fields is quite similar to that of an ordinary groundwater system. They differ from the earlier-discussed vapor-dominated geothermal fields in the fact that the hot water geothermal fields are characterized by liquid water being the continuous pressure-controlling fluid phase. Typically, the temperature of hot-water reservoirs varies from 60 to 100 °C and they occur at depths ranging from 1500 to 3000 m. Fig. 4.2 is a schematic representation of a hot-water field and the reference curve represents the variation of the boiling point of the pure water with depth. As shown in Fig. 4.2, a hot water geothermal field could develop in the absence of a cap rock, if the thermal gradients and the depth of the aquifer are adequate to maintain a convective circulation. When the cap rock is absent, the temperatures in the upper part of the reservoir cannot exceed the boiling point at atmospheric pressure, since with the convective rise, the water loses pressure and also becomes mixed with the cool groundwater.

Depending upon the temperature, chemistry and the structure of the reservoir, hot-water systems have been classified into several subtypes (White, 1974). The following is a brief description of various subtypes:

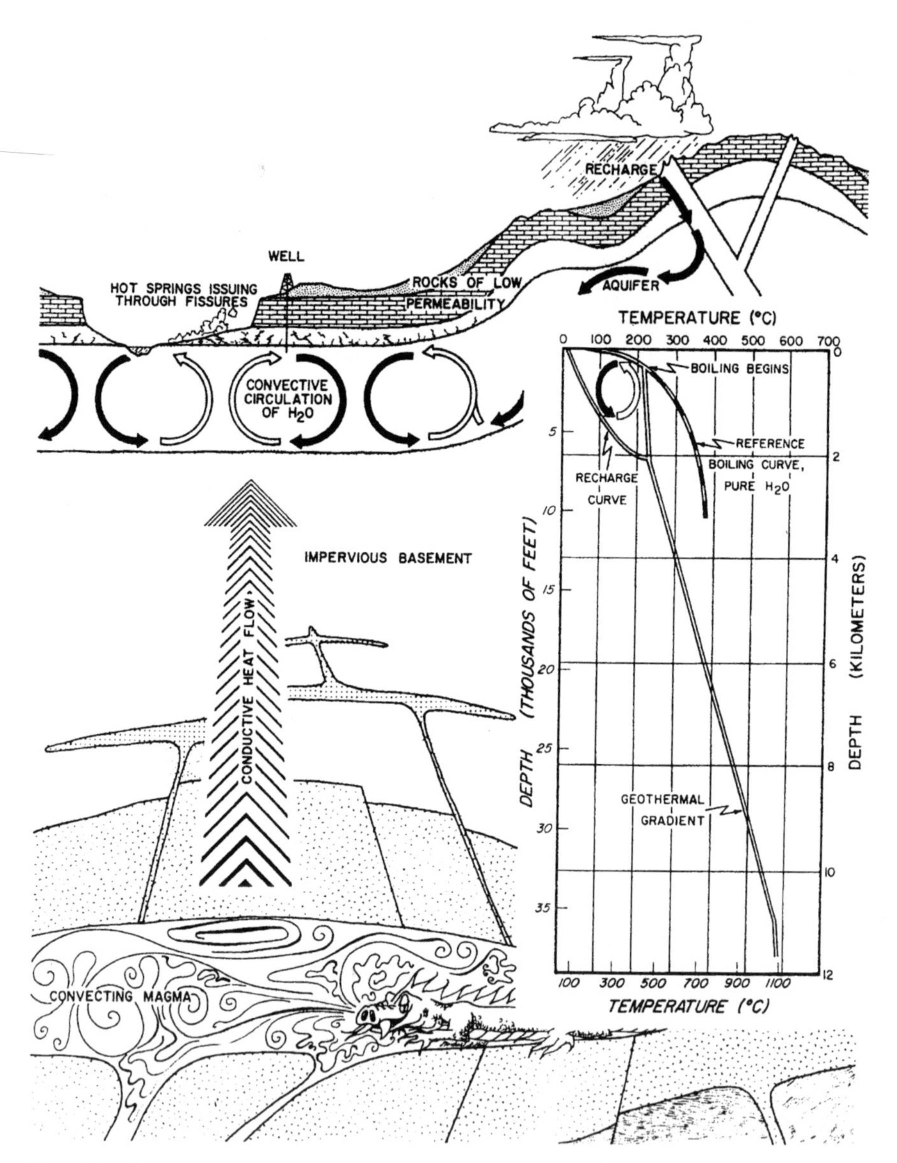

Fig. 4.2. Conceptual model of a hot water geothermal system.

(1) Systems characterized by low-to-moderate temperatures (say, 50–150 °C) and producing water with a chemical composition similar to the regional surface and shallow groundwaters.
(2) Systems characterized by the presence of partly non-meteoric water. Such systems usually occur in deep sedimentary basins.

(3) Systems characterized by the presence of brine of very high salinity. The chemistry can vary considerably from one field to another. The Salton Sea (California) and the Red Sea brine pools belong to this subtype and have very differing bulk chemistry of the sediments and the associated rocks, probably attributable to the difference in brine composition.
(4) Systems characterized by the presence of natural cap rocks. Geothermal fields at Cerro Prieto (Mexico) and Salton Sea (Califomia) have cap rocks constituted by fine-grained, low-permeability sediments.
(5) Systems characterized by the creation of their own self-sealing cap rocks. As explained earlier in discussing vapor-dominated geothermal systems, these cap rocks are formed through chemical alteration and deposition of sediments near the surface where the temperature decreases suddenly. Wairakei (New Zealand) and Yellowstone Park (Wyoming) are typical examples.

Geopressured Geothermal Resources

A type of hydrothermal environment whose hot water is almost completely sealed from exchange with surrounding rocks is called a geopressured system (Jones, 1970; Duffield and Sass, 2003). Such systems typically form in a basin in which very rapid filling with sediments takes place, resulting in higher than normal pressure of the hydrothermal water. Geopressured geothermal systems were first identified in the deep sedimentary layers underneath the Gulf of Mexico at a depth between 6 and 8 km with pore pressures of up to 130 MPa and temperatures in the range 150–180 °C.

Systematic studies on geopressured geothermal resources were carried out during the 1970s and early 1980s in the Gulf of Mexico geosyncline. This region served as a natural laboratory for studies aimed at understanding the mechanics and geology of the formation of the geopressured geothermal systems because of the wealth of geological information available from field surveys conducted for petroleum exploration. The region has witnessed large-scale subsidence in geosynclines involving downward transformation of enormous amounts of sediments to depths of 5–10 km in relatively short geological time intervals. As evidenced in the Gulf of Mexico, normal faulting with throws of up to 1 km is the most important structural feature. This process of subsidence subjects the poorly consolidated sediments to intense heat and enormous pressure. As pointed out by Jones (1970), the process of geosynclinal subsidence and rapid sedimentary deposition is analogous to igneous intrusion in an opposite sense: the motion is downward, the intruding material is cold and the rocks intruded are hot. If the sediments undergoing subsidence are constituted of clay of a swelling variety, endothermic diagenesis of clay minerals can consume much heat and thereby reduce the heat flow to the Earth's surface. This phenomenon is very noticeable in the Gulf of Mexico Basin where sediments are mostly of the swelling variety of clay and consequently the geothermal gradient map of the area shows it as a negative geothermal area (Fig. 4.3). Based upon measurements in holes less than 2 km deep, this map shows a decrease in geothermal gradient along the axis of the

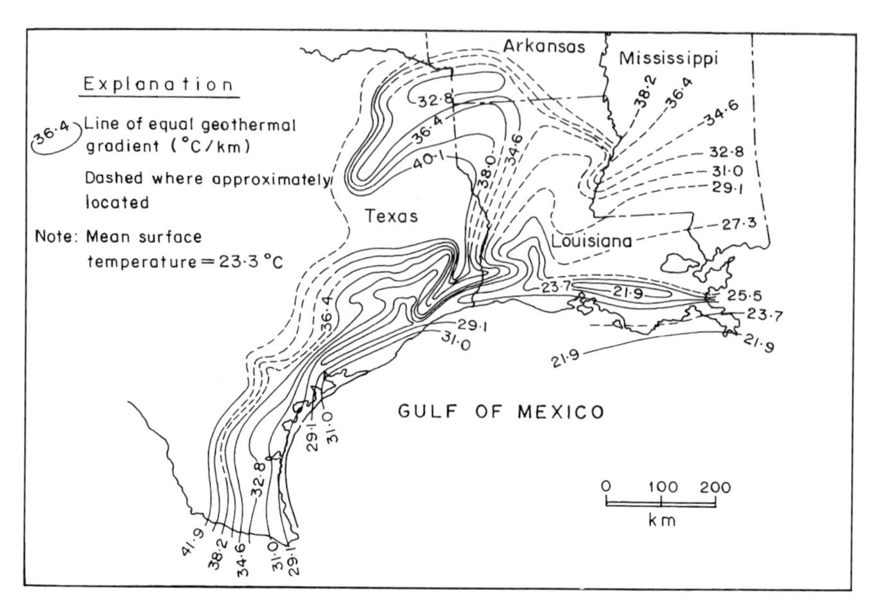

Fig. 4.3. Geothermal gradients in the Gulf Coast of the United States (from Jones, 1970). For details, see text.

geosyncline. Obviously, being dependent on shallow heat flow measurements, the map does not indicate (or conceals) the true magnitude of the existing geothermal resource potential in the northern Gulf of Mexico Basin. It has been observed that the geothermal gradient in the hydropressured zones at depths ranging from 1 to 2 km varies from 20 to 40 °C km^{-1}, and it undergoes a very sharp increase in the geopressured zone; being two to three times greater than that existing in the overlying hydropressured zone.

The deeply buried clays in the geosynclines undergo thermal metamorphism, releasing petroleum hydrocarbons and geothermal waters, the former being solute and the latter solvent, and a solution is formed. The strength of the load-bearing clay bed is drastically reduced by the process of metamorphism, resulting in the expulsion of the pore fluids to share the overburden load. Since the confining pressure decreases most rapidly in the upward direction, and the pressure applied to a confined fluid is transmitted equally in all directions, the expelled pore fluid moves out of the clay bed and moves upwards through sand beds and along the fault planes. The expelled solution of hydrocarbons and water, as it moves upward, encounters much lower pressures and temperatures. Finally, a stage is reached when the temperatures and the pressures are so low that the solute hydrocarbons can no longer remain in the solvent water. Fluid hydrocarbons and geothermal waters then appear as separate phases, and the hydrocarbons being lighter than the water seek higher points in the regional flow system and structural traps. Detailed geological mapping of structures and knowledge of sedimentary history, when combined with the isothermal, isosaline and isopressure maps are helpful tools in prospecting fluid hydrocarbon and geothermal regimes.

To date, potential geopressured geothermal fields have been discovered mainly in the Texas-Louisiana Gulf Coast region. Similar systems may exist in other hydrocarbon bearing deep sedimentary basins elsewhere. Although the huge potential of geothermal–geopressured aquifers has been recognized, the commercial development has been considered marginally economic in only special circumstances. Availability of cheap fossil fuels has been a deterrent to further research and analysis of their energy potential. In the future, new technologies including use of binary-cycle plants may allow more efficient extraction of thermal energy from geopressured brines.

Hot Dry Rock Geothermal Systems

So far we have discussed geothermal resources associated with one or another type of fluid convection system; the fluid (mostly steam and water) being responsible for carrying the heat to the Earth's surface or to the shallower depths from which it could be exploited. Occurrences of such hydrothermal resources are restricted to countries with favourable geological conditions like in the plate boundary zones. However, there exists another category of geothermal resource where geothermal heat is stored in the hot and poorly permeable rocks at shallow depths within the Earth's crust, without any fluid availability to store or transport the heat. These resources are designated HDR. Large volumes of such rocks at high temperatures are known to exist below all major geothermal areas. Geologically young igneous intrusive bodies at shallow levels of the Earth's crust, which form potential targets for this energy resource, occur in several continental areas. After having described the *hot dry rock* geothermal resources in the above manner, three questions arise immediately: "how deep?", "how hot?" and "how dry?" Different depth and temperature ranges as well as definitions of "dryness" can be found in the literature. Here we adopt the definition given by the U.S. Energy Research and Development Administration (ERDA) in their Report No. 1 of the HDR Assessment Panel to answer these questions concerning the HDR geothermal resource. Accordingly, it is "heat stored in rocks within 10 km of the Earth's surface from which the energy cannot be economically produced by natural hot water or steam". Additionally, they cannot produce an economical volume of hot water or steam and their temperature is less than about 650 °C (so as to exclude molten lava or magma). Quite understandably, HDR may be associated at some locations and/or at some depths with hydrothermal resources and/or magma.

Depending upon the cause of the HDR geothermal resources, they can broadly be classified into three categories. These are: (1) igneous related: heat being transferred from magma or stored in dry rocks surrounding the magma bodies; (2) upper mantle related: the heat being conducted to shallow crust overlying an unusually hot upper mantle, causing the anomaly; and (3) local: heat being stored locally either due to the presence of a high concentration of radioactive minerals or due to large-scale faulting and/or fracturing.

Considering the above classification of HDR based upon the different possible sources, the obvious places to look for them are the regions characterized by: (1) recent volcanism, (2) high-heat flow, and (3) localized radiometric heat sources. All magma chambers have a surrounding hot rock environment of varying size. Similarly, geothermal resource areas characterized by hydrothermal convective systems may, at some depth, be underlain by HDR. Also, high-heat flow areas, where the basement geology is favorable for low water contents, are possible HDR resource areas.

HDR technology envisages exploitation of Earth's heat stored in the high-temperature and impermeable rocks by artificially creating a fracture system at depth (that acts as a heat exchanger) and circulating water from an injection borehole towards a production borehole. Hydraulic fracturing, which involves injection of water at very high pressure into a reservoir to create new fractures or enlarge pre-existing cracks, has been one of the successful methods in creating permeability in the rocks at depth. It has been estimated that cooling of 1 km^3 of hot rock by 100 °C will enable operation of a 30 MWe geothermal power plant for 30 years (Rummel, 2005). However, generation of large heat exchangers at depth and controlling the loss of circulation fluids present the biggest technological challenges in exploitation of the HDR geothermal energy.

Detailed research and experimentation in HDR technology was first carried out at Fenton Hill, near the edge of the volcanically young Valles Caldera, near Los Alamos in New Mexico, where temperatures of the order of 195 °C at a depth of 3 km were observed in the granitic basement. Scientists and technologists of the Los Alamos National Laboratory, New Mexico were able to obtain a thermal power output of 5 MW by drilling a pair of injection and production wells and creating a permeable rock volume by hydrofracturing the rocks surrounding the bottom parts of the wells. Although they were able to demonstrate the feasibility of the technology, further tests were discouraged by the high costs involved for commercial exploitation. Over the past decade or so, renewed interest in HDR geothermal energy led to several pilot projects in central Europe, Britain, Russia, Japan and Australia. The biggest success story in the last 30 years has been the one at Soultz-sous-Forêts area located in the Upper Rhine graben near the boundary between France and Germany, where setting up a pilot HDR project to produce 6 MW electricity is nearing completion (Klee, 2005; Rummel, 2005). Drilling up to 5 km depth in the area confirmed a temperature of ~200 °C. This project could be the forerunner for several small HDR projects in the Upper Rhine graben region. Another pilot HDR project is underway in Cooper Basin, Australia, an area of very high heat flow. The basin is underlain by high heat producing granites (7–10 $\mu W\,m^{-3}$), and temperatures of ~240 °C have been measured at depths up to 3.7 km in several oil and gas wells (Chopra and Wyborn, 2003). The first borehole (Habanero-1), drilled under the HDR project met with temperatures exceeding 250 °C at 4.4 km (Beardsmore, 2004).

The project holds promise for another successful demonstration of HDR technology in the next few years.

Magma

Magma, the naturally occurring molten rock material is a hot viscous liquid, which retains fluidity till solidification. It may contain gases and particles of solid materials such as crystals or fragments of solid rocks. However, the mobility of magma is not much affected until the content of solid material is too large. Typically, magma crystallizes to form igneous rocks at temperatures varying, depending upon the composition and pressure, from 600 to 1400 °C. At its site of generation, magma is lighter than the surrounding material, and consequently it rises as long as the density contrast between magma and surrounding cooler rocks continues. Eventually, magma either solidifies or forms reservoirs at some depth from the Earth's surface, or it erupts. Magma is the ultimate source of all high-temperature geothermal resources. Plate boundaries are the most common sites of volcanic eruptions (Fig. 2.7). At several volcanic locales, magma is present within the top 5 km of the crust. It has been estimated that the average rate of production of magma at Kilauea Volcano, Hawaii, during 1952–1971 has been about $10^8\,m^3\,year^{-1}$ (Swanson, 1972). Similar estimates have been made for the Columbia Plateau (Baksi and Watkins, 1973) and elsewhere. The heat energy available from such sources, if harvested, would constitute very large additions to the global energy inventory.

Extraction of thermal energy from magma was tested during the 1980s by drilling into the still-molten core of a lava lake in Hawaii. However, up to the present, the necessary technology has not been developed to recover heat energy from magma. Economical mining of heat energy from magma presents several practical difficulties such as locating such bodies accurately before drilling into them, the prohibitive costs of drilling and longevity of deployed plant materials in a hot corrosive environment.

Chapter 5

EXPLORATION TECHNIQUES

The contemporary economic and technological state of affairs and its foreseeable development restricts extraction of geothermal energy to the upper few kilometers of the Earth's crust. The deepest geothermal wells, to date, are less than 5 km deep. The development of drilling technology and increase in the cost of other energy resources may permit drilling deeper for geothermal resources in the near future. The framework of the plate tectonics hypothesis provides broad but useful clues about the potential geothermal regions. Once a geothermal region is identified, the next step is to use different available exploration techniques to localize potential geothermal resource areas and identify suitable drilling targets for production. It is necessary to estimate the temperature, volume and permeability at depth as well as to predict whether the well will produce dry steam, wet steam or just hot water. It is also desirable to estimate the chemical composition of the fluid to be produced. To obtain this varied information, it is necessary to employ a suite of exploration techniques. Important among these are: (1) geological and hydrological techniques, (2) geochemical techniques, (3) geophysical techniques, (4) remote sensing techniques and (5) exploratory drilling. Because the geological settings of different geothermal resource areas vary widely, the exploration tools could be different as also the sequence of investigations. In most cases, the exploration strategy for a particular area is finalized by a team of geologists, geophysicists and reservoir engineers on the basis of all available information.

Before undertaking any of these detailed surveys, it is necessary to undertake a detailed search and examination of relevant existing literature. For many areas, useful geological, hydrological, geochemical, geophysical, topographical and meteorological information already exists, and these should be fully utilized before undertaking further exploration. Analyses of existing data could provide useful information regarding the suitability or otherwise of the various exploration techniques discussed in the following.

GEOLOGICAL AND HYDROLOGICAL CONSIDERATIONS

General Background

The first step in selecting a region for preliminary reconnaissance is to look for surface manifestations of geothermal resources such as thermal springs, steaming ground, fumaroles, mud-pools and geysers. A comprehensive list of all these manifestations, including a description of the locality, geologic control, physical environment, temperature, chemistry and rate of the fluid discharged, should be compiled. Many countries have long-established geological surveys and such information is often well documented. The locations of these geothermal manifestations should next be plotted on the most detailed geological maps available so that their association with a particular rock type or structural zone can be examined. If they tend to be associated with a particular zone or district, then that zone is likely to be an area of positive geothermal anomaly. It is also likely that there are many more geothermal springs in the region, which has not been reported. A field reconnaissance is desirable, with particular attention being given to areas where existing information has indicated potential geothermal resource, and extended to all hydrothermal areas. It is necessary to carry out geochemical analysis of the water from springs.

Geothermal resources, on the basis of their origin, can broadly be classified into two main categories and various sub-categories (Muffler, 1976):

(1) Geothermal resources associated with igneous intrusions in the upper crust. These can be further divided into: (a) magma, (b) hot dry rock, and (c) convective hydrothermal systems.
(2) Geothermal regimes not related with young igneous intrusions, further classified into (a) resources associated with deep circulation of meteoric water, (b) high-porosity environment resources at hydrostatic pressure, and (c) high-porosity environments at pressures greater than hydrostatic.

In the second chapter, the concept of the plate tectonic hypothesis has been introduced. This hypothesis is extremely useful in geographically identifying broad areas in which the two primary categories of geothermal resources are found. Fig. 2.8 shows the present-day configuration of the lithospheric plates and the location of the major geothermal fields. Plate boundaries are characterized by earthquake occurrences. Generation of magma takes place at the spreading ridges and at the trenches where plates are subducted, as well as at intraplate melting anomalies. Therefore, it is to be expected that geothermal resources related with the intrusion of magma in the shallow crust would primarily occur close to the spreading ridges, subduction zones and intraplate hot spots like Hawaii. However, geothermal resources of conductive crustal regime could occur even outside such active tectonic zones.

Geothermal Resources Associated with Igneous Intrusions in the Upper Crust

The three varieties of geothermal resources associated with the intrusion of igneous rocks, i.e., magma, hot dry rock and convective hydrothermal reservoir, require emplacement of a body of magma in the upper crust. It indicates the presence of some suitable barrier or cap that could prevent a portion of magma from erupting. When magma from the lower crust or the upper mantle quickly reaches the Earth's surface, it loses all its heat to the atmosphere without producing a significant geothermal resource. However, a body of magma confined at some depth from the

Earth's surface dissipates its heat by a variety of conductive and convective processes, giving rise to geothermal resources.

As discussed earlier, magma is generated at mid-oceanic ridges and hence they are potential geothermal resource targets. However, practical problems render them less attractive and to date energy is harnessed only in Iceland, where the Mid-Atlantic Ridge is above sea level. When spreading ridges impinge on a continent, partial melting of crustal material forms silicic magma. Therefore, silicic intrusions and extrusions are most commonly observed where spreading ridges impinge on continents, compared to mid-oceanic sites (Robinson et al., 1976).

At the zones of subduction, material is thrust down into the lower crust and upper mantle. The composition of this material varies from siliceous sediments to ultramafic rocks. Its partial melting produces a variety of silicic and intermediate rocks. The material properties of the subducting plates, i.e., whether it is an oceanic plate—oceanic plate subduction, or an oceanic plate—continental plate subduction, or a continental plate—continental plate contact, govern the composition of the rocks produced.

Intracontinental melting anomalies have been postulated by Morgan (1972) to lie above narrow plumes (less than 150 km in diameter) that originate deep in the Earth's mantle. Propagating fractures and shear melting have also been suggested as being responsible for melting anomalies (McDougall, 1971; Shaw and Jackson, 1973). The best-known example of such anomalies is beneath Hawaii where large volumes of basaltic magma have erupted from depths of at least 60 km (Dalrymple et al., 1973).

Depending upon the existing geological and geohydrological conditions, the three major categories of geothermal resources develop consequent on the intrusion of large volumes of molten rocks into the shallow crust. The magma itself is a potential resource. According to the estimates of Muffler (1976), a cubic kilometer of granitic magma at 800 °C contains 3×10^{18} J of heat, which is equivalent to the heat content of 480 million barrels of crude oil. However, the technological problems in harnessing geothermal energy from magma bodies are enormous. After intrusion in the crust, magma loses heat through conduction and convection. When the permeabilities of the host rock and the intrusion are low, a hot dry rock geothermal resource is developed. One cubic kilometer of hot granite at 400 °C contains 1×10^{18} J of heat. The technological problems involved in utilizing the heat of hot dry impermeable rocks are, however, substantial. When an igneous intrusion is emplaced in rocks with adequate fractures or intergranular permeability, the cooling of the igneous intrusion is considerably affected by movement of water. If adequate permeability is present, hydrothermal convection sets in and dominates the cooling history (Cathles, 1977; Norton, 1984). Under favorable conditions of hydrological continuity, meteoric water enters through permeable rocks, comes in contact with the hot intrusion, becomes heated and moves upward, setting up strong convective currents. The heat from the intrusion is thus either convected to the Earth's surface or it forms a hydrothermal reservoir at some intermediate depth. Depending upon the local hydrological situation, this could be a vapor-dominated or a hot water reservoir.

From the above it may be concluded that any one crustal geothermal anomaly, caused by igneous intrusion, can support a variety of geothermal resources simultaneously and/or sequentially. The potential of an igneous system is directly proportional to the size of the anomaly and indirectly proportional to its age. In Fig. 5.1, empirical relations derived by Smith and Shaw (1975) are shown. They have considered the age-volume data of 54 volcanic systems in the U.S.A. and plotted the approximate present position of each system in relation to its probable solidification state and to the 800 °C isotherm. Smith and Shaw (1975) have drawn pairs of lines to represent a spectrum of cooling models that identify igneous systems which are approaching ambient temperatures (points above lines 5 and 6), systems that may now be approaching the post-magmatic stage (points between lines 3 and 4) and systems which may still have magma chambers with large portions of molten magma (points below lines 1 and 2). The pair of lines covers geometric shapes varying from cubes to slabs. Lines 1 and 2 are drawn assuming that cooling takes place by internal convection till solidification takes place. Lines 3 and 4 assume that cooling is due to conduction only, both inside and outside the magma chamber. Lines 5 and 6 represent an estimate of the time required before the central temperature of the solidified pluton has fallen to about 300 °C. Points lying above line 6 are considered to be of little geothermal potential. Smith and Shaw (1973, 1975) have used these plots for reconnaissance evaluation of silicic volcanic areas.

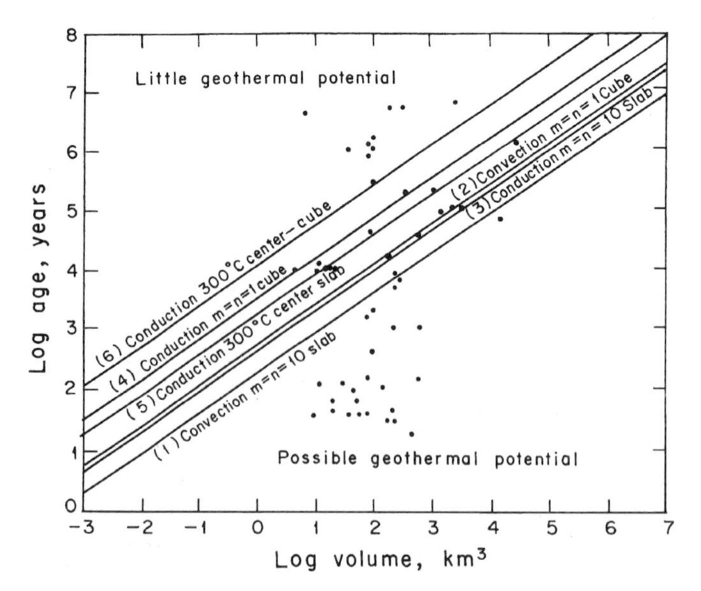

Fig. 5.1. Graph showing theoretical cooling time vs. volume of magma bodies. See text for explanation of lines 1–6 (from Smith and Shaw, 1975).

Geothermal Regimes not Related to Young Intrusions

Resources associated with deep circulation of meteoric water
Warm water springs are sometimes found in regions characterized by normal, or even low, geothermal gradients. These regions are not found to be associated with Cenozoic volcanism. Geothermometry, based on geochemical analysis, does not reveal elevated temperatures at shallow depths. Such warm springs appear to be associated with deep circulation of meteoric water. Warm springs of the Basin and Range Province of the western United States is one such example. The Basin and Range Province is characterized by extensive normal faulting and tectonic extension (Thompson and Burke, 1973). The hot springs are mostly located along faults. Another well-known example is that of the Larderello field in Italy (Marinelli, 1964), where a widely spread impermeable cap rock overlies a folded sedimentary series with a highly permeable member. The basic requirement for this kind of geothermal reservoir to exist appears to be the presence of hydraulic continuity to great depths and some sort of a barrier to form a reservoir.

Resources associated with a high-porosity environment at hydrostatic pressure
Deep basins filled with sedimentary rocks of high porosity and permeability and occurring in regions of high geothermal gradients are favored sites of geothermal resources at hydrostatic pressures. High temperatures, approaching 200 °C, can be expected at shallow depths. The Hungarian Basin, where porous Cenozoic sediments, up to 3 km thick, occur in a region of high temperature gradients of 50–70 °Ckm^{-1}, is one such example. A water reservoir comprising over 4000 km^3 of water at temperatures varying from 60 to 200 °C is estimated to exist in these porous rocks (Boldizsár, 1970). Geothermal energy from similar resources is being harnessed in Russia and France. The two basic requirements for this kind of geothermal resource to occur are the existence of high regional heat flow and a sedimentary basin with a thick pile of porous and permeable rocks. Plate margins are favored to satisfy these requirements (Muffler, 1976).

Resources associated with a high-porosity environment at pressures in excess of hydrostatic
Zones characterized by pressures greatly in excess of the hydrostatic pressure, have been revealed by drilling carried out in sedimentary basins. These zones are called geopressured zones and are associated with high temperatures. Most of the information on geopressured geothermal resources has accrued from investigations carried out in the Gulf Coast of the United States and summarized by Jones (1970) and Papadopulos et al. (1975). This has been discussed in Chapter 4. As explained by Lewis and Rose (1970), these extensive geopressured reservoirs are bounded laterally by faults and they occur in sands underlying low-permeability shales. Thermal insulation is provided by the overlying clays filled with static water.

The basic requirement for geopressured geothermal reservoirs to occur appears to be a rapid rate of sedimentation in shallow basins characterized by high regional heat flow. No geothermal energy has been yet harnessed from geopressured geothermal resources. With technological developments, and as and when it becomes possible to harness these resources, they could provide an abundant source of energy.

Choosing a Prospect Area

The above-mentioned geological and geohydrological considerations are useful in broadly selecting a promising region for geothermal prospecting. The next step is to select specific areas for detailed prospecting. In areas where geothermal manifestations are not well documented, field surveys should be undertaken to locate them and to collect information on the type of manifestation (i.e., thermal springs, steaming ground, seepage, CO_2 vent, fumaroles, mud pools and geysers). Remote-sensing surveys, which are discussed later, can be undertaken to locate the thermal manifestations over a large area in a short time.

Surface manifestations may not reflect potential geothermal conditions at depth in many situations (McNitt, 1973). This is particularly true when an overlying non-geothermal groundwater horizon masks the geothermal system. For this reason, it is not always possible to use surface heat discharge as a first approximation to the size of the system and its capacity to produce usable energy. For example, the Larderello fields in Italy, The Geysers and the Salton Sea fields in the United States and many other geothermal fields have meager surface heat discharge compared to the amount of energy obtained after drilling. At places, productive ground is found to extend many kilometers beyond a small surficial manifestation. It is particularly difficult to estimate the spread and capacity of vapor-dominated geothermal systems from surface indications. Such systems require a cap for building up of the high temperatures and pressures in the trapped vapor. This cap prevents spectacular surface indications. However, for hot water geothermal systems, a lithological or hydrological cap is not necessary. Consequently, for such systems, surficial thermal manifestations can be used to approximate their size. The Yellowstone geothermal field in the United States and Wairakei in New Zealand are excellent examples of this kind. At places impermeable lithological horizons tend to deflect the upflowing hot water laterally for several kilometers. This results in surficial thermal manifestations being remote from the center of the subsurface geothermal reservoir. These lithological barriers are commonly caused by the fine sediments of streams or lakes flowing at the top of the geothermal systems. These sediments are deposited along the fault angles and cause the upcoming hot water to move up-dip beneath them for long distances before surfacing. As mentioned by McNitt (1973), at the Kizildere field in Turkey, this kind of lithological situation causes deflection of the upcoming thermal fluid from the center of the subsurface geothermal reservoir by up to 5 km. This kind of lateral deflection should be suspected in the regions where the geothermal springs occur at the contact of alluvial beds with the bedrock rather than occurring in fault or fracture zones.

Certain hydrological conditions also considerably alter the possible surficial manifestation of geothermal reservoirs. When the top of a geothermal system comes in contact with a deep underground body of water, where the water is stagnant or slowly moving, and the heat from the geothermal system is adequate to cause the groundwater to boil, patches of gently steaming ground would appear. These patches would give no indication of the high pressures that may exist in the geothermal systems. Similarly, when the top of a geothermal system is intersected by a large cold water aquifer, the heat will be swept down gradient (McNitt, 1973). Therefore, either no evidence of the geothermal system will appear on the surface, or large-volume warm springs may appear at large distances from the center of the source of heat.

In conclusion, as summarized by Healy (1970), the following stages of investigation are suggested.

(1) Review of the existing data on thermal springs and preparation of a map showing their relation to regional geology. This could reveal areas where more attention is required.
(2) Field reconnaissance of the thermal springs, sampling of the water for chemical analysis, and study of aerial photographs where available.
(3) Detailed study of the characteristics of geothermal areas. These characteristics include the size of the area, the presence or absence of shallow and deep aquifers, subsurface temperatures, and potential for development as steam or hot water field, etc.
(4) Investigation of the hydrothermal areas and their geological characteristics to assess their relative importance from genetic, structural and regional points of view.
(5) On the basis of the above studies a decision is to be made whether production of geothermal energy appears possible or not. If there are positive indications, several areas should be demarcated in order of preference for further detailed investigations.

GEOCHEMICAL TECHNIQUES

Geochemical methods have played an important role in preliminary prospecting of geothermal resources since the early 1960s. Chemical data on hot water and steam discharges in a virgin area serve as useful indicators of the feasibility of further exploration in the area including preliminary drilling locations. Together with structural information from geological, hydrological and geophysical methods, they can guide decision making on subsurface exploration by drilling. Chemical analysis of fluids extracted from depth by drilling provides valuable information on flow patterns of subsurface fluids. Chemical geothermometers, which relate the fluid chemistry and reservoir temperature, are routinely used in assessing the energy potential of a geothermal prospect. During production, testing and subsequent utilization, monitoring the chemical changes of extracted fluids can be used to detect changes in temperature and water levels in the reservoir. Furthermore, geochemical surveys are relatively inexpensive when compared to geophysical surveys and subsurface investigations by drilling. Therefore, geochemical tools are now widely used in all stages of geothermal exploration and development (Sigvaldason, 1973; Giggenbach, 1991).

Geochemical surveys are carried out simultaneously with geological, hydrologeological and remote sensing surveys. During the reconnaissance stage, geochemical analyses are done on hot spring samples, steam samples from fumaroles and samples from hot pools. For reference purposes, cold springs if present in the area, are also studied. The application of geochemical tools to geothermal prospecting has been dealt with in detail by several authors (e.g., Fournier, 1981; D'Amore, 1992; Nicholson, 1993; Arnórsson, 2000). Here we shall briefly discuss the application of geochemistry in assessing the origin of geothermal fluids, the chemical composition of geothermal fluids, and estimation of reservoir temperatures using chemical and isotopic geothermometers.

Origin of Geothermal Fluids

The origin of geothermal fluids has been debated for a long time and a number of hypotheses have been proposed by various workers. This problem was ultimately resolved by Craig et al. (1956) and Craig (1963) through detailed studies of the isotope composition of the ratios hydrogen/deuterium (H/D) and $^{16}O/^{18}O$ of geothermal fluids. They determined these ratios for geothermal fluids from a number of widely separated localities and compared the values with the isotopic composition of meteoric waters of the corresponding localities. The H/D ratio in the meteoric and geothermal waters was found to be same for all the localities, indicating that the geothermal fluids could not have a magmatic origin. However, the $^{16}O/^{18}O$ ratio was found to differ. This difference was explained as due to exchange of the oxygen isotopes in the thermal water with the oxygen of the wall rocks through which the geothermal water circulated. A meteoric origin for most geothermal waters was thus established. All the solutes present in geothermal fluids could, however, be derived from reactions between the meteoric groundwater and the host rock formations (Ellis and Mahon, 1964, 1967). Although most geothermal fluids are of meteoric origin, chemical considerations and uncertainties in isotopic ratios indicate the possibility of a small (5–10%), but significant magmatic component (e.g., juvenile water, possibly magmatic brine) in a few cases (Nicholson, 1993; Barbier, 2002). The finding of a predominantly meteoric origin for geothermal fluids resulted in significant changes in exploration strategies with more emphasis given to hydrological considerations. Recent studies of rare geothermal gases suggest that they are mostly of atmospheric origin. However, excess of ^{3}He in some systems may come from the mantle (Craig et al., 1975). The mantle contribution of ^{3}He isotope does not necessarily indicate that other mantle-derived components are also present in the geothermal fluids.

In contrast to a meteoric origin for most geothermal waters, the geothermal fluids discharged from geothermal systems in volcanic island arcs have a large magmatic component. On the basis of a large number of studies on such geothermal systems, Giggenbach (1992) concluded that the "horizontal" ^{18}O shifts, interpreted previously by Craig (1963) as due to water–rock interaction, are rather the exception than the rule. He found significant hydrogen shifts from local meteoric water values in the case of fluid discharges from island arc volcanoes and hence proposed a common magmatic origin for these systems.

Chemical Composition of Geothermal Fluids

The chemistry of geothermal fluids are largely influenced by the fluid-mineral equilibria in hydrothermal systems (Henley et al., 1984). The waters circulating in high-enthalpy geothermal areas contain variable solute concentrations, which depend on temperature, gas content, heat source, rock type, permeability, age of the hydrothermal system and fluid source (Barbier, 2002). Cations (e.g., sodium, potassium, lithium, calcium, magnesium, rubidium, cesium, manganese and iron), anions (e.g., sulfate, chloride, bicarbonate, fluoride, bromide and iodide) and neutral species (e.g., silica, ammonia, arsenic, boron and noble gases) are the most common species constituting the waters. Accordingly, four types of waters—sodium chloride water, acid sulfate chloride water, acid sulfate water, and calcium bicarbonate water—have been proposed (Ellis and Mahon, 1977; Henley et al., 1984; Giggenbach, 1988). It must be underscored, however, that each of these waters may mix with each other giving rise to hybrid water types.

Sodium chloride water

Sodium chloride water is the most common type of fluid found at depth in large water-dominated geothermal systems. It contains 1000–10,000 mg kg^{-1} of chlorine and mainly carbon dioxide gas. It has a pH close to neutral at depth. When this type of water approaches the surface, it loses steam and carbon dioxide and consequently becomes slightly alkaline. It is believed that these waters are formed from the absorption of magmatic volatiles such as HCl, CO_2, SO_2 and H_2S into deeply circulating meteoric water. The ratio of chloride to sulfate is high. Typical examples of geothermal systems containing this type of water are Wairakei (New Zealand), Geysir (Iceland) and Ahuachapan (El Salvador).

Acid sulfate/chloride water

This type of water is commonly formed by mixing of chloride and sulfate waters at variable depths. Oxidation of sulfide to bisulfate at deeper levels causes its acidity. As the waters rise from depth, they cool down and become acidic. Geothermal waters of the Frying Pan Lake, Tarawera, New Zealand are a typical example. Corrosive acid sulfate chloride waters were also found after deepening the wells in the Miravalles geothermal field as well as some fields in the Phillipines (Sussman et al., 1993). These waters are produced due inflow of acid magmatic gases into the deepest portions of neutral NaCl systems.

Acid sulfate water

Acid sulfate waters are typically found in fumarolic areas above the upflow part of the geothermal systems, where steam rising from hot water reservoirs condenses as it approaches the surface. Boiling results in the transfer of gas species, mainly CO_2 and H_2S, into the vapor phase. This vapor phase can reach the surface in the form of fumaroles and steam jets without any interaction with shallow or surface waters. Alternatively, separated vapor may condense in shallow groundwaters or surface waters to form steam-heated waters. In this environment, atmospheric oxygen oxidizes H_2S to sulfuric acid producing acid sulfate waters. These are characterized by low chloride contents and low pH values (0–3) and react quickly with host rocks to give clay alteration products dominated by kaolinite and alunite. Dissolved cations and silica are mainly leached from the surrounding rocks, whose compositions may be approached by these acid waters. Shallow steam-heated waters may themselves boil, separating secondary steam, which reaches the surface in the form of low-pressure steaming grounds. Acid sulfate waters are highly corrosive to casing and pipelines. The waters from Norris Basin, Yellowstone National Park, U.S.A. are a typical example.

Bicarbonate water

Bicarbonate-rich waters originate through either dissolution of CO_2-bearing gases or condensation of geothermal steam in relatively deep, oxygen-poor groundwaters. Because the absence of oxygen prevents oxidation of H_2S, the acidity of these aqueous solutions is due to dissociation of H_2CO_3. Although it is a weak acid, it converts feldspars into clays, generating neutral aqueous solutions. These solutions are typically rich in sodium and bicarbonate, particularly at medium–high temperature. Sodium bicarbonate waters occur in the "condensation zone" of vapor-dominated geothermal systems and in the marginal parts of water-dominated systems. However, sodium bicarbonate waters are also present in deep geothermal reservoirs hosted in metamorphic and/or sedimentary rocks (e.g., Kizildere, Turkey). Calcium bicarbonate water often occurs as low-temperature travertine depositing springs. A typical example is the Mammoth Hot Springs, Yellowstone National Park.

Volcanic water systems are usually characterized by waters of neutral sodium chloride type at depth, which could be altered during its passage to the surface by addition of calcium, bicarbonate or acid sulfate components. The subsurface temperatures and the concentration of soluble components in the rocks affect the amount of soluble components like Cl, B, Br, Li, Cs and As observed in the thermal waters. The presence of less-soluble components like SiO_2, Ca, Mg, Rb, K, Na, SO_4, HCO_3 and CO_3 is controlled by: temperatures at depth, mineral solubility, pH value and mineral-fluid equilibria. The chloride content is usually low in the near-surface fluids of vapor-dominated systems. They normally contain H_2S, CO_2, HBO_2, Hg, NH_3, etc., which are soluble in low-pressure steam. For non-volcanic geothermal systems, the water composition varies widely from dilute meteoric waters to connate waters, oil field brines and metamorphic water. Relatively much less is known about the factors controlling the composition of waters from non-volcanic geothermal systems.

In addition to geothermal waters, analysis of gases escaping from geothermal areas is useful in understanding subsurface conditions. According to Kononov and Polak (1976), fluids related to volcanic processes are characterized either by H_2S–CO_2 gases and acid sulfate or acid sulfate/chloride waters in the oxidizing zone, or by N_2–CO_2 gases and alkaline sodium chloride waters in the reducing zone. Fluids connected with thermo-metamorphic processes have high concentrations of CO_2 gases and carbonate waters. Geothermal fluids originating from deep circulation, outside volcanic and thermo-metamorphic zones, have high concentrations of N_2 gases and dilute sodium chloride/sulfate waters.

It is not uncommon to find more than one kind of fluid occurring at the same geothermal system. It is necessary to find a logical relationship between the different kind of fluids and to interpret from the chemistry of the geothermal fluids, the chemical composition of fluids at depth as well as the physical environment of the geothermal reservoir.

Geothermometers

Geothermometers are subsurface temperature indicators derived using temperature-dependent geochemical and/or isotopic composition of hot spring waters and other geothermal fluids under certain favorable conditions. The equilibrium between common minerals or mineral assemblages and a given water chemistry is temperature dependent. White (1965, 1970), Ellis and Mahon (1967), Fournier and Truesdell (1973) and several other workers have observed that the relationship between chemical composition and temperature is predictable for certain parameters or ratios of parameters. Therefore, these parameters or ratios of parameters can serve as geothermometers. All geothermometers have certain limitations. The silica geothermometer yields reservoir temperatures independent of the local mineral suite and gas partial pressures, but may be severely affected by dilution. On the other hand, the ratio-based geothermometers (such as Na/K) are more resistant to dilution effects.

Geochemical and isotopic geothermometers rely on the following assumptions: (i) the two species or compounds co-existed and have equilibrated within the geothermal reservoir, (ii) the ratio is controlled predominantly by temperature, and (iii) re-equilibrium has not occurred during ascent and discharge. Therefore, the temperature indicated by the geothermometer is not necessarily the maxiumum temperature of the water, but the temperature at which mineral and water phases were last in equilibrium (Nicholson, 1993). Therefore, the geothermometers should be applied with caution, utilizing all preliminary information available about the geological and hydrological setting as well as the chemistry of the fluids.

Na–K Geothermometers

A commonly used geothermometer where geothermal waters are known to come from high-temperature environments (> 180 °C, up to about 200 °C) is the atomic ratio of sodium to potassium (Na/K). The ratio decreases with an increase in temperature. The cation concentrations (Na^+, K^+) are controlled by temperature-dependent equilibrium reactions with felspars and mica. The main advantage of this thermometer is that it is less affected by dilution and steam separation than other commonly used geothermometers provided there is little Na^+ and K^+ in the diluting water compared to the reservoir water.

Syntheses of available data by White (1970), Fournier and Truesdell (1973) and others have produced a number of empirical equations for reservoir temperature. The most commonly used equation is that given by Fournier (1979):

$$\theta(^\circ\mathrm{C}) = \frac{1217}{\log(\mathrm{Na}/\mathrm{K}) + 1.483} - 273.15 \tag{5.1}$$

where, Na and K are measured in ppm. According to this relation, if the Na/K ratio is found to be 10 in a certain spring water, the subsurface temperature is expected to be about 217 °C. Other empirical relationships have been proposed by later workers (e.g., Tonani, 1980; Arnórsson et al., 1983; Giggenbach, 1988). These empirical relationships yield slightly different temperature estimates, with the differences lying within uncertainty bands of the individual relationships.

The Na/K geothermometer fails at temperatures lower than 100–120 °C and gives high temperatures for solutions with high calcium contents. A drawback of this thermometer is the preferential absorption of potassium by certain hydrothermal clay minerals, such as montmorillonite, present in the top few hundred metres in geothermal areas.

Na–K–Ca Geothermometer

The Na–K–Ca geothermometer was developed by Fournier and Truesdell (1973) to overcome the problem of anomalously high computed temperatures using the Na/K method. The empirical equation is given as

$$\theta(^\circ\mathrm{C}) = \frac{1647}{\log(\mathrm{Na}/\mathrm{K}) + \beta\left[\log\left(\sqrt{\mathrm{Ca}}/\mathrm{Na}\right) + 2.06\right] + 2.47} - 273.15 \tag{5.2}$$

where, Na, K and Ca are measured in $\mathrm{mg\,kg^{-1}}$ or ppm,

$\beta = 4/3$ for Ca waters, and $\theta < 100$ °C and

$\beta = 1/3$ for Na waters, and $\theta > 100$ °C.

Changes in concentration resulting both from boiling and from mixing with cold water will affect the Na–K–Ca geothermometer. In case of boiling, loss of CO_2 occurs which can cause $CaCO_3$ to precipitate. The loss of aqueous Ca^{2+} generally results in overestimates in temperatures. Further, corrections to the computed temperatures may be necessary in case of Mg^{2+}-rich waters at temperatures <200 °C (Fournier and Potter, 1979).

Absence of magnesium in geothermal waters is an indication of high reservoir temperatures. Hydrothermal experiments show that magnesium is preferentially incorporated in clay minerals. Clay minerals, such as chlorite, are stable at high temperatures. Therefore, geothermal water at high temperatures, say above 200 °C, in contact with altered rocks is depleted in magnesium (<1 ppm).

Silica Geothermometer

The increased solubility of quartz and its polymorphs at elevated temperatures has been used extensively as an indicator of reservoir temperatures (Truesdell, 1976; Truesdell and Hulston, 1980; Fournier and Potter, 1982). Silica geothermometers are one of the oldest and most commonly used types. Geochemical studies have shown that quartz is an important secondary mineral phase present and therefore it is common to compare the silica value of the thermal waters with the quartz solubility vs. temperature curve for deducing the reservoir temperature. In systems above about 180 °C, the silica concentration is controlled by the equilibrium with quartz. At lower temperatures, the equilibrium with chalcedony becomes important.

One of the assumptions made in such calculations is that silica dissolved at high temperature at depth remains metastable in the solution and does not precipitate as the thermal waters rise to the surface. This is mostly true when the spring discharge is considerable and water rises rapidly. On the other hand, higher-temperature values will be inferred if additional silica is dissolved from the wall rocks. The silica geothermometer therefore needs to be used with caution.

The subsurface temperatures can be estimated using the following relationships for equilibrium with the various silica polymorphs in the temperature range, 0–250 °C (Truesdell, 1976; Fournier, 1981):

(i) Quartz, no steam loss:

$$\theta(^\circ\mathrm{C}) = \frac{1309}{5.19 - \log \mathrm{SiO_2}} - 273.15 \quad (5.3)$$

(ii) Quartz (max. steam loss):

$$\theta(^\circ\mathrm{C}) = \frac{1522}{5.75 - \log \mathrm{SiO_2}} - 273.15 \quad (5.4)$$

(iii) Chalcedony (conductive cooling):

$$\theta(^\circ C) = \frac{1032}{4.69 - \log SiO_2} - 273.15 \quad (5.5)$$

(iv) α-Crystobalite:

$$\theta(^\circ C) = \frac{1000}{4.78 - \log SiO_2} - 273.15 \quad (5.6)$$

(v) β-Crystobalite:

$$\theta(^\circ C) = \frac{781}{4.51 - \log SiO_2} - 273.15 \quad (5.7)$$

(vi) Amorphous silica:

$$\theta(^\circ C) = \frac{731}{4.52 - \log SiO_2} - 273.15 \quad (5.8)$$

where SiO_2 concentrations are given in ppm. In accordance with the first relation, if 200 ppm of SiO_2 are measured in a particular spring water sample, the expected temperature of the reservoir would be 180 °C.

Mixing Models

Very often the cold water from near the surface gets mixed with the thermal waters. This mixing makes the direct application of the chemical geothermometers difficult. Under favorable conditions, the original temperature of the hot water and the fraction of the cold water in the mixture can be estimated by measuring the temperature and silica content of the warm spring water, as well as the temperature and silica content of non-thermal water in the region. Fig. 5.2, from Fournier and Truesdell (1974), shows a typical situation. Hot water ascends from depth along a permeable channel, and at some point it encounters cold water. At the depth of mixing, the weight of the column of the cold water extending to the surface is greater than the weight of the hot water column to the surface. Cold water therefore enters the hot water channel and the mixture flows to the surface and is discharged as a warm spring. To determine the two unknowns, i.e., the temperature of the hot water and the proportion of the hot and cold waters, two equations can be written and solved. The first equation relates the enthalpies of the hot water (H_h), cold water (H_c) and

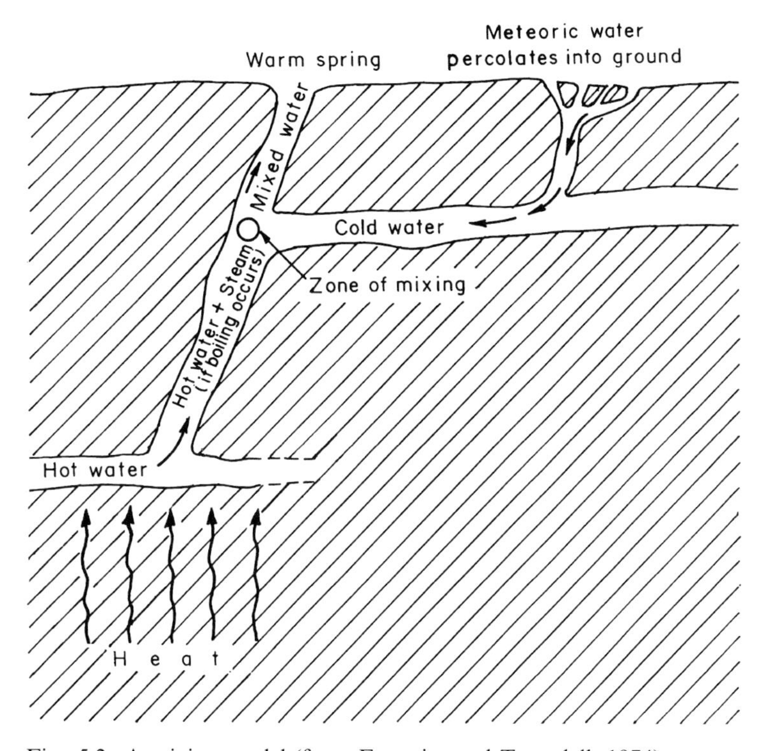

Fig. 5.2. A mixing model (from Fournier and Truesdell, 1974).

spring water (H_s) and the fractions of cold water, X, and of hot water, $(1-X)$ as follows:

$$H_c X + H_h(1 - X) = H_s \tag{5.9}$$

The second equation relates the silica contents of hot water (Si_h), cold water (Si_c), and spring water (Si_s):

$$Si_c X + Si_h(1 - X) = Si_s \tag{5.10}$$

These two equations can be solved to obtain the two unknowns. Fournier and Truesdell (1974) have suggested a graphical method for determining the unknowns as follows:

(1) A series of values of the enthalpy of hot water for temperatures listed in Table 5.1 are assumed and X_t is calculated for each value as follows:

$$X_{Si} = \frac{\text{(enthalpy of hot water)} - \text{(temperature of warm spring)}}{\text{(enthalpy of hot water)} - \text{(temperature of cold spring)}} \tag{5.11}$$

Table 5.1. Enthalpies of liquid water and quartz solubilities at selected temperatures and pressures (Fournier and Truesdell, 1974)

Temperature (°C)	Enthalpy ($cal\,g^{-1}$)	Silica content ($mg\,l^{-1}$)
50	50.0	13.5
75	75.0	26.6
100	100.1	48.0
125	125.4	80.0
150	151.0	125.0
175	177.0	185.0
200	203.6	265.0
225	230.9	365.0
250	259.2	486.0
275	289.0	614.0
300	321.0	692.0

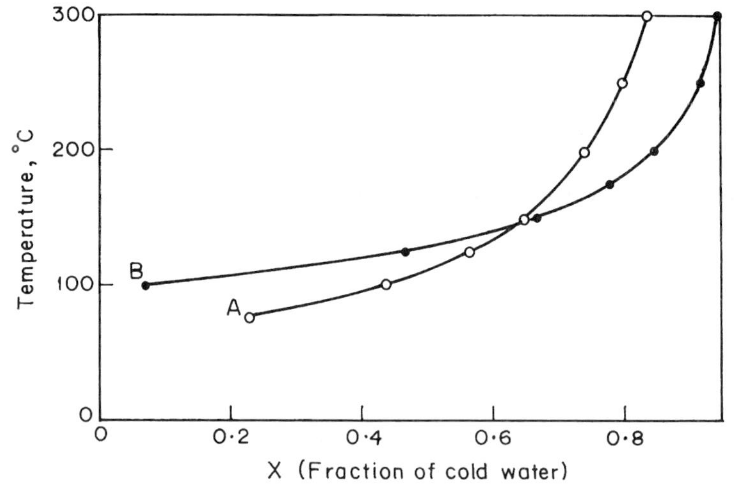

Fig. 5.3. Relations between fractions of cold water and temperature in a mixing model. Fraction of cold water based on model shown in Fig. 5.2 with enthalpy consideration (A), and with silica content consideration (B). For details, see text.

(2) X_t is plotted against the temperatures from which the assumed hot water enthalpy values are derived (curve A in Fig. 5.3).

(3) A series of values of the silica contents of hot water for the temperatures listed in Fig. 5.3 is assumed and X_{Si} is calculated for each value as follows:

$$X_{Si} = \frac{(\text{silica in hot water}) - (\text{silica in warm spring})}{(\text{silica in hot water}) - (\text{silica in cold spring})} \tag{5.12}$$

(4) X_{Si} is plotted against temperature on the same graph (curve B in Fig. 5.3). The point of intersection of these two curves gives the estimated temperature of the hot-water component and the fraction of the cold water.

In Fig. 5.3, curves A and B are drawn for a case where the warm spring water and cold water temperatures are 60 °C and 10 °C, respectively, and the corresponding silica contents are 45 and 5 mg l^{-1}, respectively. The intersection of A and B shows the cold water fraction to be 0.64 and the temperature of the hot water is inferred to be 148 °C.

At times these two curves do not intersect, or intersect at an unreasonably high temperature. These situations arise when the ascending hot water loses heat before mixing with cold water and a different procedure is suggested by Fournier and Truesdell (1974) to deal with such a situation. The accuracy of the mixing model depends to a large extent on the exactness with which the chemistry and temperature of the local cold subsurface waters are determined. Warm spring mixing models have been used in geochemical prospecting in most geothermal fields for estimating subsurface reservoir temperatures (Arnórsson, 1985).

Isotope Geothermometers

Several isotope-exchange reactions can be used as subsurface temperature indicators. The exchange of ^{18}O between dissolved sulfate and water is the most useful because of rapid equilibrium at reservoir temperatures greater than 200 °C and pH$<$7, conditions that favor sulfate ion—water exchange. Truesdell (1976) has summarized the uncertainties associated with various isotope geothermometers. Two of the commonly used empirical relationships are as follows:

$$10^3 \ln \alpha^{18}O_{SO_4-H_2O} = 3.251 \times 10^6 T^{-2} - 5.1 \qquad (5.13)$$

(after Mizutani and Rafter, 1969)

$$10^3 \ln \alpha^{18}O_{SO_4-H_2O} = 2.88 \times 10^6 T^{-2} - 4.1 \qquad (5.14)$$

(after McKenzie and Truesdell, 1977)

GEOPHYSICAL TECHNIQUES

Geophysical surveys provide the only means of delineating deep subsurface features, other than drilling. They can be used to cover large areas in a short time and at much lower cost when compared to drilling. In addition to major structural features, both shallow and deep, they address other important questions relevant to geothermal exploration. These include the source of heat, areal extent of the subsurface reservoir, zones of upflow of fluids and highly permeable pockets, and assessment of the geothermal energy potential of a resource. Certain geophysical techniques can provide valuable inputs to understanding the movement of fluids in response to withdrawal as well as re-injection during the exploitation stage of the geothermal field, and thus help in proper management of the field towards sustaining production over a long time. Therefore, geophysical investigations constitute an essential part of any exploration program, in conjunction with geological, hydrological and geochemical surveys. Geophysical anomalies associated with a geothermal prospect are usually caused by contrasts between physical properties of rocks and fluids inside or near the reservoir and those outside it. The most common physical properties that are targets of geophysical exploration are temperature, resistivity, density, porosity, magnetic susceptibility and seismic velocity. The techniques to be applied to a geothermal prospect are primarily influenced by pre-existing information such as the geological and tectonic setting, type of the hydrothermal system and hydrological characteristics of the area.

Geophysical methods in exploration of hydrothermal resources have been reviewed periodically (i.e., Banwell, 1970, 1973; Bodvarsson, 1970; Combs and Muffler, 1973; Combs, 1976; Palmason, 1976; McNitt, 1976; Meidav and Tonani, 1976; Lumb, 1981; Ward, 1983; Rapolla and Keller, 1984; Wright et al., 1985). Some of the geophysical methods such as thermal, electrical resistivity, gravity, seismic refraction and well logging, having been successfully used in several geothermal exploration projects are now well established. Other methods using controlled source electromagnetic, magnetotelluric (MT), self-potential, seismic reflection and radiometric techniques are being developed and tested. The geophysical techniques used are specific to a particular geothermal resource being explored. The choice of the technique is mainly governed by the temperature as well as the physical and chemical properties of reservoir rocks and the fluids contained in them. Therefore, before deciding on a set of geophysical surveys, it is important to have some idea about the various geophysical targets that a particular hydrothermal system may present.

Exploration work in geothermal areas is mostly aimed at delineating the geometry of shallow geothermal reservoirs with upflow zones causing geothermal manifestations. However, deeper systems exist without any upflow zones and surface manifestations. The most suitable geophysical methods for exploring shallow reservoirs differ in part from those used for deep reservoirs. The thermal state of the fluid in the reservoir adds further complications. The physical properties of porous rocks, such as resistivity and density, are different in rocks filled with steam from those filled with hot water. A critical target of exploration programs is to locate highly permeable zones (Lumb, 1981; Wright et al., 1985), which eventually control the exploitation of the hydrothermal system by locating suitable drill-holes. Primary permeability results mainly from intergranular porosity and the interconnection between the pores, and it decreases with depth due to compaction. Secondary permeability, such as fault zones, fractures, dykes and breccia zones, play a very significant role in exploitation of geothermal fluids. Leaching by hydrothermal fluids and mineral deposition often causes changes in permeability of reservoir rocks. Locating subsurface permeable zones within a reservoir is essentially an indirect procedure. Therefore, an exploration geophysicist should fully utilize available geological, hydrological and geochemical information relevant to permeability in a reservoir to decide on geophysical techniques to delineate highly permeable locales within the hydrothermal system.

The mineral assemblages produced by the thermal fluids alter the physical properties of the reservoir rocks. In most cases, the type and distribution of hydrothermal alteration depends on temperature, fluid composition, permeability, pressure, rock type and time (Browne, 1978). As described in Wright et al. (1985), a strong zonation of alteration minerals is observed in most hydrothermal systems. At temperatures of up to 225 °C, clay minerals, quartz and carbonate are the dominant secondary minerals. Chlorite, illite, epidote, quartz and potash feldspar are important at higher temperatures. At temperatures above 250 °C, metamorphism of reservoir rocks can occur leading to increase in density. Precipitation of silica may occur through cooling of hot brines. Silicification, in turn, often results in significant reduction in both porosity and permeability of the reservoir rocks, effectively sealing the sodium chloride reservoir and preventing its expression at the surface. However, steam and gas, which can travel upward, interact with meteoric water and produce near-neutral pH sodium bicarbonate-sulfate water. Therefore, a secondary geothermal reservoir is formed, which is unconnected with the primary reservoir containing sodium chloride water at depth.

Various classifications of the geophysical methods used in geothermal exploration have been adopted (Bodvarsson, 1970). The methods can be classified according to the depth of investigation. They can also be classified as direct and indirect methods (Palmason, 1976). Direct methods include thermal exploration techniques, which aim at geometrically mapping anomalous thermal zones. Indirect methods include techniques for investigating geological structures and highly permeable zones which may control the accumulation and movement of geothermal fluids. Geophysical techniques are used for exploration as well as to map the extent of the geothermal reservoir, mass changes by fluid withdrawal or injection, and changes in saturation and gas content during the production stage. Microgravity, microseismic and resistivity surveys are often used to monitor an exploited field. A broad classification of geophysical methods used in exploration of hydrothermal resources and the targets is given in Table 5.2. The most commonly used geophysical methods are discussed briefly in the following sections.

Thermal Methods

Thermal exploration techniques are extremely useful in assessing the size and potential of a geothermal system. Near-surface temperature gradient and heat flow measurements are routinely made in any geothermal exploration program and are often used as primary criteria for selection of drilling sites. The average global conductive heat flow, computed from measurements made in both continental and

Table 5.2. Use of various geophysical techniques in exploration of geothermal resources of hydrothermal origin (modified from Wright et al., 1985)

	Geophysical technique	Potential target(s)
1 Thermal	Shallow (~1 m depth) temperature surveys	High temperatures in subsurface rocks or fluids; anomalous high heat flow areas
	Thermal gradient measurements	
	Heat flow measurements	
2 Electrical	Resistivity surveys	Hot brines, fluid-induced alteration zones, faults
	Induced polarization	Mineralization zones, fluid-induced alteration zones
	Self-potential	Heat and fluid flow
	Telluric current method	Hot brines, fluid-induced alteration zones, faults
	Controlled source electromagnetics	Hot brines, fluid-induced alteration zones, faults
	Natural source electromagnetics (Magnetotelluric/ Audiomagnetotelluric)	Structure, hot brines, partial melt zones, deep magma chamber
3 Gravity	Gravity and microgravity surveys	Structure, intrusions, alteration zones, anomalous density, migration of fluids
4 Magnetic	Airborne and ground magnetic surveys	Structure, alteration zones, anomalous magnetic properties, rock type
5 Seismic	Microseisms	Active hydrothermal processes
	Microearthquakes	Active faulting and fracturing, velocity distribution and attenuation
	Telesesimics	Deep magma chamber
	Refraction surveys	Structure, velocity distribution and attenuation
	Reflection surveys	Structure, velocity distribution and attenuation
6 Radiometric	Radioelemental (K, U and Th) and heat production surveys	Anomalous radioactive zones (^{222}Radium and ^{226}Radon)
7 Borehole geophysical tools	Geophysical well logging	Anomalous temperature, porosity, permeability, rock type
	Vertical seismic profiling	Velocity distribution, fractures
	Electrical	Hot brines, alteration zones, faults

oceanic areas is 87 mW m^{-2}. In combination, or individually, factors such as high content of radioactive material, movement along faults and other similar factors can cause heat flow values slightly in excess of the normal values. On the other hand, geothermal areas, which are economically attractive at the present time have several times higher heat flow than the global average. Such anomalous heat flow values are usually restricted to fairly small areas, often around natural discharge features. It should be kept in mind that the surface expression of a geothermal activity may not necessarily be located directly above the heat source because the hydrothermal waters can flow horizontally for considerable distances.

Thermal methods of prospecting normally make use of the following three measurements:

One-meter temperature probe surveys
Temperature measurements at 1 m depth are inexpensive and quick, and can be used to detect anomalously hot areas. When a geological structure controls thermal activity, one-meter probe surveys have the same value as infrared imagery surveys. Measurements are made along profiles or in a grid pattern at distances varying from 10 to 100 m. A simple probe (Fig. 5.4C) is used. Due care must be taken of near-surface effects such as topography, precipitation, diurnal and annual solar

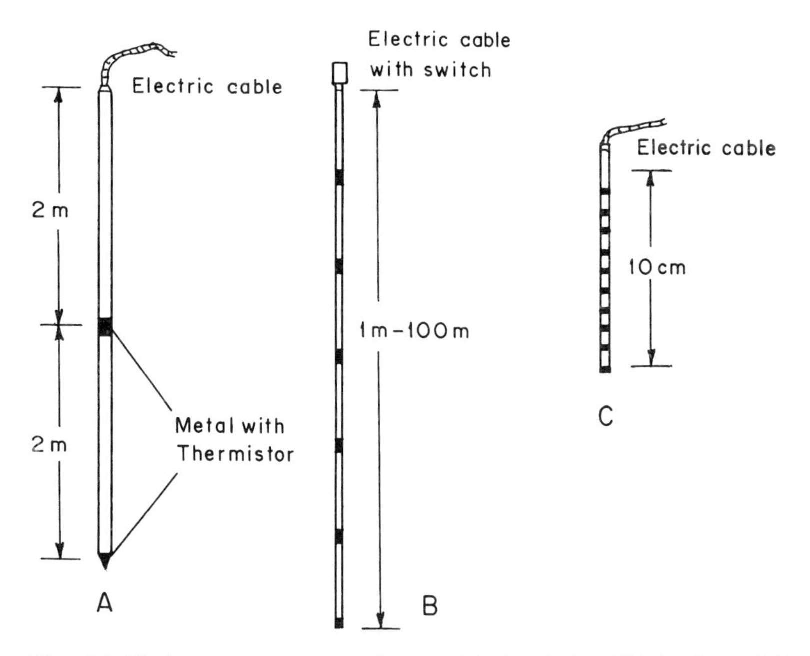

Fig. 5.4. Various temperature probes used in boreholes. (A) 4 m-long rigid probe with two thermistors. (B) Cable probe with a number of thermistors. (C) Short probe for near-surface measurements.

heating variations. Changes in surface albedo, which affect the amount of heat energy absorbed, variations in soil thermal diffusivity, and movement of groundwater, should also be taken care in interpreting the results of 1-m depth probe surveys. The effect of groundwater movement is most important. Relatively slowly moving groundwater can carry away conductive heat, even over a strong geothermal anomaly, and grossly distort the pattern of surface temperature. Except groundwater flow, most of the other effects become negligible below a depth of about 25 m. These surveys are often extended to depths of a few meters. The results obtained from the shallow temperature surveys could guide whether undertaking intermediate-depth, thermal gradient surveys are warranted.

Temperature gradient surveys

Temperature gradient surveys provide basic data about subsurface temperatures and have been often used as a primary criterion in selecting drilling sites. Therefore, these surveys form part of any systematic exploration program. A thermal probe used for such measurements is shown schematically in Fig. 5.4A. Essentially, temperatures are measured simultaneously at two different depths using thermistors embedded along the probe and gradient is computed from the ratio of temperature difference between the two measurements and the distance between them. In specific situations, this procedure can be extended to a number of thermistors fixed at regular intervals along the probe, as shown in Fig. 5.4B. Measurements made in the near-surface zone are often perturbed by recharge of meteoric water with vertical and lateral flow of cold water. Therefore, temperature measurements are commonly made in boreholes to depths ranging from 10 to 150 m in hard rock terrain and to depths of a few hundred meters in sedimentary areas so as to penetrate the near-surface hydrologic regime. A typical value of gradient at shallow levels (a few meters to a few tens of meters) is about $3\,^{\circ}\mathrm{C}\,100\,\mathrm{m}^{-1}$, whereas in potential geothermal areas it could be greater than $7\,^{\circ}\mathrm{C}\,100\,\mathrm{m}^{-1}$ (Combs and Muffler, 1973). Temperature gradient surveys are useful in defining the areal extent of thermal anomalies. However, care should be exercised in extrapolating the temperature gradients to deeper levels. The linear extrapolation of gradients is often found to be erroneous, giving high values, due to decrease in porosity with depth and the effect of convection cells at depth. Particularly in sedimentary rocks, porosity decreases considerably with depth. The thermal conductivity of most minerals is much higher (3–10 times) than that of water. Consequently, with the decrease of porosity, bulk thermal conductivity increases considerably and the gradient decreases. The convection cells have an even larger contribution in decreasing temperature gradients at depth. Fig. 5.5, from White (1973), shows that convection produces higher thermal gradients at the top of the convection cell and lower thermal gradients in the region within the convection cell.

Heat flow investigations

Geothermal gradient surveys are often adequate to outline a geothermal area in general. For a better understanding of the regional subsurface thermal regime and

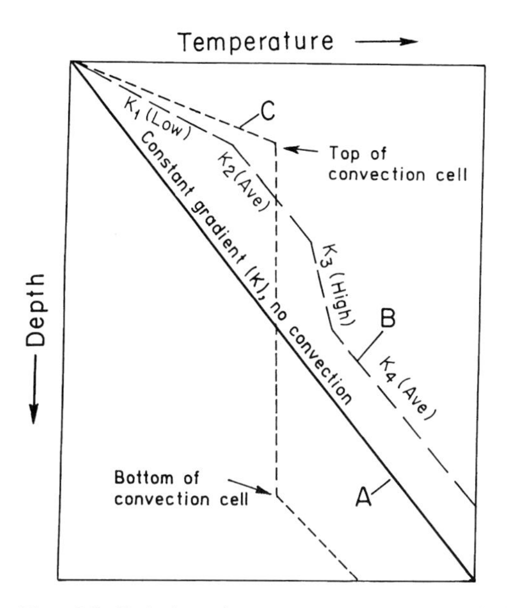

Fig. 5.5. Relations between temperature and depth when: (A) heat flow is controlled by thermal conduction in rocks of constant conductivity, (B) rocks of variable conductivity, or, by (C) major convective disturbances (from White, 1974).

delineation of the central production zone from the less productive marginal zones, heat flow determinations are useful. The most important advantage of heat flow data compared to temperature gradient data is that heat flow values are independent of the in situ thermal conductivity of the rocks. Therefore, as explained by Sestini (1970), in inhomogeneous terrain the heat flow data alone can provide accurate information regarding productive zones. One of the earliest demonstrations of the usefulness of heat flow investigations in delineating the central production zone was given by Boldizsár (1963) in the natural steam field at Larderello, Italy. A summary of the heat flow techniques commonly used in geothermal exploration is given by Rybach (1989).

The distance between boreholes used for heat flow investigations depends upon the size of the subsurface heat source. Magmatic intrusions, which are usually the sources for economic geothermal fields, cause geothermal disturbances of at least 1–2 km lateral extent. Therefore, measurements at distances of 500 m are considered to be adequate. Depths of boreholes required for heat flow investigations are at least 150 m in hard rock terrain and a few hundred meters in sedimentary areas. Heat flow is a product of thermal gradient and thermal conductivity of rock type over which the gradient has been computed. Therefore, it is necessary to measure the thermal conductivity of the formations using cores obtained by drilling.

Interpretation of near-surface temperature gradient and the heat flow data is complicated by a variety of factors, as discussed earlier. A few classical studies that provide detailed accounts of acquisition of thermal datasets in anomalous geothermal settings, analysis and interpretation, are those by Lachenbruch et al. (1976a, 1976b), Sass and Sammel (1976), Blackwell and Morgan (1976), Lachenbruch (1978), Chapman et al. (1981), Sass et al. (1981), Blackwell et al. (1982), Blackwell, 1985; Lachenbruch et al. (1985); Blackwell and Steele (1989). In a recent study of The Geysers—Clear Lake magmatic/volcanic geothermal system, Erkan et al. (2005) analysed temperature-depth data from numerous deep wells to characterize the thermal regime to depths where temperatures in the range 250–350 °C occur. This geothermal system is associated with a large intrusive center, but has a very limited surface expression. Although seismic and electrical methods did not reflect the presence of a magma chamber in the crust, the interpretation of the thermal datasets yielded valuable information on thermal effects of large intrusions in the upper and middle crustal levels. A close correspondence between the thermal anomaly and large negative gravity anomaly has been observed. Studies carried out by Williams and Grubb (1998) also demonstrated the usefulness of thermal investigations in constraining the lateral extent of the geothermal reservoir. Erkan et al. (2005) concluded that gravity and thermal measurements are more effective in locating and characterizing many types of upper crustal magma chambers than are seismic and electrical studies. Sass and Walters (1999) have elucidated the thermal regime of the Great Basin, U.S.A. from numerous heat flow determinations in the region, and indicated possible localities of enhanced geothermal systems.

The importance of subsurface temperature and heat flow measurements in detection of groundwater flow, on both regional and local scales, has been well recognized. However, further researches are needed to improve our understanding of regional and local hydrologic effects on temperature measurements, so that the observed thermal gradient and heat flow patterns in basins and geothermal fields can be converted into quantitative estimates of flow parameters. Studies are underway on 2-dimensional and 3-dimensional modeling of coupled mass and heat transport in complex geologic situations including extension, faulting, intrusions, uplift and erosion and sedimentation and burial, which would result in a better interpretation of thermal signatures in tectonically active terrains and further widen the scope of geothermal measurements (see, e.g., Furlong and Chapman, 1987; Williams et al., 1997; Wisian et al., 1999; Wisian, 2000; Clauser, 2003 and references therein).

Other Thermal Techniques

Other temperature-sensitive methods such as *snow-melt photography* and *airborne thermal infrared imagery* have been sparingly used in reconnoitery geothermal exploration for identifying "warm or hot spots" in large areas. In a novel study conducted many years ago, White (1969) estimated heat flow in Yellowstone National Park, U.S.A. utilizing individual snow falls as calorimeters. Snow melts faster over thermally anomalous areas relative to normal areas. Therefore, after a favorable snow fall—characterized by being of brief duration, heavy, and occurring when the air temperature was close to 0 °C with no wind—contacts between snow-free and snow-covered ground were mapped at a suitable scale. Each mapped contact, as time passed, represented a heat flow contour. Airborne thermal infrared imageries have been used to map the occurrence of warm ground and distribution of hot springs in Kenya (Noble and Ojiambo, 1976) and hot springs in Hawaii (Fischer et al., 1966). However, the thermal infrared technique has seen very limited use as an exploration tool for hydrothermal resources in recent times.

Thermal Conductivity Measurements

A number of experimental methods for measuring thermal conductivity of rocks have been in use over the past few decades (Roy et al., 1981; Beck, 1988; Jessop, 1990). They can be classified into two broad groups: (a) one-dimensional, steady state, comparative type, and (b) two-dimensional (cylindrical), absolute type (Roy et al., 1981). The steady state, comparative, divided-bar apparatus has been the most widely used of all existing methods in view of its simplicity of construction and operation, reasonable accuracy, and an equilibrium time that is compatible with routine measurements (Jessop, 1990).

A schematic diagram of the divided-bar apparatus is shown in Fig. 5.6. The rock samples are in the form of cylindrical discs, usually 25–50 mm diameter and 10–25 mm thick, sliced from cores obtained from boreholes. Each disc is inserted between two identical reference discs (e.g., polycarbonate), whose thermal conductivity is calibrated against a standard material such as fused quartz, or crystalline quartz cut so that the heat flow is perpendicular to the optic axis. The reference discs are of the same diameter as the rock sample but of appropriate thickness so that its thermal resistance is comparable to that of the rock sample to be tested. The advantage of a low-conductivity reference material such as polycarbonate is that the stack length does not become too large. Cut surfaces of the reference as well as rock discs are polished until the thickness variation is less than 0.01 mm. The flat and polished surfaces, together with applied pressure at both ends of the stack, minimize the contact resistances within the stack during measurement. A constant temperature differential is maintained across the stack comprising the rock disc and the two reference discs. A suitable insulated housing for the stack ensures that lateral heat losses are minimal and flow of heat takes place in the vertical direction only. At steady state, i.e., when heat flow across each segment of the stack is constant, temperature differential across each segment is measured using thermocouples. To facilitate measurement of temperature differential across each disc, the thermocouples are often inserted within thin and highly polished discs of very high conductivity such as copper having same diameter as the reference and sample discs. Each sample disc is sandwiched between two such copper discs. This procedure eliminates the need for drilling fine holes into the reference and sample discs and also

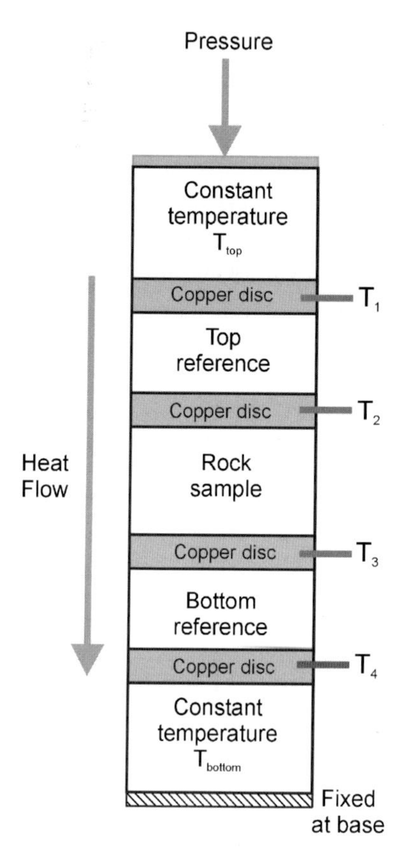

Fig. 5.6. Schematic diagram of steady-state divided-bar thermal conductivity apparatus.

helps minimize subjective errors while inserting them in the stack prior to measurement. The temperature differentials are combined with the individual disc thickness and conductivity of reference material to compute the thermal conductivity of the rock sample.

At steady state, heat flow across the reference material is the same as that across the rock sample, i.e.,

$$q_{\text{ref}} = q_{\text{rock}} \tag{5.15}$$

$$\lambda_{\text{ref}} \left(\frac{\partial T}{\partial l} \right)_{\text{ref}} = \lambda_{\text{rock}} \left(\frac{\partial T}{\partial l} \right)_{\text{rock}} \tag{5.16}$$

where, λ is the thermal conductivity and $(\partial T/\partial l)$ the thermal gradient, l being the thickness of the sample. The conductivity of the rock sample is therefore given by:

$$\lambda_{\text{rock}} = \lambda_{\text{ref}} \frac{\left(\frac{\partial T}{\partial l}\right)_{\text{ref}}}{\left(\frac{\partial T}{\partial l}\right)_{\text{rock}}} \tag{5.17}$$

The combined systematic and random errors in the measurement of a single rock disc are $<5\%$, much lower than the variation from specimen to specimen in most rocks.

Methods using the two-dimensional (cylindrical) type employ both steady-state as well as transient techniques. In both cases, a line source of heat (e.g., a long heating probe, with length $L \gg$ diameter) is used to supply energy at a known rate Q' per unit time per unit length. In the steady-state technique, temperature sensors (usually, thermocouple or thermistor) are placed at known radii, r_1 and r_2, and from the temperature difference, $(T_1 - T_2)$, between them the conductivity, λ, can be determined as follows (Roy et al., 1981):

$$\lambda = \frac{Q'}{2\pi} \frac{\ln(r_2/r_1)}{T_1 - T_2} \tag{5.18}$$

In the transient case, the growth of temperature is measured as a function of time, and the conductivity, λ, is determined using the following equation (Roy et al., 1981):

$$\Delta T = \frac{Q'}{4\pi\lambda}[2h + \ln(Dt) - (4h - B)(2B\tau)^{-1} + (B - 2)(2B\tau)^{-1}\ln(Dt)] \tag{5.19}$$

where $D = 4\alpha/e^{g}a^2$, g being the Euler's constant (0.5772), $\tau = \alpha t/a^2$, $h = \lambda/ac$ (dimensionless), B twice the ratio of thermal capacities of the medium and probe material (dimensionless), a the probe radius, t the time, α the thermal diffusivity, and c is the thermal conductance per unit length of the contact layer.

If the radius of the probe, a, very small, so that τ is always very large, even for relatively small times, Eq. (5.19) reduces to the line source equation:

$$T = \frac{Q'}{4\pi\lambda}(\ln t) + A \tag{5.20}$$

where the constant $A = [(Q'/4\pi\lambda)\ln D] + 2h$. A plot between temperature rise versus $(\ln t)$ is a straight line in the early part of the experiment. The slope of the straight-line plot yields a value of thermal conductivity

$$\lambda = \frac{Q'}{4\pi} \frac{\ln(t_2/t_1)}{(T_2 - T_1)} \tag{5.21}$$

A relatively recent variant of the transient technique for measuring thermal conductivity of rocks is the optical scanning technique (Popov et al., 1985, 2001). The method is based on scanning a surface of the rock sample with a focused, mobile and constant heat source in combination with a temperature sensor. The heat source and sensor move together with the same speed relative to the sample and at a constant distance to each other. A laser source is commonly used for heating, and infrared detectors are used as temperature sensors. The sensor displays the value of the maximum temperature rise, θ, along the heating line behind the source. The maximum temperature rise is given as

$$\theta = \frac{\bar{Q}}{2\pi x \lambda} \tag{5.22}$$

where $\bar{Q}$ is the power of the heat source, x is the separation between heat source and sensor, and λ is the thermal conductivity of the rock sample. The method provides several advantages such as characterization of inhomogeneity and identification of three-dimensional anisotropy in conductivity of rock samples, high speed of operation, and freedom from constraints on sample size, shape and laborious machining of sample surfaces (Popov et al., 1999). The equipment, although relatively more expensive than those using other techniques, has contributed to several-fold increase in volume of conductivity data acquired for geothermal researches over the past few years.

Measurements on Rock Fragments and Unconsolidated Sediments

Owing to the nature of rocks in geothermal areas, cores are not available and the rocks cut by the drill bit are taken out of the hole in the form of chips or cuttings only. In the case of unconsolidated sedimentary rocks met with during geothermal exploration in sedimentary basins, even cores disintegrate quickly upon retrieval from the hole. Sass et al. (1971) have developed a method for measuring the conductivity of such rock samples using a steady-state divided-bar apparatus. The material in the form of chips, cuttings or unconsolidated sediments are first crushed to the size of coarse sand and then tightly packed into a specially designed cell. A typical cell is shown in Fig. 5.7. The base and the lid of the cell are made of a good heat conductor such as copper or brass and the walls are made of polycarbonate or other plastic material of very low thermal conductivity. The walls are thin to ensure that the necessary correction for the plastic is as small as possible. The material inside the cell is saturated with water before closing the lid of the cell. The cell is then introduced between two reference disks in the stack and thermal conductivity of the cell is measured in the usual way. A set of corrections is made to account for the conductivities of the plastic wall and the fraction of water in the cell before the thermal conductivity of the rock can be computed. The details of the method are

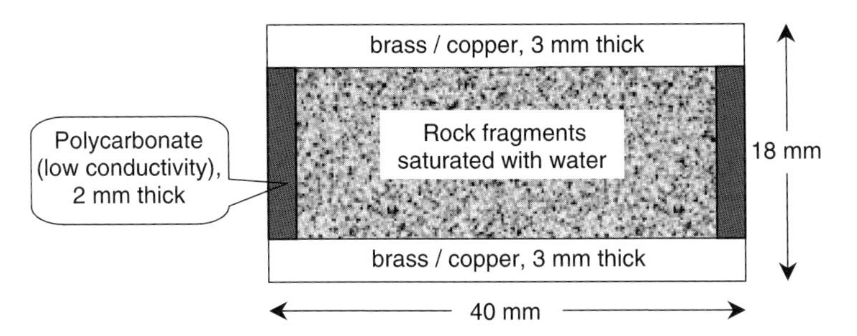

Fig. 5.7. Typical dimensions of a cell used for measuring thermal conductivity of rock fragments using a steady-state divided-bar apparatus.

given in Sass et al. (1971). With good measurement practice, the method can yield results accurate to about 10%.

Measurement of thermal conductivity of sedimentary rocks requires additional caution. Measurements are conventionally made in the laboratory on drill core samples and more often on cuttings after saturating the samples with water so that in situ conditions are simulated. The best measurements, however, have large uncertainties resulting from poor sampling, inadequate correction for porosity, lack of control on the orientation of cuttings along which thermal conductivity is determined for anisotropic samples, and inadequate information about the relationship between in situ and laboratory measured thermal conductivity (Rao et al., 2003). Shales, which constitute a large proportion of the rocks in sedimentary basins, are particularly problematic because they are (a) highly porous and disintegrate easily making measurements difficult, (b) are anisotropic, and their thermal conductivity is not a function of compaction (or depth), and (c) show structural variations associated with transformation to kaolinite (Blackwell and Steele, 1989). These uncertainties should be kept in mind while developing a thermal model.

Temperature Dependence of Conductivity

The thermal conductivities of several rock types show significant variations with increase in temperature. In case of boreholes deeper than 1–1.5 km, particularly in regions of high temperature gradients such as found in geothermally anomalous areas, the temperature effect on conductivity becomes substantial. This necessitates a temperature correction to thermal conductivity prior to computation of temperatures at deeper levels on the basis of heat flow measurements. Large uncertainties in temperature computations caused due to inadequate characterization of the temperature effect can result in (a) inaccurate estimation of depth to a particular reservoir temperature and (b) masking of changes in heat flow with depth, which could lead to misinterpretations of heat and fluid flow conditions within a hydrothermal system (Williams and Sass, 1996). Both steady-state comparative divided-bar as well as transient needle probe techniques have been suitably modified to carry out conductivity measurements at high temperatures of about 800 °C, and the temperature dependence of thermal conductivity has been established for a few rock types (Birch and Clark, 1940; Zoth and Haenel, 1988; Pribnow et al., 1996; Williams and Sass, 1996; Seipold, 2001).

Electrical Resistivity Methods

Electrical resistivity methods have been very successfully used in geothermal exploration as is evident from the numerous reported case histories in the literature (e.g., Keller, 1970; Meidav, 1970; Risk et al., 1970; Zohdy et al., 1973; Stanley et al., 1976; Tripp et al., 1978; Ward et al., 1978; Razo et al., 1980; Ross et al., 1996; and others). Change in electrical resistivity of the rock-fluid volume is the most important physical property change due to the presence of a hydrothermal system, other than elevated temperature and heat flow (Moskowitz and Norton, 1977). Ionic mobility increases with increase in temperature up to about 300 °C, resulting in an increase in conductivity. Ionic conduction in rocks also increases with increased porosity, increased salinity and increased amounts of certain minerals such as clays and zeolites. Hydrothermal systems, which are generally characterized by one or more of these features, are therefore associated with a high-conductivity (or low-resistivity) anomaly (Wright et al., 1985).

The three classical configurations of electrodes used in resistivity surveys are shown in Fig. 5.8. In the Wenner array, four electrodes are equally spaced at intervals a. C_1 and C_2 are the current electrodes and P_1 and P_2 are potential electrodes. For

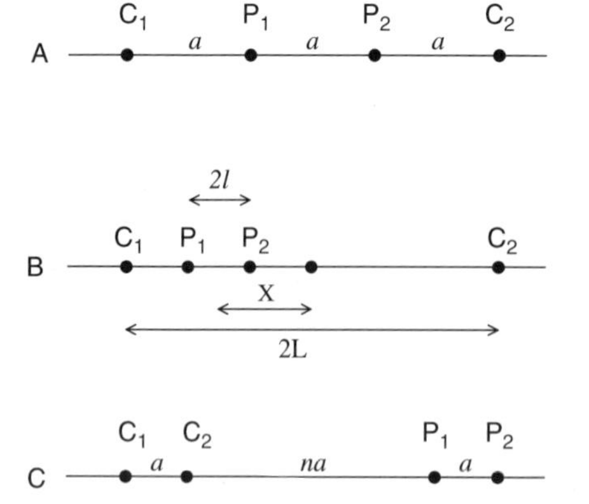

Fig. 5.8. Commonly used electrode configurations in electrical resistivity surveys: (A) Wenner, (B) Schlumberger, and (C) Dipole–Dipole.

this configuration, apparent resistivity ρ_a is given by

$$\rho_a = 2\pi a \frac{\Delta V}{I} \tag{5.23}$$

where ρ_a is the potential difference between P_1 and P_2 when current I is sent through the current electrodes. The distance between potential electrodes, $2l$, in the generalized Schlumberger array is small compared to the distance, $2L$, between the current electrodes. When $L \geq 5l$, resistivity can be estimated using the relation:

$$\rho_a \cong \frac{\pi}{2l}\left[\frac{(L^2 - X^2)^2}{L^2 + X^2}\right]\frac{\Delta V}{I} \tag{5.24}$$

where X is the distance between the center of the current electrodes and the center of the potential electrodes. In the symmetrical Schlumberger array arrangement, $X = 0$, and apparent resistivity is estimated using the relation:

$$\rho_a \cong \frac{\pi L^2}{2l}\frac{\Delta V}{I} \tag{5.25}$$

In the dipole–dipole configuration of electrodes, the potential electrodes are placed outside the current electrodes, each pair having a constant mutual separation, a. When the distance between the two pairs, na, is large (Fig. 5.8), the current source can be treated as a dipole and the apparent resistivity is estimated using the relation:

$$\rho_a \cong \pi n(n+1)(n+2)a\frac{\Delta V}{I} \tag{5.26}$$

The above mentioned configurations are the standard ones. Various others have been tested in recent years to increase the depth of penetration when the near-surface resistivities are low—a situation mostly met with in hydrothermal areas.

Resistivity sounding and profiling are two commonly employed procedures for estimating underground resistivity. The object of resistivity sounding is to estimate variation of resistivity with depth below a given point. Such measurements are required when the ground consists of a number of more or less horizontal layers and vertical variation in resistivity is to be estimated. With increasing separation of the current electrodes, the current penetrates deeper. While using the Wenner configuration for resistivity sounding, the electrode separation, a, is increased in steps, for example $a = 4$, 8, 16, 32 m. The object of resistivity profiling is to detect the lateral variation of resistivity of the ground. This kind of survey is undertaken to delineate underground intrusions. In electrical profiling with the Schlumberger method, the current electrodes are fixed at a large separation and the potential electrodes are moved at a constant small separation within the current electrodes (Fig. 5.8). Profiling is normally conducted at right angles to the strata of the structure such as faults and dykes. Standard master curves are available in the literature to interpret field observations (Compagnie Generale de Geophysique, 1963).

Various factors contribute in increasing the electrical resistivity contrast between the geothermal systems and the surrounding rocks. First, the electrical resistivity decreases with the increase of temperature. Second, with the increase of temperature, solubility increases and consequently there is an increase in salinity and decrease in resistivity. Third, the pore space in the central portion of a geothermal field increases with increase in temperature, thereby further decreasing the resistivity. All these factors contribute in decreasing the resistivity of geothermal fields relative to the surrounding rocks. In a classical geothermal system, typically the electrical resistivity is about 5 Ω m, while that of the surrounding rocks could be in excess of 100 Ω m (Meidav, 1972).

As summarized by Meidav and Tonani (1976), there are three major factors, which introduce ambiguity so that electrical resistivity methods cannot always be used as a perfect geothermal exploration tool. These are (1) the effect of dry steam, (2) the effect of brine, and (3) the presence of highly porous but non-permeable rocks like clay and shale.

In liquid-dominated geothermal reservoir systems, the resistivity decreases with the increase of temperature. This is not the case with dry-steam-dominated geothermal reservoirs. Investigations carried out at vapor-dominated geothermal systems at Larderello, Italy (Batini and Menut, 1964), The Geysers, California (Stanley et al., 1973) and Matsukawa, Japan (Hayakawa, 1966) have failed to show low resistivity. When a non-condensable gas cap overlies a boiling water table, resistivity soundings have revealed a high-resistivity layer sandwiched between two layers of low resistivity. Fig. 5.9, adapted from Meidav and Tonani (1976) shows a model of the type of resistivity-depth sounding curve observed in an area where a dry steam layer exists, compared to a region where temperatures increase moderately with depth and the sandwiched vapor-dominated layer is missing.

Cold brine or seawater has an electrical resistivity of less than 1 Ω m. Therefore many basins, where brine is accumulated, exhibit low resistivity without any connection with elevated subsurface temperatures. Meidav and Furgerson (1972) have reported a classical example from the Imperial Valley, U.S.A., where the resistivity value decreases regionally from 30 Ω m at Colorado River near Yuma, Arizona, to 1 Ω m near the Salton Sea, California. These observations indicate an increase in groundwater salinity, in general, in a northwestward direction.

Large sedimentary basins which are rich in clay sand sequences, such as the Gulf Coast of Mexico, are generally characterized by low resistivities. The decrease in resistivity has been caused by the increase in clay or shale content. Therefore, in sedimentary basins, unless there is some other direct evidence, decrease in resistivity cannot be interpreted in terms of the existence of a geothermal reservoir. Meidav and Tonani (1976) have presented a schematic plot of resistivity vs. temperature (Fig. 5.10), which is useful in diagnosing the subsurface rocks in terms of their geothermal energy potential. Only a few combinations of high temperature gradient

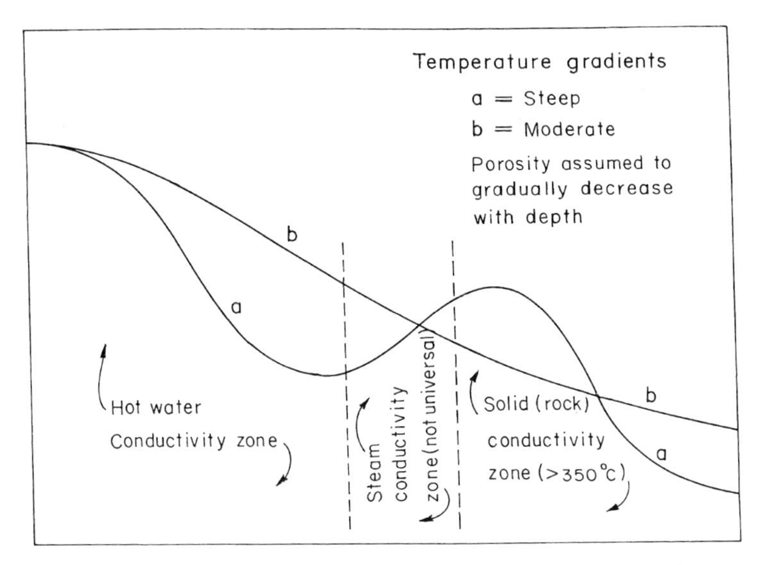

Fig. 5.9. A schematic representation of resistivity-depth graphs for areas where temperature increases moderately with depth (b), and for areas where temperature increases rapidly with depth (a) where a gas or dry steam layer may exist (from Meidav and Tonani, 1976).

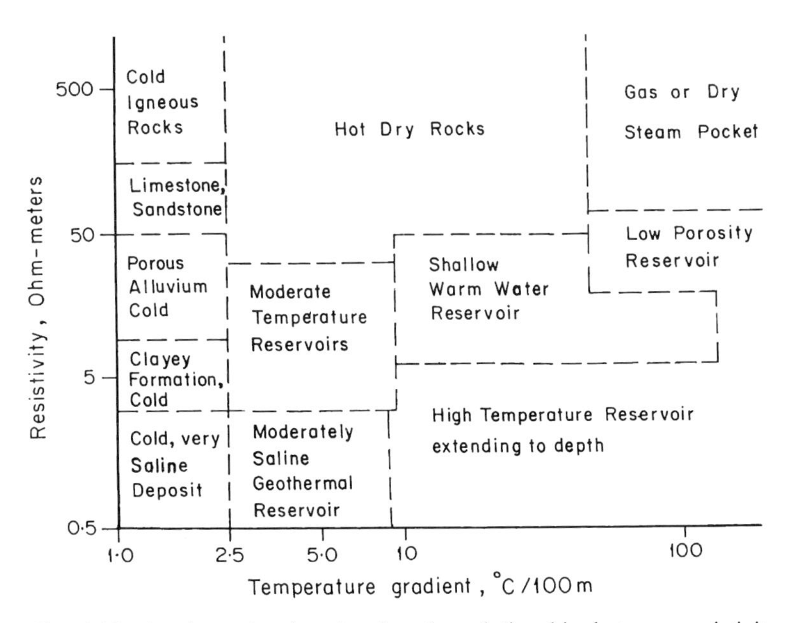

Fig. 5.10. A schematic plot showing the relationship between resistivity and temperature gradients. Such plots are useful in roughly estimating the nature of the subsurface rocks in terms of their geothermal energy potential.

and resistivity are indicative of favorable geothermal conditions. As pointed out by Meidav and Tonani (1976), ambiguity in interpretation of resistivity data can also be minimized by relating the resistivity data with the lithology and salinity of the groundwater. Knowledge of local geological and hydrological conditions is useful in interpreting resistivity data.

Electromagnetic Methods

The effectiveness of electromagnetic methods for geothermal prospecting has been demonstrated in a large number of areas (Keller, 1970; Keller and Rapolla, 1974; Keller et al., 1981; Kauahikaua, 1981; Goldstein et al., 1982; Ward, 1983; Hutton et al., 1989; Volpi et al., 2003; and others). In principle, the electromagnetic inductive method involves generating a magnetic field, which varies in time, and detection and measurement of either the electric field or the magnetic field arising from the current induced in the Earth. Theoretically, electromagnetic methods have a few clear-cut advantages over electrical resistivity methods. Unlike the direct current resistivity methods, the depth of penetration can be varied by changing the signal frequency without changing the geometry of the field configuration. Second, in the electromagnetic inductive methods signal size increases with decreasing resistivity, making them more suitable for geothermal areas. Third, unlike electrical resistivity methods, electromagnetic inductive methods are not adversely affected by the presence of a high-resistivity top layer. Time-domain electromagnetic methods are particularly successful in volcanic areas of high surface impedence (e.g., in the volcanic islands of Kilauea in Hawaii), where grounded electrical resistivity surveys are inadequate. Several configurations of the electromagnetic method have been used since the 1960s. In the following, we discuss two-loop profiling, two-loop sounding and the audio-frequency magnetotelluric (AMT) method, which have been used in geothermal exploration most successfully. Keller (1970) has made an exhaustive review of these methods. Subsequently, Ward and Wannamaker (1983) have reviewed the application of MT methods to geothermal prospecting.

Two-loop profiling method

The system consists of two small loops used as a transmitter and a receiver. The transmitter is powered from a small oscillator with selectable frequencies and a power generation capacity of a few tens of watts. Most commonly used frequencies are from a few hundred to a few thousand hertz. The receiver consists of a tuned amplifier and a ratiometer. The signal received through the ground is compared with the reference signal transmitted along a wire connecting the transmitter with the receiver. The relative amplitudes of the in-phase and out-of-phase components are measured as a percentage deviation from the amplitudes of these components when no conductive ground is present between the transmitter and the receiver. The voltage, which is 90° out of phase with the current in the transmitter, is considered to be a cent-percent response, and the voltage which 180° out of phase with the current in the transmitter is taken to be zero percent response. Typically, the separation between the transmitter and the receiver ranges from 20 to 50 m. The transmitter and the receiver are moved along a profile and readings are taken every few tens of meters. The depth to which conductivity is estimated is about one-half the distance between the transmitter and the receiver. The relative readings from a two-loop profiling system can be converted into apparent ground conductivity using some simple relations. When both coils are horizontal, the voltage at the receiver coil is given by

$$V_c = \frac{\mu\omega A_r M}{2\pi r^2 R^5}\left[9 - \left\{9 + 9\nu R + (2\nu R)^2 + (\nu R)^3\right\}e^{-\nu R}\right] \qquad (5.27)$$

where V_c is the complex voltage detected at the receiver coil, ω the frequency in rad s^{-1}, R the separation between two coils in meters, A_r the effective area of the receiver coil, M the moment of the source coil, i.e., the product of current and effective area, μ the magnetic permeability of the Earth, usually taken to be $4\pi \times 10^{-7}$ in MKS units, and ν the complex wave number for uniform Earth, usually taken to be $(i\omega\mu\sigma)^{1/2}$, σ being the conductivity in ℧ m^{-1}.

The above-mentioned Eq. (5.27) cannot be solved explicitly for conductivity. A curve is drawn from the in-phase and out-of-phase components of V_c, as a function of νR, such as the one shown in Fig. 5.11, and the value of ground conductivity is read from it. Keller (1970) has commented on various features of such a curve. Usually two values of ground conductivity are given from the out-of-phase component. A unique value of ground conductivity is provided from the in-phase component, provided the in-phase response is less than the 100% free air response. When the in-phase response is more than 100%, two values are given. When the response is close to 100%, the value of conductivity cannot be determined and a maximum possible conductivity is inferred. When the ground under investigation is uniform, one of the out-of-phase values will agree with the in-phase values. Deviation from this implies that the Earth in the vicinity of one of the coils is not uniform. The spacing between the coils and the frequency used should be so chosen that the in-phase response drops to 95% for the lowest conductivity to be resolved. Keller (1970) has suggested a number of combinations of frequencies and spacings for geothermal areas. These are given in Table 5.3.

Two-loop sounding method

In the two-loop sounding method, measurements are made at many frequencies and the data thus generated are interpreted in terms of a number of layers with differing resistivities lying one over the other. With the two-loop profiling method, it becomes difficult to carry the bulky generator used for powering the source, which is necessary to generate the required power at lower frequencies for deeper penetration. This difficulty is somewhat resolved in the two-loop sounding procedure.

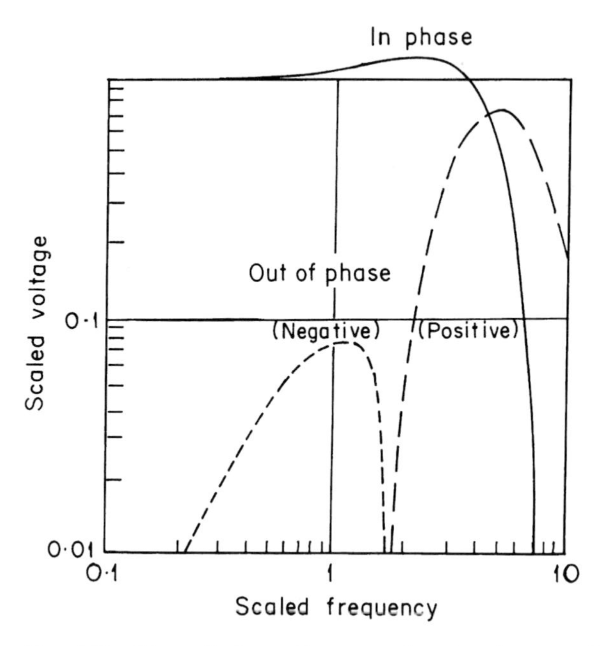

Fig. 5.11. Theoretical coupling curve for a two-loop profiling system (after Keller, 1970). For details, see text.

Table 5.3. Frequency and spacing between the coils as suggested by Keller (1970) for the two-loop profiling method in geothermal areas

Frequency (rad s^{-1})	Spacing (m)
10	3,800
100	1,200
1,000	380
10,000	120
100,000	38

Keller and Frischknecht (1966), Vanyan (1967) and others have discussed the theory of the two-loop electromagnetic sounding method. When the Earth consists of a number of horizontal layers, the voltage at the receiver end can be expressed by a relation of the type:

$$V_c = \frac{A_r M \mu}{2\pi R} \frac{\partial}{\partial R} \left[R \frac{\partial}{\partial R} \int_0^\infty \frac{m}{n_o + n_i/Q} J_o(mR) \mathrm{d}m \right] \tag{5.28}$$

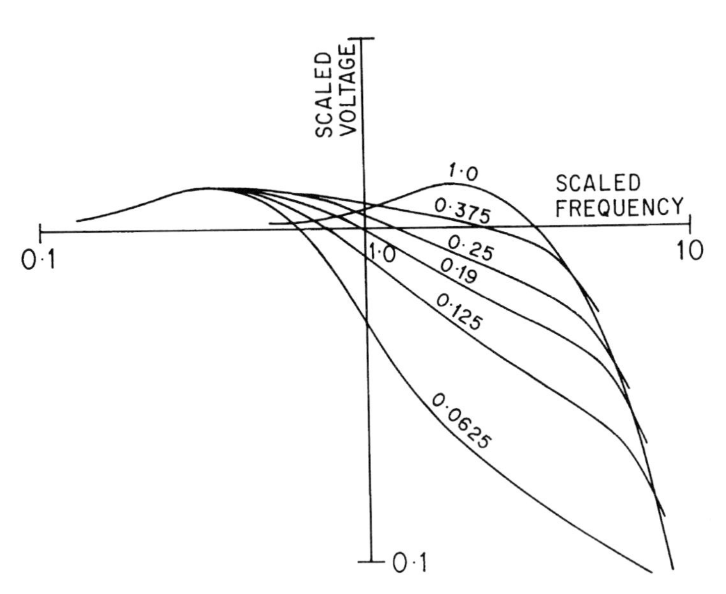

Fig. 5.12. A set of reference curves for electromagnetic coupling between two loops (from Frischknecht, 1967). The curves are valid for a case when a layer rests on a half-space which is 30 times more conductive. For details, see text.

where A_r is the effective area of the receiving loop, M the moment of the source, R the separation between the two loops, J_o the Bessel function, n_o the modified wave number in air, given by $n_o^2 = m^2 + \lambda\omega^2\mu\varepsilon$, where ε is the dielectric constant of air, n_i the modified wave number in the surface layer given by $n_i^2 = m^2 + \lambda\omega^2\sigma_i$, m is a dummy variable of integration, and Q the function, which can be written in the form of a recursive algorithm (Keller, 1970).

Evaluation of the above-mentioned expression is quite time consuming. However, tables of results for 1, 2 and 3 layers have been published by Frischknecht (1967), which can be conveniently used. Fig. 5.12 gives a typical set of curves for a two-layer case where the top layer has a 30-times higher resistivity than the underlying medium. Each curve given in Fig. 5.12 corresponds to a different thickness of the top layer expressed as a ratio to the distance between the two loops. For interpretation, field measurements are plotted on bi-logarithmic graph paper on the same scale as the reference curve and compared to select a reference curve agreeing closely with the observations. The desired value of conductivity is inferred from the relative positions of the origins on the two plots, and the depth is obtained from the parameters associated with the selected reference curve.

Audio-frequency magnetotelluric method

The two-loop profiling and sounding methods require a somewhat cumbersome and bulky source to establish the required moment before measurements can be made.

The MT method makes use of natural electromagnetic fields (Cagniard, 1953). The AMT method makes use of natural signals in the frequency range of 8–20 Hz and it has been found useful in geothermal reconnaissance surveys.

Strangway et al. (1973) have described the AMT method as applied to mineral exploration. AMT exploration makes use of the natural energy originating from worldwide lightning storms. Tropical storm cells, most frequent during the summer months, are the principal contributors to the natural energy. The scalar apparent resistivity is given by

$$\rho_a = \frac{1}{5f}\frac{|E_x|^2}{|H_y|^2} \tag{5.29}$$

where f is the frequency in hertz, E_x the horizontal electrical field component in the x-direction in $10^{-6}\,\mathrm{V\,m^{-1}}$ and H_y the horizontal component of magnetic field in the y-direction in gammas. Apparent resistivity at a given frequency is determined by measuring the mutually orthogonal electric and magnetic fields of the distant lightning. The depth of measurement, δ, in meters, is estimated from the relation (Hoover and Long, 1976):

$$\delta = 503\sqrt{\frac{\rho}{f}} \tag{5.30}$$

The equipment and procedural details of the AMT method in geothermal exploration have been given by Hoover and Long (1976) and Whiteford (1976).

Whiteford (1976) has compared the apparent resistivity measured by the AMT method with those determined by several electrical resistivity surveys carried out in the Broadlands geothermal field in New Zealand. The resistivities obtained by the AMT method were found to be systematically higher. However, the sharp boundaries of the Broadlands geothermal field could be delineated and matched with those deciphered from electrical resistivity surveys.

In general, electromagnetic induction methods are found to be useful in measuring resistivity in hydrothermal areas. The AMT method provides a good and easy to use reconnaissance tool. Controlled source methods, using a coil or loop, are useful in estimating resistivity at depths ranging from a few tens of meters to a few kilometers. One of the most important applications of induction methods is in areas characterized by surface layers of high resistivity, where direct electrical resistivity measurements are difficult for the deeper layers. The major handicaps of induction methods are the cumbersome equipment and not very direct procedures adopted in determining the resistivity.

Wide-band magnetotelluric method

The MT method utilizes naturally occurring, broadband electromagnetic waves over the Earth's surface to image subsurface resistivity structure. The electromagnetic waves originate from regional and worldwide thunderstorm activity and from the interaction of solar wind with the Earth's magnetosphere. Due to the remote nature of the sources and the high refractive index of the Earth relative to air, the electromagnetic waves are assumed to be planar and to propagate vertically into the Earth. However, the scattering of electromagnetic waves by subsurface structure can be arbitrary in polarization, necessitating a tensor description (Wannamaker et al., 2005). Accordingly, two components of electric field (E_x and E_y) and three components of magnetic field (H_x, H_y and H_z) are measured. A schematic diagram of a typical MT site is shown in Fig. 5.13. The frequencies of the waves (signals) range from about 1 Hz to a fraction of milli Hertz, which allows to image a wide depth range. A detailed account of the MT method is given in Vozoff (1991). Ward and Wannamaker (1983) have described the data acquisition, processing and interpretation for the method, including a discussion of the problems associated with exploration for geothermal resources. A recent study by Wannamaker et al. (2005) has demonstrated the applicability of MT technique in exploration and monitoring of the Coso geothermal field.

Data acquisition comprises simultaneous measurements of orthogonal components of magnetic and the induced electric field variations of the natural electromagnetic signal. The relationship between the orthogonal electric and magnetic components yields a measure of impedence over a frequency range. Impedence is usually expressed in terms of apparent resistivity and phase, and provides information about the subsurface conductivity distribution. A model for the subsurface geoelectric structure is obtained from analysis of the apparent resistivity and phase over a range of frequencies (and hence over a range of depths). The method has been

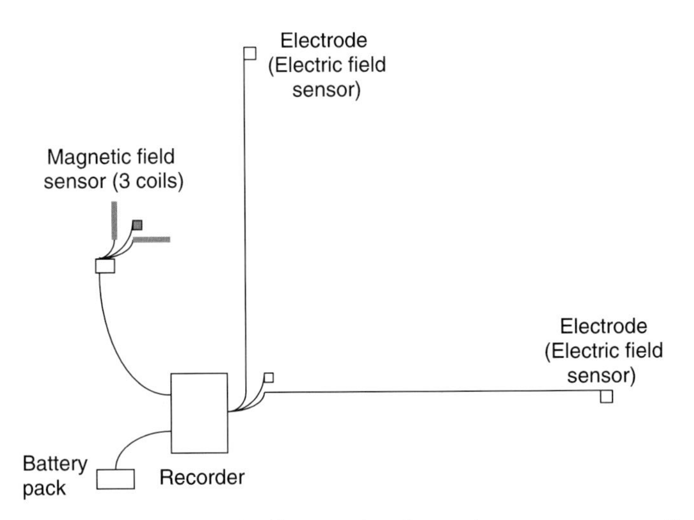

Fig. 5.13. A schematic diagram showing various components of a MT data acquisition system (after Wannamaker et al., 2005).

used successfully at several geothermal locations worldwide. Researches are underway to construct a three-dimensional (3D) conductivity model of a hydrothermal system by 3D MT inversion and modeling. Such studies are, at present, quite demanding in terms of computational resources and time, and development of efficient modeling tools could result in a new quantitative tool for locating geothermal wells (Newman et al., 2005).

Self-potential method

Self-potential anomalies of the order of a few tens to a few hundreds of millivolts have been observed over a number of convective hydrothermal systems, e.g., Long Valley, California (Anderson and Johnson, 1976), Kilauea, Hawaii (Zablocki, 1976), Cerro Prieto, Mexico (Goldstein et al., 1989), East Mesa (Goldstein et al., 1989), Mori, Japan (Ishido and Pritchett, 2000), Sumikawa, Japan (Matsushima et al., 2000), Yanaizu-Nishiyama, Japan (Tosha et al., 2000). The anomalies can be either positive or negative, and are often dipolar. The exact mechanism of development of these anomalies is not known with certainty, and several causes such as electrokinetic, thermoelectric and even chemical diffusion effects associated with the flow of fluids and heat have been proposed by earlier workers (Zohdy et al., 1973; Corwin and Hoover, 1979; Sill, 1983). It is now believed that the main mechanism is related to the electrokinetic effect, wherein streaming potentials are generated by subsurface fluid flow (Ishido et al., 1987; Goldstein et al., 1989).

Data acquisition for the self-potential method is carried out using either "fixed reference" technique or "leap frog" electrode configuration. In the former, one electrode remains fixed at a base station during the entire survey and the second electrode is moved along the survey line. The base electrode is connected to a multimeter using a very long wire. Usually, multiple base stations are set up and loop errors are distributed to minimize cumulative errors. The station separations vary between 100 and 350 m, but smaller station separations generally yield better resolution and data quality. In the "leap frog" technique, both electrodes move in an alternating fashion along the survey line. The length of the dipole remains fixed. Alternating the positions of the leading and following electrodes in leap frogging helps reduce the cumulative error caused by electrode polarization. It is also necessary to monitor the potential generated by telluric currents during the survey to check for periods of strong noise, and to carry out frequent checks of electrode polarization and drift. Additional sources of noise in such surveys arise due to topographic effects, variations in soil moisture, cultural noise, vegetation potentials, and electrokinetic potentials due to flowing surface or subsurface waters. In many geothermal areas, the cumulative noise due to one or more of these sources can potentially mask the data, and complicate the interpretation process. Therefore, it is necessary to scrutinize self-potential data carefully before assigning a geothermal origin to them.

In addition to geothermal exploration, self-potential methods have found applicability in reservoir monitoring studies during the last couple of decades. Changes associated with large-scale fluid withdrawal during production and pumping back of the fluids into the reservoir during re-injection are reflected in self-potential measurements, as for example, in the Cerro Prieto and East Mesa geothermal fields (Goldstein et al., 1989), Mori geothermal field (Yasukawa et al., 2001) and Hachijojima geothermal field (Nishino et al., 2000). The method is applied for monitoring of both liquid- and steam-dominated geothermal systems.

Seismic methods

Applications of seismic methods in geothermal exploration have been reviewed by Iyer (1978). They can be broadly classified into active and passive methods. Active methods make use of seismic signals generated by an artificial source such as explosives and vibrators, whereas passive methods make use of natural sources such as earthquakes, microseisms and seismic noise.

Active methods
These methods include seismic reflection and refraction surveys, and travel time residual and attenuation studies. Although, active seismic methods have been very successfully used in exploration for oil and the theory and field procedures are well established, their use in geothermal exploration has been rather limited.

The seismic refraction technique has been used mainly as a reconnaissance tool for mapping velocity distributions in the top few kilometers of the crust, from which faults, fracture zones, intrusions, rock types and other structural features are inferred. Several studies have demonstrated the usefulness of the technique in geothermal areas (Hill, 1976; Majer, 1978; Gertson and Smith, 1979; Hill et al., 1981 and others). Refraction profiles have been useful in estimating the depth of the geothermal reservoir in the Reykjanes geothermal field in Iceland (Palmason, 1971). At Reykjanes, it is interesting to note that geothermal aquifers are more abundant in deeper, higher-velocity ($V_p \approx 4.2\,\mathrm{km\,s^{-1}}$) layers and are not so often located in the shallower, low-velocity material ($V_p \approx 3.0\,\mathrm{km\,s^{-1}}$). It has been also discovered that the highest porosity rocks are not necessarily always the most productive (Macdonald and Muffler, 1972).

Seismic refraction surveys were found useful in evolving an average crustal model in Long Valley, California (Hill, 1976). Strong attenuation of high-frequency seismic waves was also found to be associated with surface geothermal phenomena in the area (Hill, 1976). Evidence for the presence of material with high-velocity and high-attenuation characteristics in the top 1 km layer at The Geysers geothermal field in California have been found by Majer and McEvilly (1979) using an array of closely spaced seismometers. Combs et al. (1976a) have reported results of investigating lateral variation in compression wave velocity and attenuation in the East Mesa geothermal field, California, using explosive charges from 9 to 90 kg at depths varying from 15 to 60 m. They delineated a low-velocity layer with a horizontal extent of a few kilometers beneath the East Mesa geothermal field. A few other surveys using active seismic methods have been reported in the literature (Iyer, 1978); however, their use in geothermal exploration is quite limited.

The seismic reflection technique has been used as an effective structural-stratigraphic mapping tool at several geothermal locations (Hayakawa, 1970; Denlinger and Kovach, 1981; Ross et al., 1982; Blakeslee, 1984). In case of a layered subsurface structure, where reflectors can be traced horizontally and interruptions in reflectors can be used to identify faults along which displacement has taken place, seismic reflection technique is very effective for studying a potential reservoir before drilling. Batini and Nicholich (1984) have described the application of seismic reflection surveys to geothermal exploration. A detailed discussion of the data acquisition and processing methods are beyond the scope of this book.

Although the data acquisition is quite expensive when compared with refraction method, it provides much better resolution of horizontal and shallow-dipping layered structures located at deeper levels, such as the Mexicalli Valley and the Imperial Valley. The reflection method does not work well where the subsurface structures are highly faulted or folded due to diffraction of seismic waves at the discontinuities. Wherever the budgets permit, coincident reflection and refraction experiments are often the preferred mode of survey.

Passive methods

Seismic ground noise surveys. A high level of seismic noise is invariably found to be associated with geothermal areas. The seismic noise surveys are carried out using a closely spaced group of seismic stations recording for at least 48 h (Iyer and Hitchcock, 1976). The process is repeated in different parts of the geothermal area. It is desirable to operate one reference station continuously. Analysis techniques include computation of average noise level in several frequency bands, using carefully selected noise samples, and plotting their spatial variations. Very often power spectra, cross spectra, azimuth and velocity waves are also calculated. Iyer and Hitchcock (1976) have summarized the results obtained at four geothermal fields in the United States: The Geysers, Imperial Valley and Long Valley in California, and Yellowstone National Park in Wyoming. The basic results are:

(a) All four seismic areas have high noise levels in the 1–5 Hz frequency band.
(b) At the Yellowstone geyser basins, the noise is generated at depth and it is not a surface phenomena.
(c) The cultural noise present at The Geysers and the East Mesa area in the Imperial Valley makes it difficult to interpret the geothermal noise.
(d) The noise anomaly found at Long Valley is partly caused by the amplification of seismic waves by soft alluvial basins.

Iyer (1978) concludes that although tens of noise anomalies have been reported from geothermal areas, attempts at discriminating geothermal noise from cultural noise have not been very successful. More experimental work under varying geological conditions needs to be carried out to develop the seismic noise survey as a useful geothermal exploration tool.

Teleseismic P-wave delays. Seismic waves arriving from earthquakes occurring at far-off distances (larger than 1,000 km) are called teleseismic waves. The teleseismic P-wave delay technique, as applied in geothermal areas, involves measuring delays in the relative travel time of teleseismic P-waves that have traversed the geothermal area. Low P-wave velocities and hence delays in travel times could arise from partial melts or magma chambers located beneath geothermal systems. Usually, an array of seismographs with accurate time keeping is deployed and one of the stations in the array is chosen as the reference station and relative differences in arrival times at the other stations are calculated. These delays are then interpreted in terms of subsurface structure and low-velocity layers. Fig. 5.14 shows the delays observed by Steeples and Iyer (1976) at the Long Valley Caldera, California, and Fig. 5.15 shows a model of subsurface structure interpreted by them along section AA′ of Fig. 5.14. It is estimated, according to one of the models, that a volume of about 1000 km^3, with

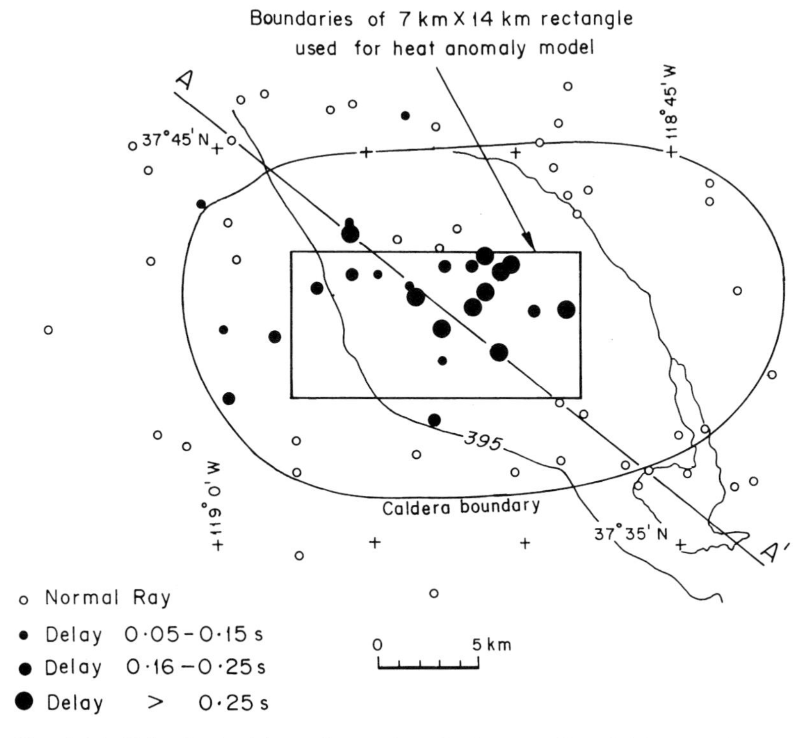

Fig. 5.14. Teleseismic delays observed at the Long Valley Caldera, California (from Steeples and Iyer, 1976). Rectangle in the middle was used to estimate heat anomaly.

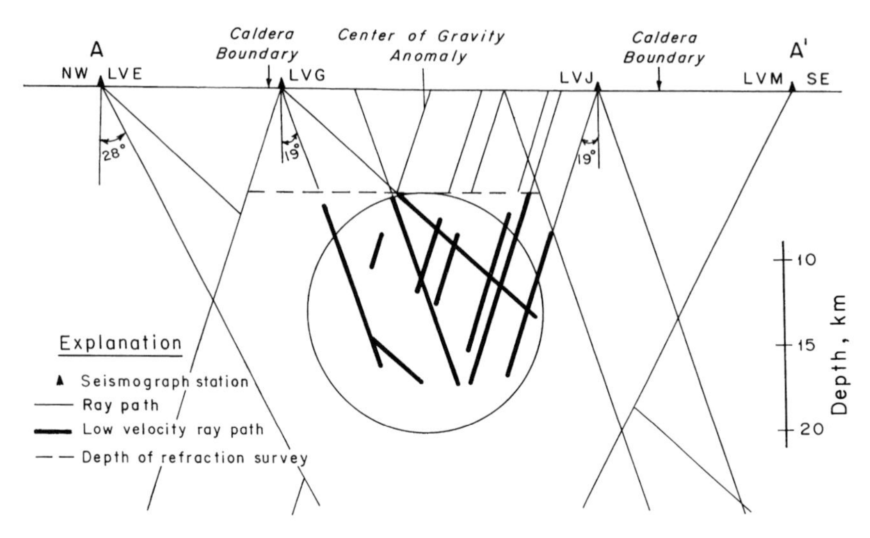

Fig. 5.15. Sectional view of Long Valley Caldera along AA′ of Fig. 5.14. A velocity contrast of 15% along ray paths shown by thick lines with the ray path length proportional to delays seen along those paths, is represented in the model (from Steeples and Iyer, 1976).

temperatures in excess of about 300 °C above normal exists at depths of 10–20 km below the surface. Iyer et al. (1979) have reported large teleseismic delays, exceeding 1 s, from Mount Hanna, near the Clear Lake volcanic field and from The Geysers in California. They postulate that a molten magma chamber under the surface volcanic rocks of Mount Hanna and a highly fractured steam reservoir at The Geysers are responsible for the observed delays. Other sources of P-wave delays include large-scale alteration, compositional differences, lateral temperature variations, or locally fractured rock (Iyer and Stewart, 1977). Therefore, it is necessary to analyze the results of teleseismic surveys together with information obtained using other geological and geophysical techniques before inferring a geothermal cause for the changes in P-wave travel times from one station to another.

P-wave attenuation and delay. Anomalous P-wave travel time delays and attenuations in the high-frequency range have been reported for a number of geothermal fields (Iyer, 1978). Combs et al. (1976b) used an array of nine short-period, high-gain three-component seismographs at the East Mesa geothermal field in California to investigate travel time and attenuation anomalies. Records of several well-located earthquakes from the Brawley earthquake swarm of 1975, with epicentral distances varying from 20 to 50 km were examined. They discovered significant P-wave travel time delays for ray paths passing through the zone of high heat flow. Spectral analyses of the observed seismic waves from the swarm showed that the relative attenuation of body wave amplitudes increased in the frequency range of 10 Hz and higher along the paths through the East Mesa geothermal field.

Young and Ward (1980) have developed techniques to estimate the attenuation of teleseismic P-waves and have discovered a zone of large attenuation coinciding with a zone of large delays at The Geysers in California. Similar results have also been reported by them for the Coso geothermal area.

Surface wave dispersion. In contrast to seismic body waves whose depth of penetration in a given medium is governed by the distance between the source and the observation point, the depth of penetration of surface waves is controlled by wavelength, and the corresponding depth resolution is well established. This property makes surface waves particularly suitable for a 3D estimate of shear wave velocity structures. In spite of the extensive application of seismic methods in petroleum exploration, however, surface wave analysis has played virtually no part in the study of shallow structures. This fact can be traced to the use of explosive sources by which the surface waves are poorly excited and seldom observed (Dobrin et al., 1954). Geothermal systems, however, are characterized by an abundance of natural seismicity (e.g., Ward, 1972; Hill et al., 1975; Combs and Rotstein, 1976), which produces relatively greater surface wave excitation than explosive sources (Richter, 1958). The utilization of surface waves in geothermal areas is dealt with here in some detail.

Seismic stations have been installed at a number of geothermal areas around the world. This makes it possible to investigate interstation surface waves and interpret them in terms of shallow crustal structure. We shall briefly examine the results of an investigation carried out at the Coso geothermal fields in California, U.S.A. (Fig. 5.16). The area, characterized by the occurrence of Pleistocene volcanic rocks, exhibits a high heat flow exceeding 600 mW m^{-2} within the ring-type caldera. A number of seismic stations (Fig. 5.17) are in operation in the area. The interstation impulse response method of measuring dispersion (Gupta et al., 1977) accurately estimates surface wave dispersion and attenuation for paths between pairs of stations. For two stations which are in line with an epicenter, the surface wave which arrives at the farther station has passed through the nearer station. The signal at the nearer station is considered to be the input to a filter, the output of which is the signal at the farther station. The impulse response of the filter then represents the effect of the path between the stations, independent of the source, the path to the nearer station, and the seismometers when they are identical. Interstation group velocity, phase velocity and attenuation can be calculated from frequency-time analysis (e.g., Dziewonski et al., 1968; Nyman and Landisman, 1977) of the impulse response. The impulse response $x_1(t)$ is the inverse of the Fourier transform of the signal at the farther station, x_2, divided by the Fourier transform of the signal at the nearer station x_1 (Gupta and Nyman, 1977a):

$$x_i(t) = \frac{1}{2\pi}\int_{-\infty}^{\infty}[F_2(\omega)/F_1(\omega)]\exp(i\omega t)d\omega \qquad (5.31)$$

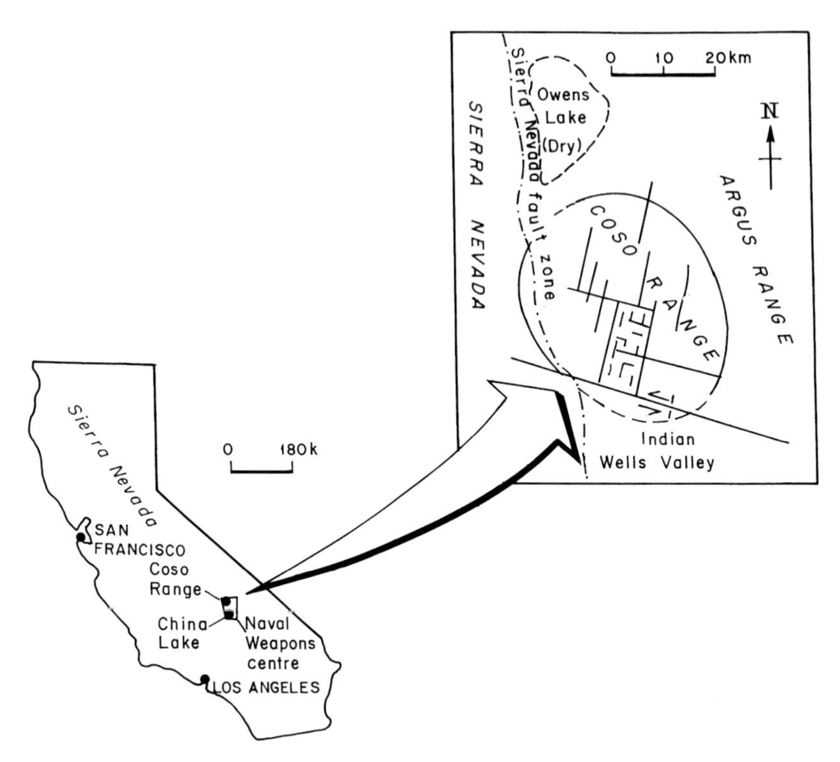

Fig. 5.16. Location of the Coso geothermal system with index map of California. The more detailed map shows faults in the Coso Range area (after Duffield, 1975).

where

$$F_j(\omega) = \int_{-\infty}^{\infty} [x_j(t)]\exp(-i\omega t)dt, \quad j = 1,2 \tag{5.32}$$

Gupta and Nyman (1977b) have applied the interstation impulse response method to investigate the subsurface velocity structure in the Coso geothermal area, California. One of the highlights of their investigation is the confirmation of a low Poisson's ratio in this area, which was earlier inferred by Combs and Rotstein (1976).

Fig. 5.18 shows seismograms at two stations located at distances of 6.7 and 3.3 km from the epicenter of a shallow earthquake. The epicenter and the seismic stations lie approximately on a straight line. Using display-equalized frequency-time analysis as described earlier, a Rayleigh wave group velocity of about 2.8 km s^{-1} is inferred in the period range of 0.07–0.15. This implies that Rayleigh wave velocities to a depth of about 300 m are 2.80 km s^{-1}. It is well known that Rayleigh waves propagate with a velocity of approximately nine-tenths that of S-waves. Therefore, the S-wave velocity would be 3.11 km s^{-1} to the depths of 200–300 m from the surface. Detailed

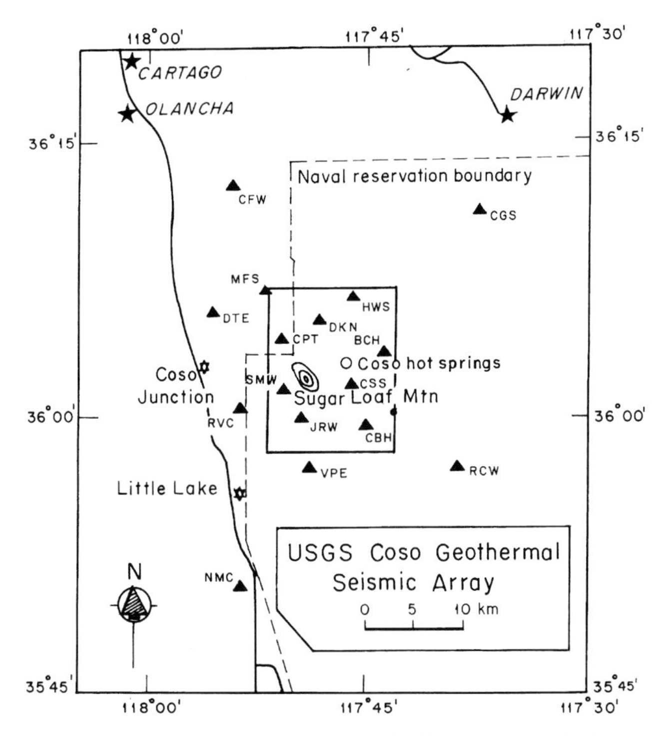

Fig. 5.17. Location map of the U.S. Geol. Surv. seismic network in the Coso geothermal area. Station locations are indicated by filled triangles.

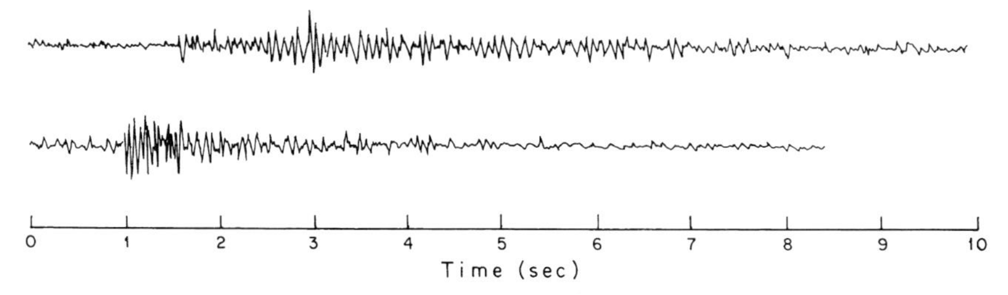

Fig. 5.18. Seismograms at two seismic stations in the Coso geothermal area for a small earthquake. The epicentral distances are 3.3 km (lower seismogram) and 6.7 km respectively.

P-wave velocity data obtained from nine calibration blasts show a constant P-wave velocity of 4.75 km s^{-1}. These P- and S-wave velocities infer a Poisson's ratio of 0.13 compared with the 0.25–0.30 normally observed.

The low Poisson's ratio observed for the Coso geothermal field implies that the shallow subsurface could be deficient in water content, or more likely the cracks and void spaces are filled with steam. Similar results from other considerations have been reported by Combs and Rotstein (1976). These data help in inferring that the Coso geothermal field is a vapor-dominated rather than a hot water system.

Microearthquake surveys. Investigation of microearthquakes (magnitude range: 1–3) in tectonically active and volcanic areas has shown that major hydrothermal convection systems are often characterized by a high level of microearthquake activity (Ward et al., 1969; Lange and Westphal, 1969; Ward and Bjornsson, 1971; Hamilton and Muffler, 1972; Combs and Rotstein, 1976; Ward et al., 1979; and others). One of the main uses of monitoring and precisely locating microearthquake hypocenters is their application in estimating the depth of fluid circulation in hydrothermal systems, which is difficult to estimate otherwise, and locate active faults that act as channels for upflow of hot fluids. Microearthquake surveys have potential to contribute to locating drilling targets, particularly in the case of blind hydrothermal systems. Several microearthquake studies are reported from geothermal areas, e.g., Bjornsson and Einarsson (1974) in the Reykjanes Peninsula in Iceland, Combs and Hadley (1977) in East Mesa, California, Combs and Rotstein (1976) in Coso, and, Hunt and Lattan (1982) in Wairakei.

Analysis and interpretation of microearthquake data can yield estimates of P- and S-wave velocities in the area. However, a reliable subsurface velocity distribution model, such as obtained from high-resolution refraction surveys, is necessary to constrain the hypocenter locations. In a study of The Geysers geothermal area, Gupta et al. (1982) have derived the regional P- and S-wave velocities using analysis of microearthquake data. Absorption coefficients determined from microearthquake data can indicate the presence of fluid-filled, steam-filled or silica-filled fractures (Wright et al., 1985). Determination of Poisson's ratio from microearthquake surveys helps in distinguishing between water and vapor-dominated reservoirs. At The Geysers geothermal field, the Poisson's ratio ranges from 0.13 to 0.16 within the production zone relative to much higher values exceeding 0.25 outside the production zone (Majer and McEvilly, 1979; Gupta et al., 1982). The lowering of Poisson's ratio has been partly explained due to a decrease seismic P-wave velocity.

Gravity surveys

The gravity method exploits the density contrasts among different rock formations to map subsurface structure, delineate faults and intrusive bodies below the surface and determine the thickness of alluvium in sedimentary basins. It has been most widely used in prospecting for hydrocarbons, often in conjunction with seismic methods, to delineate favorable structures for deposition of oil. In the context of geothermal prospecting, gravity surveys are used to define lateral density variation related to deep magmatic body, which may represent the heat source. Residual gravity highs have been interpreted to reflect structural highs, anomalous geometry of fault zones, densification of sediments due to metamorphism and silica deposition (Muffler and White, 1969; Biehler, 1971; Isherwood, 1976; Isherwood and Mabey, 1978; Razo and Fonseca, 1978; Ross and Moore, 1985; Salem et al., 2005).

Gravity surveys are rather simple and inexpensive. However, a good elevation control is necessary and in rugged terrain, a considerable amount of effort is required to incorporate the necessary terrain correction. Therefore, gravity surveys are most useful in plains or areas of smoothly varying terrain. Exhaustive discussions of gravity meters, data acquisition and reduction procedures are extensively available in literature (e.g., Nettleton, 1971; Dobrin and Savit, 1988). As in any potential field method, the interpretation of gravity data is ambiguous—it does not provide independent estimates of the depth to the causative body, but yields a product of the density contrast and depth. Here we shall give a brief account of the successful application of the method to a few geothermal areas only.

Local gravity anomalies in geothermal fields are generally associated with metamorphism or increase in density of the sediment due to deposition of minerals from the rising plumes of thermal water. Meidav and Rex (1970) report that every known geothermal field in the Imperial Valley, California, is associated with a measurable residual gravity anomaly varying from 2 to 22 mgal. These anomalies are believed to have originated from the combination of (i) intrusive bodies which may have been the original source of the geothermal anomaly and (ii) precipitation of minerals out of the thermal water at a shallower depth. In New Zealand geothermal areas, the observed positive residual gravity anomalies are considered to be caused by rhyolitic domes and hydrothermal alteration of reservoir rocks (Hochstein and Hunt, 1970; Macdonald and Muffler, 1972). On the other hand, a large gravity anomaly has been observed over the Mt. Hannah area at the Geysers, most likely due to the presence of hot silicic magma beneath this area (Isherwood, 1976). In regions where other geological and geophysical considerations indicate the presence of economic geothermal reservoirs, gravity surveys are useful in providing a clue to identifying localized subareas where geothermally related processes may have taken place. The reverse is not true. Gravity anomalies alone cannot necessarily be indicative of a geothermal region.

A relatively recent application of gravity surveys is to monitor subsidence in geothermal areas and temporal changes in water level in the reservoir due to production. Repeated high precision, microgravity measurements coupled with simultaneous high-precision leveling surveys have been used to investigate subsurface mass changes during production of steam/hot water from the reservoir as well as reinjection of wastewater back into it (e.g., Grannell, 1980 for the Cerro Prieto geothermal field; Allis and Hunt, 1986 for the Wairakei field; Allis et al., 2000 and Gettings et al., 2002 for The Geysers; Apuada and Olivar, 2005 for the Leyte geothermal field, Phillipines). The gravity changes in such studies are of the order of a few microgals to a few tens of microgals. A major problem in these surveys, therefore, is to extract very small signals from large background noise. Stringent precision requirements during data acquisition and availability of accurate reservoir data inputs for modeling are a key to increase the confidence that the gravity changes are tracking real signals.

Magnetic surveys

The magnetic method exploits the contrasts in magnetization of subsurface rocks. As a reconnaissance tool, it has been extensively used in hydrocarbon exploration for structural and lithologic mapping. The procedures for data acquisition, processing and interpretation are discussed extensively in literature (see e.g., Dobrin and Savit, 1988). The method has come into use for identifying and locating masses of igneous rocks that have relatively high concentrations of magnetite. Strongly magnetic rocks include basalt and gabbro, while rocks such as granite, granodiorite and rhyolite have only moderately high magnetic susceptibilities. Therefore, the method is useful in mapping near-surface volcanic rocks that are often of interest in geothermal exploration. Other uses in geothermal prospecting include mapping of hydrothermal alteration zones that exhibit decrease in magnetization relative to the host rock.

Magnetic anomalies in the New Zealand geothermal fields have been interpreted as having been caused by conversion of magnetite to pyrite (Studt, 1964). Distinct magnetic anomalies are found to be associated with the high-temperature geothermal fields. Namafjall and Krafla geothermal fields in northern Iceland are good examples (Palmason, 1976). Ground magnetic surveys are also extensively used in the low-temperature fields of Iceland for tracing hidden dykes and faults which often control the flow of geothermal water to the Earth's surface (Palmason, 1976; Flovenz and Georgeson, 1982). The magnetic method has not been successful in detection of geothermal anomalies in every case. Often, the pattern of magnetic anomalies is complex due to its dipolar nature, and interpretation is ambiguous in detail. However, in conjunction with other geophysical methods such as gravity, thermal and seismic, it provides valuable constraints for determining the subsurface structure and detecting anomalies associated with hydrothermal alteration zones in many cases.

AIRBORNE SURVEYS

Airborne surveys constitute invaluable reconnaissance prospecting tools in geothermal exploration, providing quick coverage of large and often, inaccessible areas. Although a few airborne geophysical as well as thermal infrared imaging methods have been used in the past, the aeromagnetic surveys are the most commonly used airborne technique in geothermal exploration at the present time. Moreover, during the recent years, there has been a shift in emphasis from surface features to geothermal anomalies concealed below the surface, for which aeromagnetic, and to some extent aeroelectromagnetic, methods are quite useful. Aeromagnetic as well as airborne and spaceborne remote sensing surveys have been made over several geothermal areas worldwide. Salient features of both types of surveys are discussed in the following.

Aeromagnetic Investigations

Detection of geological features such as faults, horsts and grabens is important in geothermal exploration. Under favorable circumstances, these features can be successfully picked up by high-precision aeromagnetic surveys. Dupart and Omnes (1976) have listed the technical requirements for such surveys. These include (1) constant barometric flight level, (2) small average ground clearance, (3) tight grid of profiles, (4) sensitivity of 0.01 or 0.02 gamma, ensuring a tenth of a gamma as an usable measurement, (5) digital recording at an interval of half a second, (6) accurate navigation, and (7) removal of diurnal variations.

Aeromagnetic surveys are quite expensive, and interpretation of data is often complex. Magnetic data reflect the variations in the Earth's magnetic field that arise from the distribution of magnetic materials in the ground. In addition to ambiguity in mapping this distribution, there exist additional problems in unambiguously determining their geologic association. Grauch (2002) has discussed some of these limitations. Nevertheless, high-precision aeromagnetic surveys have been successful in estimating the sedimentary cover overlying crystalline basement quite accurately. Deguen et al. (1974) report that surveys carried out over a shallow basin in Canada were very successful and the depth of basement estimated from aeromagnetic surveys was found to be accurate, the error being less than 10% of the depth obtained through drilling at 8 sites out of a total of 12 drill holes. McEuen (1970), for Imperial Valley in California, suggested that the magnetic intensity deduced from airborne survey is solely dependent on depth of basement. He found it possible to obtain useful information regarding temperature distribution at depth by comparing the apparent resistivity values obtained from long-spacing resistivity profiles with magnetic intensity values obtained from similar profiles generated from an airborne magnetic map. Another interesting example of the application of airborne magnetometry to geothermal exploration has been presented by De la Funte and Summer (1974) for the Colorado River delta area, Baja California, Mexico. Carefully planned, high-resolution airborne magnetic surveys can delineate large inter-basin faults.

Conventional aeromagnetic surveys are designed to focus on mapping magnetic rocks, such as igneous and metamorphic basement rocks underlying sedimentary cover. In contrast, high-resolution surveys are flown at lower altitudes and with narrower line spacing. This allows better resolution of subtle magnetic contrasts, such as those arising from the juxtaposition of basin-fill sediments having different rock types. It also allows for better definition of sources with limited lateral extent, and overall better resolution of details in map view. High-resolution aeromagnetic surveys flown over the Albuquerque basin of the Rio Grande rift, New Mexico, have recently demonstrated that aeromagnetic methods can successfully map concealed and poorly exposed faults in a basin environment (Grauch, 2001). The surveys revealed the overall pattern of faulting and yielded estimates of the attitudes, depths and geometries of many of the hidden faults lying at shallow levels below the surface (Grauch et al., 2001). In the Dixie Valley geothermal area, Nevada, aeromagnetic surveys were carried out by the United States Geological Survey employing a helicopter-borne cesium vapor magnetometer. The helicopter was flown at a height of 120 m with a traverse-line spacing of 200 m and tie-line spacing of 1000 m (Grauch, 2002). The study demonstrated the efficacy of high-resolution aeromagnetic surveys in revealing the distribution of shallow, hidden faults and their extensions at depth. This basic information is critical in understanding the relationship between the faults and flow of geothermal fluids within the basin.

Remote Sensing Techniques

Several efforts in the past have been made to use thermal infrared imagery, multiband black and white photography, color photography and infrared color-shift photography for detecting and mapping geothermal activities. New geothermal areas have been reported to be discovered in New Zealand (Dickinson, 1973) and Italy (Hodder et al., 1973) with the help of aerial infrared surveys. Dupart and Omnes (1976) recommend that high-sensitivity aeromagnetic surveys should also be considered in geothermal exploration programs. Marsh et al. (1976) report the use of satellite data for geothermal exploration.

It should be borne in mind that both photographic and thermal infrared sensors can make measurements only to depths of a few microns from the surface. This limitation is due to the normal opacity of soil at these wavelengths. Therefore, only surface effects can be detected. Photographic sensors (visible and infrared) are usable in identifying hydrothermal alteration associated with thermal seep or fumaroles. Moreover, photographic sensors can operate only during the time when the reflection of incident sunlight to detect thermal manifestations is available. These methods have become less popular in recent times due to the development of multispectral and hyperspectral imaging tools. Nevertheless, for the sake of completeness, the fundamentals of infrared radiation and atmospheric transmission windows and the application of infrared aerial surveys in geothermal exploration are discussed briefly, followed by a discussion of the recent airborne and satellite-based remote sensing tools.

Infrared Radiation and Atmospheric Transmission Windows

The electromagnetic spectrum between the wavelengths 0.7 and 1,000 μm is known as the infrared radiation. The lower wavelength limit coincides with the upper limit of visible radiation. Radiations with wavelengths in excess of 1,000 μm are called the microwave spectrum. Both these limits, lower and upper, are arbitrary and no change in characteristics occurs as they are passed. Conventionally, the 0.7–1.5-μm band is called the near infrared region, and from 20- to 1,000-μm band the far infrared region.

All bodies above absolute zero in temperature emit energy as electromagnetic radiation. In the infrared region, this radiation is governed by the Stefan–Boltzmann law, as discussed in Chapter 3. Emissivity, the ratio of energy emitted by a material to that emitted by a black body, for the Earth materials ranges from 0.70 to 0.98 compared to a black body emissivity of 1.0.

A large portion of the infrared radiation is absorbed in the atmosphere and consequently certain wavelengths, for which absorption is low, can only be detected by airborne sensors. Fig. 5.19 shows the atmospheric transmission at different wavelengths of the infrared spectrum.

Infrared aerial surveys

Aerial infrared surveys furnish useful information that can be related to the heat discharge at the Earth's surface. Several airborne surveys with varying degrees of success in the reconnaissance of geothermal resources have been reported in the literature. Hochstein and Dickinson (1970) have used infrared scanning equipment operating in the 4.5- to 5.l-μm range in the Karapiti area, near Wairakei, New Zealand, to determine whether infrared surveys could monitor changes in the boundaries of discharge areas over a period of a few years. They conducted a trial run at noon on an overcast day over a selected strip at an altitude of about 1000 m. A comparison with panchromatic and reflective infrared photographs and near-surface

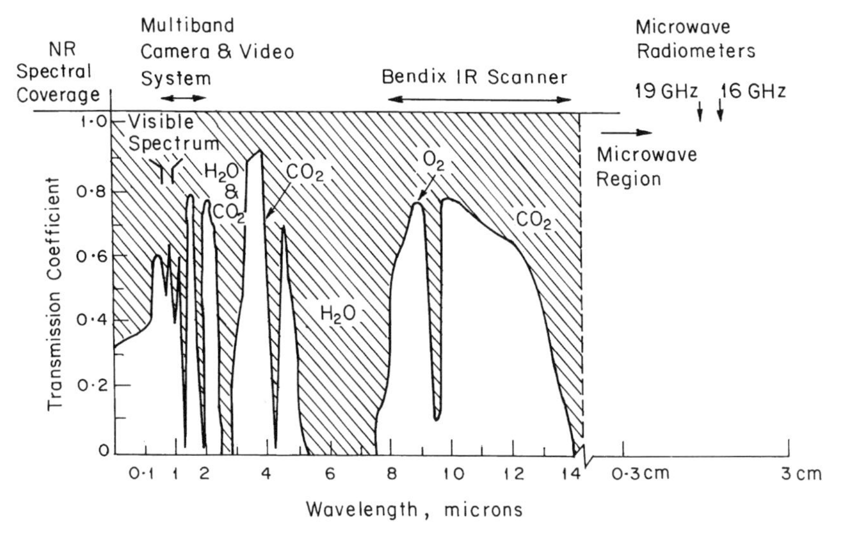

Fig. 5.19. Atmospheric transmission and sensor coverage (from Hodder, 1970). For details, see text.

temperature measurements revealed that the infrared scanner picture displayed all the essential thermal features in relatively great detail.

Palmason et al. (1970) have reported results of aerial infrared surveys at the Reykjanes and Torfajokull thermal areas in Iceland. The surveys were conducted with an infrared scanner system in the 4.5- to 5.5-μm wavelength band. To interpret features depicted on infrared imagery and to relate zones of high heat flux to tectonic structure, supplementary geological studies were undertaken. These included preparation of a shallow ground temperature map at a depth of 0.5 m. Aerial infrared surveys were found useful in outlining the surface thermal patterns and in relating them to possible geological structures controlling the upflow of hot water. Palmason et al. (1970) conclude that in addition to their use in the preliminary study of high-temperature areas, infrared surveys, when conducted at regular intervals over geothermal areas under production, can provide useful information on changes occurring in surface manifestations with time.

Hodder (1970) has presented an excellent summary of a laboratory and field research program undertaken to evaluate the application of remote sensing techniques in geothermal exploration. Multiband photography in the visible and near infrared band, passive infrared imagery and radiometry in the 8- to 14-μm band, as well as passive microwave radiometry at 16 and 19 GHz (18,750 μm and 15,800 μm wavelengths, respectively) were used. The field experiments were carried out in known geothermal areas of California with various geological backgrounds. The effects of varying flight altitudes on thermal anomaly signatures were also investigated. Fig. 5.20 shows the relationship between flight altitude and the corresponding applications. The major result of this research program was the identification of temperature anomalies in the microwave and the Rayleigh-Jeans region of spectrum in addition to the commonly used infrared region.

Dickinson (1976) has reported aerial infrared surveys in the 4.5- to 5.5-μm band, flown over urban, industrial and undeveloped land in the Tauhara and other geothermal fields in New Zealand. Maps have been prepared from the thermograms dividing the surface temperatures into three categories. These are at ambient temperature, 1–3 °C above ambient, and more than 3 °C above ambient. The 3 °C above ambient temperature contour coincides with the boundary between regions of conductive and convective heat flow, as determined from surface measurements. Dickinson (1976) succeeded in identifying some previously unknown areas of warm ground and in identifying and removing thermal anomalies of non-geothermal origin. He proposed repeating similar surveys at intervals of approximately three years to monitor any changes in hydrothermal activity.

Del Grande (1976) described a highly sensitive, aerial reconnaissance, method for identifying and evaluating potentially valuable geothermal resource areas. He designed a geothermal energy multiband (GEM) detection system, which could resolve 0.05–0.5 °C temperature enhancements for areas larger than 1 km^2 by ratioing narrow infrared spectral bands at 2.2, 3.5, 3.9, 4.8 and 13.2 μm. These signal ratios have

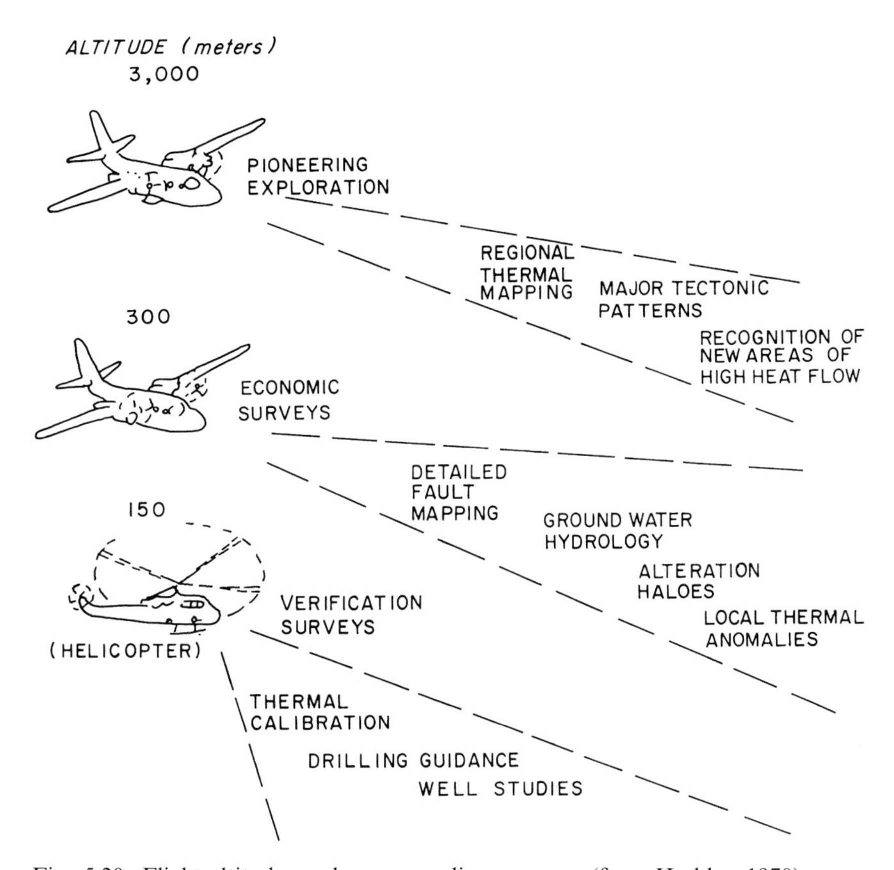

Fig. 5.20. Flight altitudes and corresponding coverage (from Hodder, 1970).

the advantage of being insensitive to surface emissivity for natural terrains. The signal ratios vary as a high power of the absolute surface temperature, and they avoid the 6- to 13-μm region which has interpretive uncertainties.

Banwell (1970), in his review of geophysical techniques in geothermal exploration, concludes that infrared surveying is useful in detecting and mapping relatively strong geothermal anomalies under favorable conditions using optimum equipment and data-handling facilities. However, its application is constrained by various random noise effects and other uncertainties. Therefore, careful considerations of various factors are necessary before undertaking airborne infrared surveys for geothermal exploration.

A number of recently developed remote sensing tools are available today, which provide better resolution and coverage. These tools are discussed in the following.

Recent remote sensing tools

Remote sensing techniques are emerging as useful reconnaissance tools for mapping the geology, detecting anomalous surface temperature anomalies and identifying geothermal indicators including hydrothermal alteration minerals (clays, sulfates), sinter and tuff in prospective geothermal areas (Calvin et al., 2005). Significant enhancements in wavelength coverage, spectral resolution and image quality over the past two decades have resulted in the development of state-of-art optical, near-infrared and thermal infrared imagery tools such as ASTER, MASTER, AVIRIS, HyMap and SEBASS (Fig. 5.21; Table 5.4). Data sets can be acquired over several wavelength channels (multispectral) as well as over hundreds of wavelengths (hyper-spectral). ASTER data are acquired using spaceborne mappers while the remaining four datasets are obtained through airborne sensors. When compared to LANDSAT imagery, the ASTER datasets provide better spatial resolution in the visible region and more spectral channels in the near-infrared region.

A combination of both day-time as well as night-time spectral imagery are used for detecting geothermal anomalies. The former provides information on albedo, while the latter reflects the geothermal anomalies as distinct from the cooler surroundings. Comparisons between the two can reveal the temperature differences. In the case of mineral mapping, optical and infrared spectroscopic technique distinguishes between the absorption characteristics of rock-forming elements like iron, and molecules and anions such as water, hydroxyl, carbonate and sulfate.

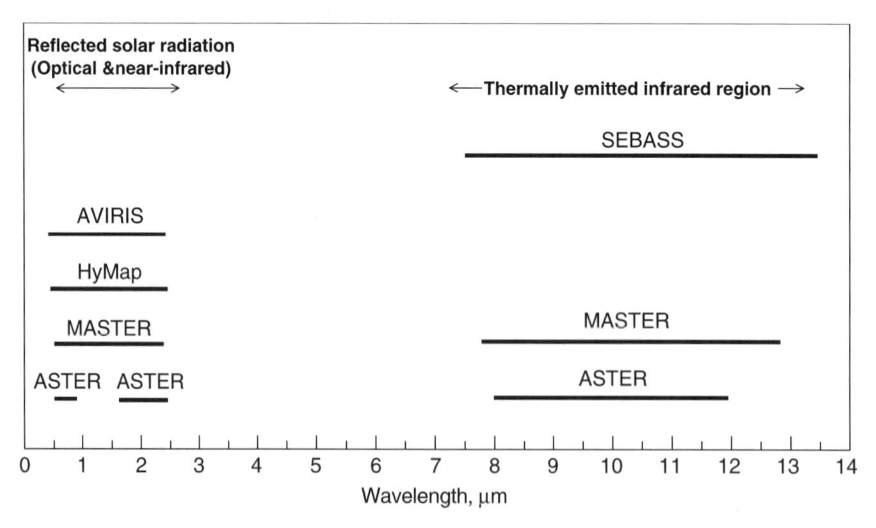

Fig. 5.21. Wavelength coverage by recent remote sensing tools. The datasets span parts of both the reflected solar radiation (optical and near-infrared region) and the thermally emitted infrared region. Wavelength ranges for the various tools, as given in Calvin et al. (2005), are used.

Table 5.4. Features of commonly used remote sensing tools (after Calvin et al., 2005)

Tool	Airborne/ spaceborne	Data acquisition mode	Wavelength coverage (μm)	No. of channels of data	Spatial resolution (m)
ASTER Advanced spaceborne thermal emission and reflection radiometer	Spaceborne	Multispectral	0.5–0.9 1.6–2.45 8–12	3 6 5	15 30 90
MASTER MODIS and ASTER airborne simulator	Airborne	Multispectral	0.46–2.39 7.76–12.88	25 10	5
AVIRIS airborne visible and infrared imaging spectrometer	Airborne	Hyperspectral	0.45–2.5	224	3
HyMap Hyperspectral mapper	Airborne	Hyperspectral	0.45–2.5	126	3
SEBASS Spatially enhanced broad array spectrograph system	Airborne	Hyperspectral	7.5–13.5	128	2

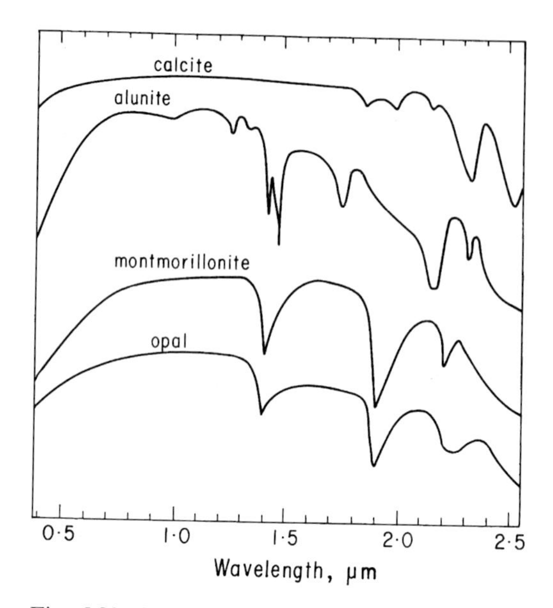

Fig. 5.22. Spectral characteristics of common geothermal indicator minerals in the visible—to infrared—region (modified from Calvin et al., 2005)

Sinter, carbonate and hydrothermal alteration minerals such as clays and sulfates have characteristic spectral signatures in the visible to infrared region, as shown in Fig. 5.22. Spectral signatures of common minerals in the visible and near-infrared regions are discussed in Farmer (1974) and Hunt (1977).

Details of data acquisition and processing methods are available in recent literature (Coolbaugh et al., 2000; Coolbaugh, 2003; Vaughan et al., 2003; Calvin et al., 2005). Recent applications demonstrating the utility of these techniques in delineating surface thermal anomalies and mapping of hydrothermal alteration minerals include those in Steamboat Springs and Bradys Hot Springs, Nevada (Kruse, 1999; Coolbaugh et al., 2000; Calvin et al., 2005), Brady-Desert Peak, Nevada (Kratt, 2005), Long Valley Caldera, California (Martini et al., 2003), Dixie Valley, Nevada (Martini et al., 2003; Nash et al., 2004) and hot springs of Yellowstone National Park, Wyoming (Hellman and Ramsey, 2004).

EXPLORATORY DRILLING

The culmination of an exploration program is marked by exploratory drilling of borewells. The locations of these borewells are decided on the basis of geological interpretation of subsurface conditions as revealed by various geological, hydrological, geochemical and geophysical surveys. The wells are used to confirm the anomalous temperature conditions in the subsurface, determine the lateral and vertical extent of the thermal anomaly, delineate the natural fracture pattern in the subsurface reservoir and assess its permeability, and carry out production tests to assess the sustainability for power production.

CONCLUDING REMARKS

Geophysical surveys have been quite successful in exploration and assessment of geothermal prospects over the past several decades. However, neither a single method nor a pre-determined combination of two or more methods has been found to be successful in every case. Individual surveys should be designed for each geothermal field keeping in mind its geological, tectonic and hydrological setting, and the results obtained should be verified from other available data or surveys. Geophysical methods provide valuable information even after drilling and during exploitation of the resource, e.g., in monitoring mass changes due to fluid withdrawal or injection, and changes in saturation and gas content in the reservoir.

Ambiguities in interpretation of potential field datasets, uncertainties in representative physical properties of reservoir and surrounding rocks, progressive degradation in resolution of geophysical anomalies with increasing depth, and complexities associated with terrain effects and variable subsurface hydrological regimes in rugged terrains should be accounted for before undertaking expensive drilling. As a rule, well locations for drilling should be selected after the results of geophysical studies have been integrated with those obtained from geological, hydrological, geochemical and remote sensing investigations, and internal ambiguities and inconsistencies have been resolved as far as possible.

Chapter 6

ASSESSMENT AND EXPLOITATION

DRILLING TECHNOLOGY

Drilling is one of the areas where geothermal resource development has gained considerably from the expertise of the oil and gas industry. Drilling for geothermal energy is quite similar to rotary drilling for oil and gas. The main differences are due to the high temperatures associated with geothermal wells, which affect the circulation system and the cementing procedures as well as the design of the drill string and casing. Geothermal steam is at some places encountered at shallow depths of 50–200 m, while at others drilling to depths in excess of 3,000 m has been necessary. During the last few decades, drilling technology has witnessed a great advancement. In 1947 only about a dozen holes deeper than 5,000 m existed, but by 1972 the number of such holes reached the 500 mark. Today, several wells are drilled to depths of about 5,000 m, and temperatures often exceed 250 °C. At the same time, through superior metallurgy, the bit life has increased by twenty-fold. The hoisting capacity of deep-well rigs has also increased 4 to 5-fold through the use of heavier and more efficient equipment. Besides overall improvements being made in different aspects of rotary drilling, research is presently being conducted to develop novel drilling techniques. These techniques operate using laser-powered basic rock destruction mechanisms like thermal spalling and chipping, mechanically induced stresses, chemical reactions and melting and vaporization. At low laser power, spalling or chipping of the rocks occurs. Increase in the laser power, with a fixed beam diameter, results in phase changes and reactions in rock, like dehydration of clays, releasing of gases and inducing thermal stresses. Progressively increasing laser power induces melting and vaporization. Laser technology, when perfected, has the potential to reduce drilling time several fold and impact the final cost of producing geothermal power.

In conventional rotary drilling for geothermal energy, a normal oil-well rig such as the one shown in Fig. 6.1 is used. A typical rig to drill a 1,000–2,000-m deep hole and to provide adequate hoisting capabilities would have a 400–500-hp (horsepower) engine to operate a 300-hp draw-works, and an independent 500-hp mud pump to supply the hydraulics needed to drill efficiently; the mast and substructure would have a capacity of over 100 tons to provide an adequate margin of safety in running the liner. A drill string of $3\frac{1}{2}''$ drill pipe and $6''$ drill collar would be sufficient. To facilitate mobility, all the components would preferably be trailer mounted. Mud is

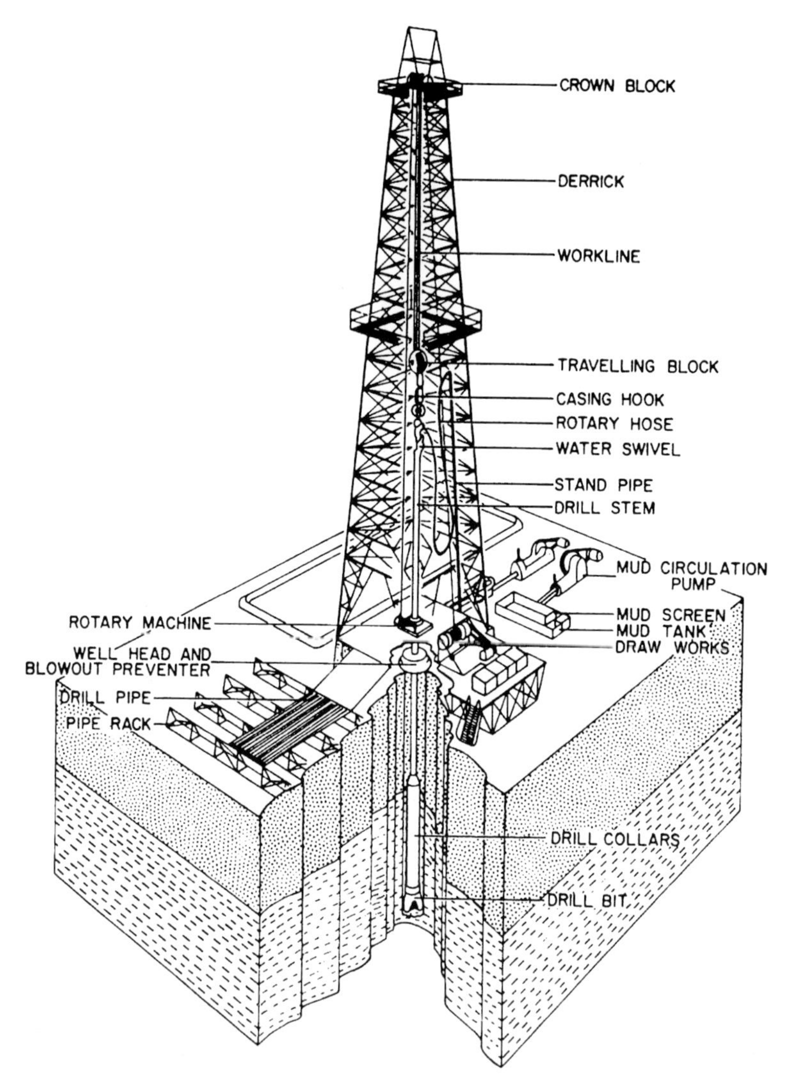

Fig. 6.1. Schematic representation of a rotary drilling rig and the peripherals used in geothermal drilling.

usually used as the drilling fluid. However, the use of air instead of mud makes drilling much faster and cheaper and has drawn increasing attention during the recent decades. One major obstacle in applying air drilling in geothermal areas is its unsuitability in formations bearing excessive water or in formations that tend to slough.

Table 6.1. Estimates of drilling time in geothermal areas under favorable conditions (Matsuo, 1973)

Depth (m)	Actual drilling (days)	Finishing and testing (days)
500	15–30	10
1,000	25–45	10
1,500	35–55	10
2,000	50–70	10

Drilling with Mud

The time required for drilling depends upon the ease of access and the topography of the site, the hardness of the formations and the number of meters to be drilled in the productive zone where mud losses are high, and on the amount of casing required. The estimates of drilling time made by Matsuo (1973) are given in Table 6.1.

Drilling mud

Clay-based muds, such as bentonite, are the most commonly used muds in geothermal drilling at low temperatures (~150 °C). At higher temperatures, ordinary clay-based muds tend to gel and therefore chrome-lignite–chrome-lignosulphonate (CL–CLS)-based mud is generally used. CL–CLS-based muds are particularly effective in formations that cave in easily. They can withstand temperatures of up to 250 °C. For more efficient drilling, it is desirable to keep the temperature and the solid content of the drilling mud as low as possible. The temperature is often reduced through 10–15 °C by installing cooling towers at the surface. Further cooling could be achieved with the help of large fans. The solid content of the mud is sometimes mechanically reduced by a large high-speed shaker, a desander and a desilter.

Drilling bits

Roller bits with hardened-steel teeth or tungsten-carbide inserts are commonly used in geothermal drilling. Since the steel used in roller bits is drawn at temperatures ~200–250 °C, these bits lose much of their strength when operated at temperatures in excess of 250 °C. This causes rapid failure of bearings and steel teeth as well as loss of inserts with the insert bits. Expensive roller bits are provided with sealed lubrication systems with rotating rubber seals to hold the grease in the bearings. Since these rubber seals have a temperature limitation of about 200 °C, improved seals and improved high-temperature lubricants are required in high-temperature geothermal drilling. Diamond drills can drill at temperatures in excess of 500 °C. However, since their drilling rate is much slower compared to roller bits, they do not provide a very acceptable solution to the problem of high-temperature drilling.

It is always important to know how a drill bit has been behaving with the passage of time. This should be estimated without removing the drill bit from the borehole.

Often decrease in the rate of penetration of the drill bit is correlated with the wearing of the tooth of the drill bit. However this is not adequate if the strength of the rock is not known. The decrease in the rate of penetration could be due to increase in the hardness of the rock formation, or the accumulation of debris below the bit with chips of the hard rock. Cooper (2002) has come up with an approach to estimate real-time measurement of the drill-bit tooth wear. This could be achieved by comparing the actual drilling performance and compare it with the theoretical calculations based on the mechanical properties of the drill bit, the drilling conditions and the strength of the rock being drilled. Cooper (2002) has stressed on making measurements on the properties of the rocks soon after the rocks have been penetrated.

Casing

To obtain a sustained flow of steam from a reservoir, it is necessary to choose an appropriate diameter for the production well. Additionally, it is necessary to provide adequate casing at correct depths to prevent hot water from higher formations from entering into the well. When the pH value of the hot water produced with the steam is low, it tends to corrode the casing and reduce the life span of the well.

Slotted liners are recommended when the wall of the hole tends to slough and caving in of the hole is suspected. Sloughing can be recognized by the presence of sloughed-off particles among the drill cuttings when passing through the production zone. If the wall of the hole collapses, it would take much longer to re-drill when an ordinary liner is used.

The recommended hole diameters and casing specifications under different steam-production volumes (after Matsuo, 1973) are given in Table 6.2.

For the sake of longevity, the casing to be used must be capable of withstanding wear, corrosion and attrition due to friction and vibration. A sudden change in the diameter of the casing pipe at the joints causes turbulence in the high-speed steam flow and results in the erosion of the upper corners of the joints (Fig. 6.2A) as well as the inside surface (Fig. 6.2B). This can be avoided by the use of internal flush-butt joints (Fig. 6.2C), in which the inside diameter does not change suddenly. At times, casing is worn by the fine sand carried with the steam; this can be reduced by regulating the speed of the steam with a flow regulator installed at the wellhead.

Circulation losses

To prevent circulation losses of drilling mud, cementing is sometimes done at the horizons where losses occur. This is usually a permanent measure against such losses. At times the formations are plugged with materials such as nutshell powder, cottonseed hulls, cellophane and fiber scrap. In the course of time these materials carbonize and block the fractured zones.

Cementing

Proper cementing of geothermal wells requires that the cement slurry should rise uniformly and continuously from the casing shoe to the ground level. If it is

Table 6.2. Recommended borehole diameters and casing specifications for different steam-production volumes (Matsuo, 1973)

(1) *Steam volume: 10–25 tons* h^{-1}		
17″ open hole	$13\frac{3}{8}$″ surface casing	Fully cemented
$12\frac{1}{4}$″ open hole	$9\frac{5}{8}$″ intermediate casing	Fully cemented
$8\frac{5}{8}$″ open hole	7″ production casing	Fully cemented
$6\frac{1}{4}$″ open hole	$4\frac{1}{2}$″ slotted liner	
(2) *Steam volume: 25–50 tons* h^{-1}		
18″ open hole	16″ surface casing	Fully cemented
$14\frac{3}{4}$″ open hole	$11\frac{3}{4}$″ intermediate casing	Fully cemented
$10\frac{5}{8}$″ open hole	$8\frac{5}{8}$″ production casing	Fully cemented
$7\frac{5}{8}$″ open hole	$6\frac{5}{8}$″ slotted liner (outer diameter of coupling skimmed by 1/16″)	
(3) *Steam volume: 50–80 tons* h^{-1}		
22″ open hole	18″ surface casing	Fully cemented
17″ open hole	$13\frac{3}{8}$″ intermediate casing	Fully cemented
$12\frac{1}{4}$″ open hole	$9\frac{5}{8}$″ production casing	Fully cemented
$8\frac{5}{8}$″ open hole	7″ slotted liner	

envisaged that the formation cannot be adequately cemented while drilling, or that it will be difficult to cement the entire height of the hole in one stage, multistage cementing should be planned from the beginning. For proper cementing, the formation of thick filter cakes should be prevented. When thick cakes cannot be prevented, they may be scraped using scratchers. Sufficient clearance between the casing and the wall of the hole should be ensured in order for the void to be uniformly filled with cement. If the annular void from the well head to the bottom has not been properly filled with cement, it is usually preferable to provide a rigid support to the head section of the casing instead of attempting to re-inject cement.

Ordinary cement is adequate for temperatures up to 150 °C, but to resist higher temperatures, silica is often mixed with it. In geothermal wells where steam is accompanied by low-pH hot water, it is necessary to use acid-resistant cement.

Drilling with Air

Rotary drilling with air is quite similar to drilling with mud, but with air replacing the mud. The technique has been tried in many countries for several years and has been particularly successful at The Geysers geothermal field in California. Two important advantages of air drilling over mud drilling are: (1) the production zone is not damaged by the circulating mud and (2) compared to mud drilling, the drilling

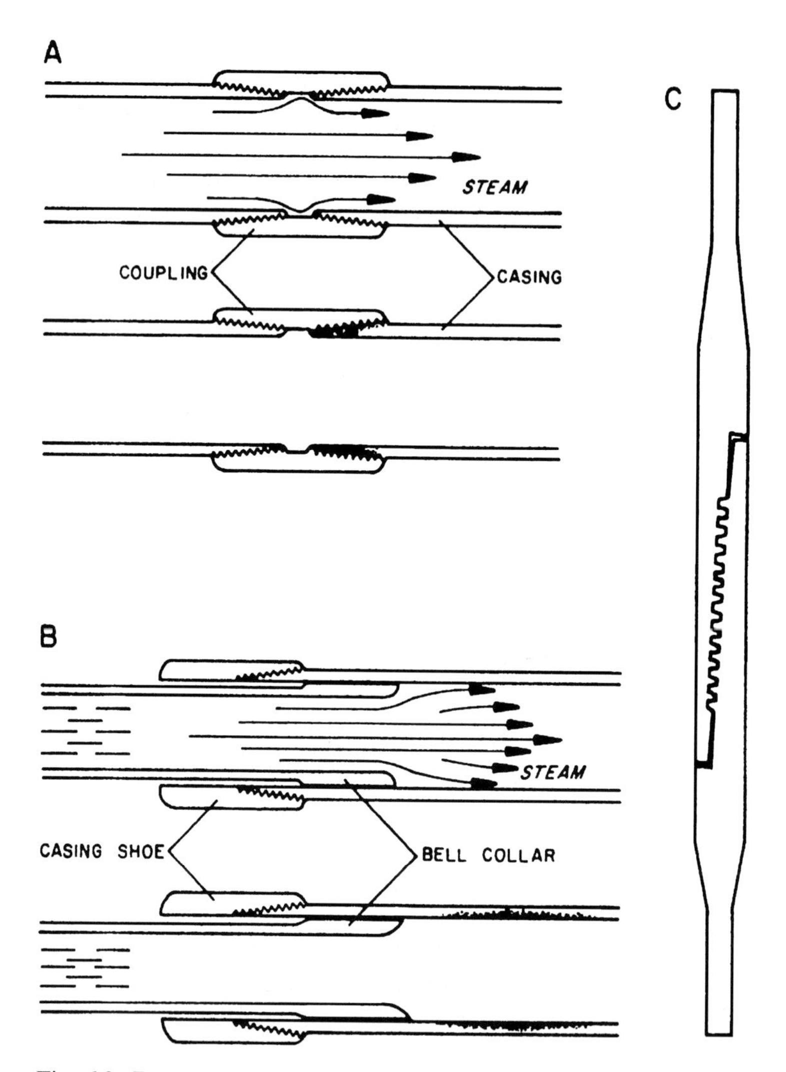

Fig. 6.2. Damage to casing due to variation in the inside diameter at the joints, and its prevention (modified from Matsuo, 1973). (A) Turbulence in high-speed steam flow caused at the joint (top), and the consequent damage at the corner (bottom) shown by shading. (B) Same as A for a bell collar. (C) Flush-butt joint, which helps in preventing such damages.

speed is 3–4 times faster and the bit life is 2–4 times longer. However, air drilling is not suitable in formations bearing excessive water and in those that tend to slough. Under suitable subsurface conditions it is not unusual to drill a hole employing both air and mud drilling.

Equipment
The equipment used in air drilling is quite similar to that used in mud drilling. Fig. 6.3 shows the former process diagrammatically. The mud pump is replaced by an air compressor and a booster compressor. The air compressor provides an air velocity of 10–25 m s^{-1} in the annular section between the drill pipe and the hole, and the booster compressor provides the additional pressure necessary to blow out the water accumulating in the hole.

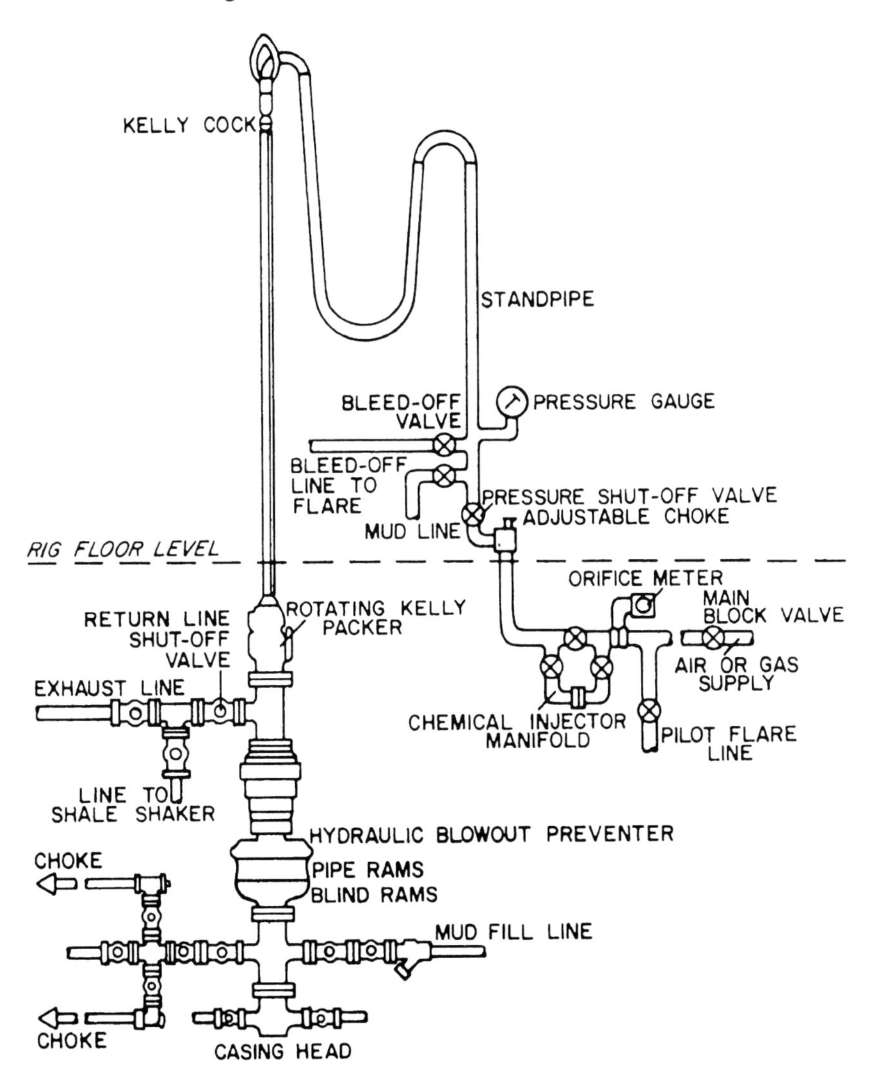

Fig. 6.3. Schematic representation of the air-drilling setup and the peripherals (modified from Matsuo, 1973).

The space between the wellhead and the drill pipe is closed with a rotary packer while lowering and lifting the drill pipe. This prevents the air from flowing back to the rotary table and the floor of the derrick. The packer is made of circular strip rubber. It can withstand temperatures of the order of 250 °C and may need as frequent replacement as the bit. An inside blow-out preventor is attached to the lower part of the drill pipe. This prevents the steam from flowing out of the drill pipe during raising and lowering. A discharge pipe is used for blowing off air, dust and cuttings. The area of the discharge pipe must be equal to or smaller than that of the annular space between the casing and drill pipe to stop the cuttings from accumulating at the drill collar. Sometimes the discharge pipes are provided with a water sprayer to ease the removal of cuttings and the dust compared to the bits used in drilling with mud, the air-drilling bits have enlarged nozzles and air-cooled bearings.

Prevention of water intrusion

The efficiency of air drilling is rapidly reduced with an increase of the amount of water intruding into the well from the surrounding formations. When the water seepage is very small (e.g., 500 l or less per hour), it can be successfully stopped by injecting finely ground silica gel and calcium stearate through a pump into the air stream. The quantities injected may be of the order of 2% by weight of the quantity of the cuttings ejected from the well. Calcium stearate forms a coating, which repels water from the cuttings, and silica gel is helpful in reducing the torque on the drill pipe. When the water seepage is of the order of 500–10,000 $l\,h^{-1}$, a foaming agent such as lithium stearate is used. In case of excessive water seepage (over 10,000 $l\,h^{-1}$), some suitable material such as plastic or cement is used for sealing off the water-bearing horizons. In places silicon tetrafluoride gas has also been used effectively for sealing the formations. When using the gas to prevent seepage, the upper and lower parts of the horizon requiring sealing are sealed with packers and gas is injected in the sealed-off area. On contact with water, a precipitate is formed which blocks the pores of the formation. When water seepage problems are insurmountable, it becomes necessary to switch over to conventional mud drilling.

Well Spacing

In spite of numerous studies, no definite criterion for well spacing in geothermal fields has been formulated. Even when a good cap rock exists, it is difficult to determine the contribution of different underground faults and fissures to the steam produced by a well. Consequently, it cannot be determined how the overall production is going to be affected by siting a well at a certain distance from an existing well. However, in shallow steam fields (500–2,000 m deep), a well spacing of 100–300 m is common. Large distances are usually kept to avoid the possibility of pressure interaction between adjacent wells, which could lead to a reduction in discharge. It has been pointed out that the costs of geothermal field development can be reduced by decreasing the distance between production wells and thus curtailing the length of steam transmission lines, which are quite expensive. It is suggested that for geothermal systems where the flow is through fissured formations and the crack permeability is good, the spacing could be somewhat less and successful wells with fairly large discharges could be spaced at distances of the order of 50 m without interaction effects.

Safety Measures

Safety measures to be undertaken and their relative importance to one another vary from one location to another. A few common problems and their remedies are as follows:

(1) Gas detectors and masks should be readily available. Gas emitted by a well may be poisonous and could produce giddiness and eye injury. If emitted in a high concentration, it could even become lethal.
(2) An adequate water supply should be available at the drilling site. At times it takes a long time for the preventor to control the steam gushing out during a drilling operation. Water is useful in averting emergencies. Also, the functioning of the preventors should be checked regularly.
(3) The blow-out preventor is held in position by the casing. To stop the steam from gushing out when the preventor is closed, the space around the casing must be perfectly sealed with cement. The casing shoe must be placed and held firmly in a competent formation and the void outside it must be uniformly and completely filled with cement.
(4) Sometimes it becomes difficult for workers to reach the derrick escape ladder in the event when the hot water and steam start gushing out. Hence, it is essential to provide an escape cable and seat.

Repairing of Wells

With the passage of time, the pressure in a reservoir declines and consequently the production of steam at the same pressure cannot be sustained with the original casing diameter. It then becomes necessary to insert a smaller-size casing in the well to correspond with the available pressure. The production of a well could also decline or stop due to casing failure, collapses and fractures. The exact procedures followed to clean the well and restore production differ from one location to another. As an example, the following procedure was in vogue at the Cerro Prieto geothermal field in Mexico (Dominguez and Vital, 1976):

(1) The wells were permitted to cool. The cooling was ascertained by taking several temperature readings.
(2) Pressure, temperature and caliper logs were carried out.
(3) After determining the obstruction, lead impressions were taken to determine its type and size.

(4) Sand plugs were eliminated by mud circulation. The depth of fracture was determined through the flow of sand. Blocked portions of the hole were cleaned by milling to permit the lowering of new production casing and other tools.
(5) Cement grout was injected through the fractures and milled windows to obtain good seals. The seals were checked by hydrostatic tests.
(6) The windows and fractures were covered using $7\frac{5}{8}''$ and $5\frac{1}{2}''$ diameter casings. The lower end of the casing was placed as close as possible to the upper end of the liner and the casings were cemented along their entire length using modified cement. At the end of these operations, the plugs, couplings, shoes, etc., were scoured until the well was clean.

Cyclic Production Behaviour

A very good case study has been provided by Iwata et al. (2002) to mitigate cyclic production behaviour at Uenotai Geothermal Field in Japan. The Uenotai Geothermal Field was commissioned in 1993 and has been producing 25 MW (net) since then. It has been provided geothermal steam by Well T-52. Throughout the production history, this well has exhibited characteristics, which are commonly seen in the wells that intersect multiple production zones. This well has a water dominated production zone in the depth range of 1,675–1,780 m, and a steam dominated production zone in the depth range of 1,450–1,460 m. The two zones produced in an alternating pattern, with the contribution from the deeper located water zone periodically resulting in a strong surge. This surge caused sudden fall in wellhead pressure. Sometimes this fall in pressure rendered the wellhead pressure below the production pressure, suspending the production of the power. A mitigation plan for such situations was developed by plugging off the lower water dominated production zone. The plugging operation was completed in about a week's time and it took another month for the well to recover. Iwata et al. (2002) have reported that after this operation the cyclic behavior of the well T-52 was successfully eliminated and the generation capacity of the well increased by more than 30%. This kind of situation is quite common in geothermal production fields and the approach adopted at the Uenotai Geothermal Field can suitably be adopted to overcome the cyclic behavior of the wells.

Cutting Drilling Costs

Drilling is an essential and expensive component of geothermal exploration, power production and maintenance. High-temperature corrosive fluids and hard, fractured formations increase the cost of drilling, logging and completing geothermal wells compared to exploration and exploitation for oil and gas. Often, drilling and completing the production and injection well accounts for one-half of the capital cost for a geothermal power project. Therefore, cutting down drilling costs becomes critical because it ultimately brings down the cost of producing geothermal power.

Cost reduction in geothermal drilling can be accomplished in one of the following ways: (a) faster drilling rates, (b) increased bit or tool life, (c) minimizing production problems such as occurrences of twist-off, stuck pipe etc., (d) higher per well production through multi-laterals and others.

Slim-hole drilling techniques

During the 1970s, slim-hole drilling technology was introduced for gas and oil exploration. This made it possible to reduce the infrastructure cost considerably, and makes smaller oil and gas fields attractive from the financial angle. Ciptomulyono (2002) has overviewed the application of slim-hole drilling technology for geothermal prospecting. These holes are usually less than 15 cm in diameter. Table 6.3, after Ciptomulyono (2002), provides a technical comparison of the slim-hole drilling and conventional drilling. Slim-hole drilling has found a wide application in geothermal exploration and development of small geothermal resources in Indonesia and elsewhere.

Slim holes are usually drilled using diamond-coring rigs as against the large-diameter wells that are drilled using rotary rigs. Slim holes cost less than large-diameter wells because (a) the smaller rigs require less transportation and site preparation, (b) the smaller-diameter holes have less expensive drilling tools, casing and cement jobs and, (c) if the hole is core-drilled, there is no need to repair lost circulation zones before drilling ahead (Finger and Jacobson, 2000).

Table 6.3. Comparison between slim-hole drilling and conventional drilling (after Ciptomulyono, 2002)

Technical specification	Slim-hole drilling	Conventional drilling
Drilling depths (m)	1,000–2,100	1,000–2,100
Hole diameter, at final depth (mm)	50–75	150–216
Rig weight (metric tons)	14–22	40–65
Installed power (KW)	74–90	280–350
Drill-string weight (tons)	3.5–5.5	30–42
Mud-pump power (KW)	44–73	180–300
Hole volume (m^3)	3.5–6.0	18–37
Mud-tank capacity (m^3)	8–10	60–75
Circulated mud volume (m^3)	8–10	60–75
Additive cost of drilling mud (%)	20	80–110
Casing weight ($kg\,m^{-1}$)	4.4	100
Casing cost (%)	30	25
Drill area (m^2)	25 × 32	150 × 64

RESERVOIR PHYSICS AND ENGINEERING

Reservoir physics and engineering deal with making the necessary physical measurements, modeling and laboratory studies for estimating subsurface reservoir conditions. The measurements usually carried out in the reservoirs pertain to their size, temperature, pressure, permeability and fluid composition. Additionally, flow rates of the various fluid constituents, i.e., steam, gas and hot water, in the wells are measured. Pressure-transient analyses are being increasingly used to determine critical parameters such as drainage volume, porosity, permeability and mean formation pressure. Mathematical modelling has been helpful in solving two-phase flow, convection patterns and other problems concerning dynamics of the reservoir under certain simplified assumptions.

The properties of steam and water under varying conditions are known quite accurately and are published in steam tables. These, and other derived properties, such as flash and dryness, are extensively used in geothermal well measurements. Some of these properties are shown in Fig. 6.4.

Subsurface Well and Reservoir Conditions

Because of the disturbances resulting from drilling, it is very difficult to measure the natural reservoir conditions. The reservoir temperatures are temporarily lowered by the cooling action of the circulating mud (Fig. 6.5A). The mass permeability of the reservoir is permanently affected by the introduction of the hole. The measured temperatures and pressures are closer to reservoir conditions in the sections where formation permeability is high. In the uncased sections or the sections with perforated casing, the pressures measured are close to the reservoir values. However, the temperatures may still vary considerably in the sections with fluid flow between permeable zones. The subsurface temperatures can vary unpredictably and inversions are often observed (Fig. 6.5B). Fig. 6.5C depicts some field observations. Hence it is not advisable to interpolate or extrapolate the temperatures in such cases. Unlike the temperatures, pressure inversions are never observed and pressure gradient changes drastically only when there is a phase change. Therefore, interpolation and extrapolation of pressures is often done. Measurement of pressure gradient helps in estimating fluid density, and when coupled with knowledge of the temperature, it is possible to determine the fluid type. With the help of steam tables, it is easily determined whether the fluid in question is water or steam. The presence of gases like hydrogen sulfide and carbon dioxide, which are most commonly encountered in the geothermal fields, can be easily detected since they are much heavier than steam under the same temperature and pressure.

The time taken in stabilization of the down-hole condition, after the drilling has been stopped, mostly depends upon the permeability of the formations and the amount of disturbance caused by drilling (Jaeger, 1961). Drilling particulars also affect the stabilizing time. For example, in the case of diamond drilling, where the

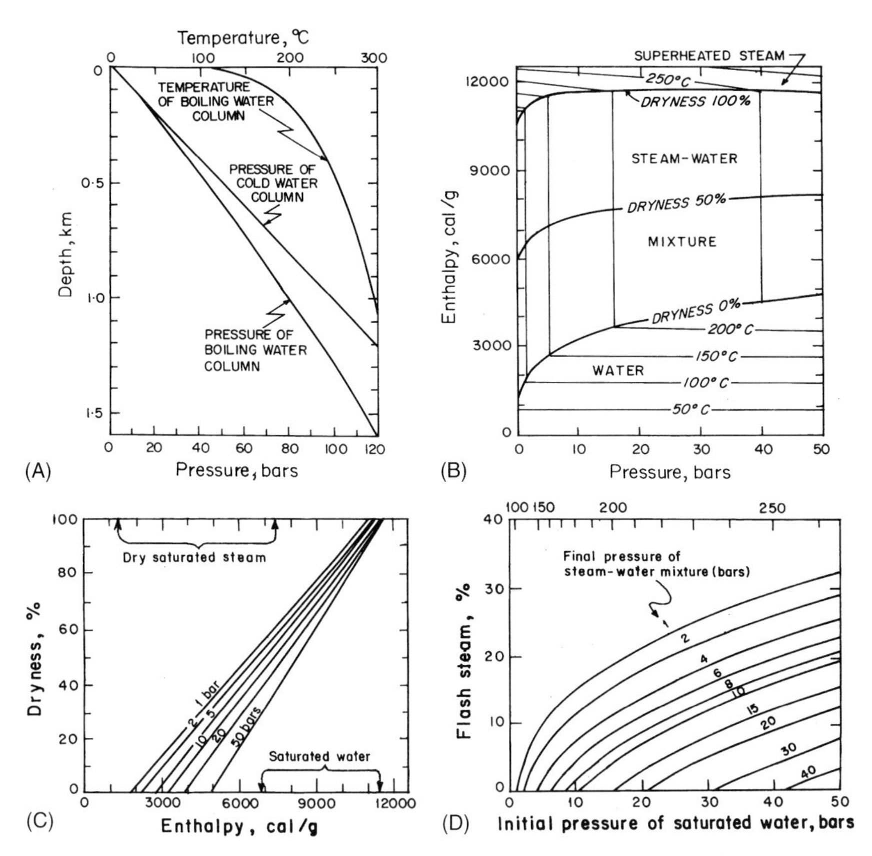

Fig. 6.4. Water and steam properties (Modified from Dench, 1973). (A) Theoretical temperature and pressure of a water column. (B) Temperature, pressure and enthalpy of steam. (C) Dryness, pressure and enthalpy of steam. (D) Flash steam and pressures for steam.

water circulation is relatively small, the regions at the top and at the bottom of the hole are the most affected. Therefore, temperature stabilization in the borehole is quick. On the other hand, in rotary drilling high fluid velocities are used and the temperature in the hole is governed by the input temperature of the fluids for a long time. In extreme cases, stabilization may be achieved within a few hours or may take a few months. In Fig. 6.5D (modified from Jaeger, 1961), recovery in the temperature of a borehole, whose temperature has been changed by θ_0 for time t_0 and then left undisturbed, is depicted for varying hole radii and diffusivities of the surrounding rocks. Kappelmeyer and Haenel (1974) have given useful dimensionless empirical relations for estimating stabilizing time under varying borehole conditions.

Flow measurements, involving physical and chemical methods, are employed to identify and measure the rate of discharge of various constituents being produced at

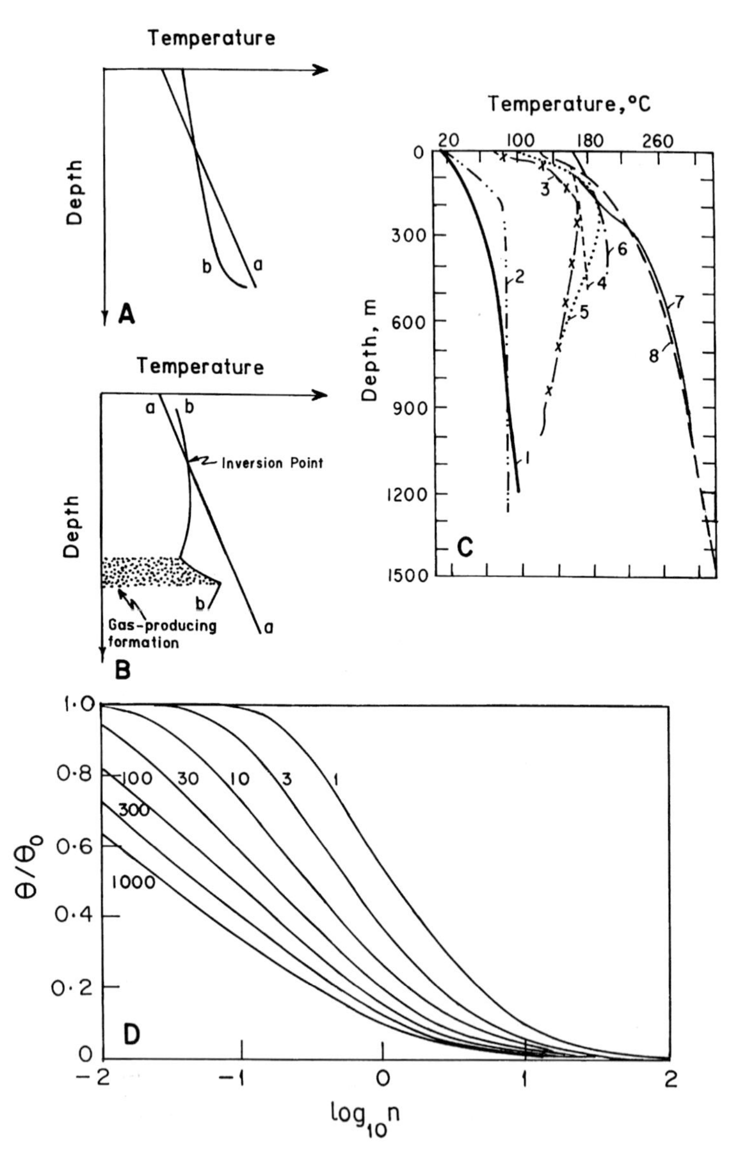

Fig. 6.5. (A) Schematic representation of rock temperatures before drilling (a) and immediately after drilling (b) under the ideal conditions of uniform thermal conductivity of rocks, normal drill-hole conditions and continuous drilling. (B) Temperature inversion due to a gas-producing zone: a = original temperatures with depth, and b = temperatures when a gas-producing horizon exists. (C) Temperature–depth profiles of wells at Kamchatka geothermal field (1–7) and the theoretical curve (8) for changes in boiling point with increasing normal hydrostatic pressure (from Vakin et al., 1970). (D) Recovery in temperature of a borehole whose temperature has been changed by θ_o for time t_o and then left undisturbed (modified from Jaeger, 1961). The temperature θ at a time nt_o after the beginning of the undisturbed period is plotted against $\log_{10}n$. The numerals on the curves are $(\chi t_o/a^2)$, where χ is the diffusivity and a the radius of the hole.

a well. The geothermal fluids—water and steam—are at times contaminated with various other gases like H_2S and CO_2, and solid impurities like sand and salt. The energy content of the fluid arriving at the wellhead is practically the same as that when it enters the borehole. Water or mixtures of water and steam occurring at high temperatures boil during upward transmission in accordance with the saturation conditions, whereas dry steam and water below 100 °C temperature reach the wellhead without any phase conversion. Formation permeability and the physical restrictions at the wellhead control the flow rate. When a well passes through zones producing fluids at the same temperature, the enthalpy of the fluid produced is found to be fairly constant over a wide range of flow rates.

A well-flow pressure maximum is reached when the well is choked to limit by a complete closure of the well. The rate of closure affects the maximum discharge pressure. The lower limit of the flow pressure at which a well continues to be operative depends upon the reservoir permeability and the friction encountered in the surface pipework. When the fluid produced by a well is drawn from two or more horizons at different levels, changes in the wellhead throttle influence the proportion of the fluids being drawn from each level.

Under the circumstances of two-phase flow, small and rapid pulsations are most commonly observed. These pulsations are caused by the continuous boiling which takes place as the fluid rises to zones of lowering pressures in the well. Another characteristic of a two-phase flow is the difference in the velocities with which the fluid moves in the pipe section. Water flows with relatively slower velocities along the walls of the pipe while steam and water move much faster in the central part of the pipe cross-section.

Temperature Measurements

Until the 1960s, thermometers used for measuring temperatures in boreholes were basically of two categories: (1) self-recording expansion type and (2) electric signal transmission type. The thermometers belonging to the first category either operate on the principle of change in volume of a liquid or on the curving of a bi-metallic sensor with changes in temperature. These thermometers are simple and convenient in operation and do not require an expensive heat-resistant cable. However, they require repetition of the manual operation of lowering and raising in a borehole to investigate temperature changes with time. Further, they have limitations in accuracy and resolution for both temperature and depth. These tools were usually employed in wells immediately after completion of drilling, i.e., in non-equilibrium conditions, which resulted in further deterioration of data quality (Blackwell and Spafford, 1987).

The thermometers belonging to the second category, i.e., the electric-line tools, make use of the thermoelectric effect and change in electrical resistance of metals with temperature. The signal is mostly transmitted through an electric cable, and can be recorded continuously by an analog or digital recorder. These tools provide better accuracy and precision in both temperature and depth. Continuous temperature-depth logs could be obtained rapidly. However, use of these tools are limited by the maximum temperature that the cable can withstand, usually about 150 °C and up to about 250 °C at the most, and costs of maintaining very long electric cable lengths.

The accuracy, ease and cost of making temperature measurements in high-temperature geothermal wells have undergone significant improvements over the past two decades. Wisian et al. (1998) provide a succinct summary of the temperature logging tools currently in use, namely, the slick-line computer tools, and the Distributed optical-fiber Temperature Sensing (DTS) system. The slick-line tools employ self-contained, battery-powered, computer and temperature sensor housed in a Dewar flask assembly, which is lowered into the well on a solid wire. The Dewar flask protects the sensor under high-temperature environments inside the well and has been tested up to temperatures of 400 °C continuously for about 10 h. Because slick-line reels are often available at most drilling sites, an additional advantage of this tool is that the sensor assembly can be directly hooked to it and logs obtained without the necessity of carrying the slick-line to the field site. The DTS tool, developed by Hurtig et al. (1994), has the potential of withstanding well temperatures of up to 550 °C. The tool works using Raman effect backscattered laser light in an optical fiber. Observations of the intensity of backscattered light with time can be used for determining the temperature along the entire length of the optical-fiber cable instantaneously. Although this tool is less accurate in temperature and depth by an order of magnitude relative to the electric-line and computer-based slick-line tools, it can be gainfully employed to monitor transient events in a well by keeping the entire cable lowered inside the well for several days without perturbing the water column due to repeated lowering and raising of the tool. Improvements such as precision and resolution in temperature and depth, reduction in the weight as well as cost of the sensor, and development of sophisticated filters for the extracting valuable additional information from temperature logs are necessary before this tool can be used universally.

Pressure Measurements

Along with development of new temperature measuring tools, pressure sensors for down-hole measurements have undergone substantial improvements with time. Until the 1960s, pressure gauges equipped with multiple-coil Bourdon-tube-type pressure elements (e.g., Kuster gauge) were mostly used for measuring geothermal fluid pressure in boreholes. The bellow-Bourdon tube fluid-filled system sensed the pressure from the well and a change in pressure rotated the Bourdon tube. These pressure sensors are now replaced with pressure transducers, which can accurately monitor changes in pressure over a wide range. The transducer is lowered into the borehole by means of a cable to a level below the water level which will remain submerged during changes in the water level. The transducer measures the depth of water as pressure, which can be converted into depth of submergence. The water level is obtained by subtracting the depth of submergence from the distance the transducer was lowered into the hole. Broadly, two types of pressure transducers are used: (a) strain-gauge bridge type for temperatures less than about 80 °C and (b) digiquartz pressure transducer for temperatures up to about 110 °C. The working principles for the two types are summarized in Lienau (1992).

In a strain-gauge pressure transducer, the sensor is made by diffusing a fully active four-arm strain-gauge bridge into the surface of single crystal silicon diaphragm. The active diameter and thickness of the diaphragm can be varied according to pressure range and application. In cases where the pressure media are not compatible with the silicon diaphragm, stainless steel or other materials are used as isolating diaphragms such that the sensor is totally immersed in oil with negligible degradation in performance. These instruments require a reference pressure such as atmospheric pressure; therefore, the connecting cable to the surface includes a vent tube in addition to four conductors. They are available in a series of operating pressure ranges, such as 1, 5, 10, 15, 20, 30, 50, 100, 150, 200, 500 and 900 psi gauge. The pressure range is selected according to the amount of variation in water level in the well.

The digiquartz transducer operates on the principle of a quartz crystal resonator whose frequency of oscillation varies with pressure-induced stress (Lienau, 1992). Frequency outputs allow interfacing with counters, computers or other digital data acquisition systems. The transducers also provide a quartz crystal temperature signal for full thermal compensation over a wide temperature range. The resolution of the digiquartz pressure transducer is $<0.1\%$ of full scale.

Flow Measurements

During the process of drilling a well, a log of the losses of the drilling fluid and of temperature is normally made. The logs are useful in estimating the approximate output of a well, as well as in deciding a suitable method for the measurement of flow. Observations of the variation of wellhead pressure with time provide useful indications of any changes in quality and quantity of flow. Usually, the flow is very erratic within the first few hours to a few days of completion of a well.

For a better understanding of the flow characteristics of a well, it is desirable to measure the rate of flow of total mass; the temperature and chemical constituent of single phases. These measurements are made at several wellhead pressures. The wellhead pressures are controlled by throttling the flow with valves or orifices with different diameters. Instead of trying to achieve a specific pressure for flow measurement to be made, it is more efficient to use a convenient choke setting and obtain a graph between resulting pressures and flows. In the event that the well exhibits a smooth relationship, flows at desired pressures can be accurately obtained through graphical interpolation. It is advisable to repeat flow measurements for both increasing and decreasing stages of throttling. However, when changes in output with time are being investigated, it is best to repeat the same process.

Measurement through an orifice
Single-phase flows under pressure are suitably measured by observing the pressure difference across an orifice in a straight run portion of the flow pipe. The pressure difference is measured using a manometer. Standard engineering textbooks carry the necessary information in regard to orifice geometry, flow calculations, possible errors and their mitigation. Since flow measurements using an orifice require single-phase flow, saturated water must either be cooled or pressurized to avoid boiling at the orifice.

Measurements with a calorimeter
Calorimeters are used to measure the volume and heat produced during a certain time interval. The flow is passed through a tank partly filled with water whose thermal capacity and weight are known before the beginning of the experiment. With the measurement of the increase of temperature and volume of water during known time duration, the heat and flow rates are calculated. Such calorimeters are particularly suitable for small flow measurements. In the case of large flows, a small (say, 0.01 m) diameter sampling tube can be inserted to collect the fluid to be measured by the calorimeter. The tube is made to traverse the entire radius of the well. The timing for sampling at each radius should be proportional to the area of flow it represents. This method is inexpensive, but the results obtained are not very reliable.

Cyclone separator vessel
Large steel cyclone separator vessels, like the one shown in Fig. 6.6 (from James, 1976), are often used to measure the mass flow and enthalpy of a two-phase discharge. The steam and water phases are separated in a vessel with the help of horizontal pipes. Throttle valves are used to control the flow, and orifice meters are used to measure the rate of discharge. The method is quite simple and straightforward as long as one phase only passes through each orifice, but the presence of a second phase could cause serious errors. The presence of steam in the pipeline containing the orifice meter being used to measure the flow of saturated hot water is detected by a significant rise in differential pressure across the orifice. James (1976) has given empirical relations to estimate the percentage of steam quantities from the percentage increase in manometer differential pressure.

New tools are now available that can measure and record temperature, pressure and water flow continuously inside a geothermal well (Stevens, 2000). This tool can be lowered into the well using mono-conductor cable in both surface readout mode as well as slick-line down-hole memory mode. The temperature sensor is a resistance thermometer such as a platinum RTD (resistance temperature detector), while the pressure sensor is a strain gauge. Impellers are provided for flow measurements, which can be used for a range of water flow velocities and well sizes. These help detect up-flow and down-flow zones within a well as also zones of water loss during re-injection.

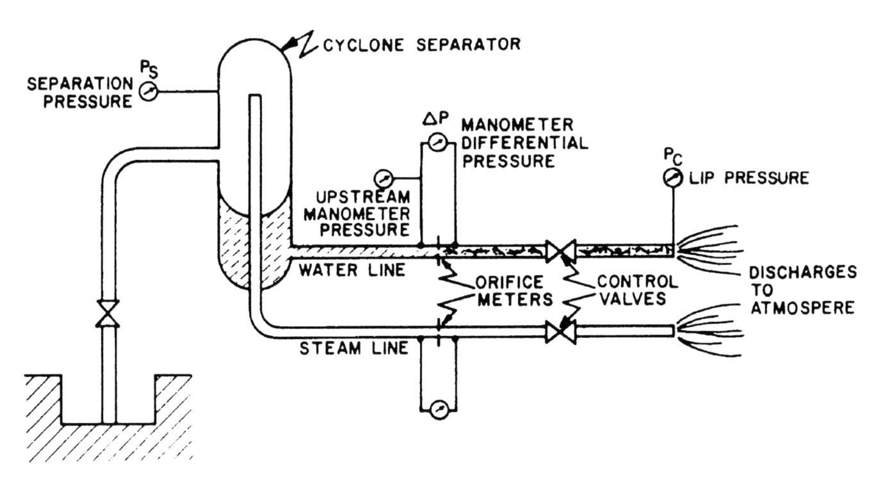

Fig. 6.6. An arrangement for measuring separated water and steam from a geothermal well (modified after James, 1976).

Pressure-transient Analysis

The effect of precipitation of solids (and consequent blocking of the drainage in the borehole face) into the well as well as the periphery of the entire geothermal reservoir is important. The precipitation occurs when the cold recharge fluids contact the hot geothermal fluids. This results in decrease in production with time. Steam wells in the Larderello geothermal field in Italy are known to have an active life of about 12 years. The pressure-transient analysis is helpful in determining the reasons for the decline in production often observed in oil, gas and geothermal fields. In the simplest pressure-transient analyses, the static bottom-hole pressure is measured at increasing intervals of time after closing the well. The lower the permeability, the longer it takes for the pressures in the well to stabilize. This information is used in determining formation permeability. Fig. 6.7A represents a Miller–Dyes–Hutchinson graph (after Ramey et al., 1976) for a well in the center of a reservoir whose boundaries could be approximated by a square. Further, a full recharge at the boundaries is hypothesized. In the graph, dimensionless, static build-up pressures are plotted against dimensionless shut-in time with the following definitions:

$$P_{DS} = \frac{k\ d\ (p_i - p_{ws})}{0.4568\ V_{sc}\ q\ B\ \mu}, \quad \text{for liquid flow} \tag{6.1}$$

$$P_{DS} = \frac{M\ k\ d\ (p_i^2 - p_{ws}^2)}{0.2789\ q\ Z\ \theta}, \quad \text{for steam or gas flow} \tag{6.2}$$

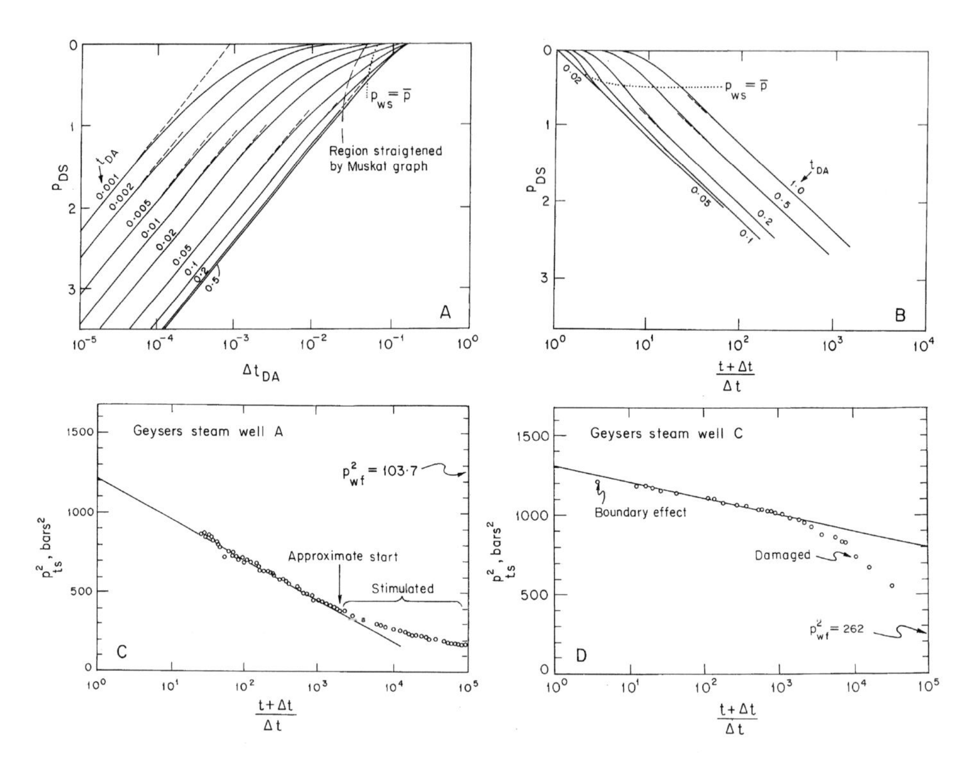

Fig. 6.7. Graphs commonly used in pressure-transient analyses, and their application to detect stimulated and damaged conditions at the Geysers. Miller–Dyes–Hutchinson graph (A) and Horner graph (B) for a well located at the center of a constant-pressure square. Detection of stimulated (C) and damaged conditions (D) in steam wells at the Geysers. (Modified after Ramey et al., 1976; Ramey, 1976). See text for details and the notations used.

$$t_{DA} = \frac{0.3604\, k\, t}{\phi\, \mu\, C_\theta\, A} = t\, D\, (r_w^2/A) \tag{6.3}$$

The relationship between slope b and effective permeability k is

$$k = \frac{0.5258\, V_{sc}\, q\, B\, \mu}{b\, d}, \quad \text{for liquids,} \tag{6.4}$$

$$k = 0.3210\, \frac{q\, \mu\, Z\, \theta}{b\, d}, \quad \text{for steam and gases} \tag{6.5}$$

where:

P_{DS} = dimensionless static pressure
k = effective permeability to the flowing phase (darcy)
d = net formation thickness (m)
p_i = initial pressure (kg cm^{-2})

p_{ws} = bottom-hole pressure (static) (kg cm^{-2})
V_{sc} = specific volume under standard conditions (cm^3 g^{-1})
q = production rate
B = formation volume factor (reservoir volume/standard volume)
μ = viscosity of flowing fluid (10^{-3} P)
M = molecular weight (g/g-mol)
Z = real gas law deviation factor
θ = absolute formation temperature (K)
ϕ = porosity in fraction of the bulk volume
C_θ = total system effective isothermal compressibility (cm^2 kg^{-1})
r_w = well radius (m)
A = drainage area (m^2)
b = slope of the semilog graph, (kg cm^{-2})/log cycle for liquids, and (kg cm^{-2})^2/log cycle for gases
t = time (h)
t_D = dimensionless time
t_{DA} = dimensionless time based on drainage area (A)

In Fig. 6.7B, a Horner graph for a well in the center of a constant-pressure square, i.e., fully recharged, is shown (from Ramey, 1976). An infinitely long shut-in time (Δt) would make $[(t+\Delta t)/\Delta t]$ approach unity and a dimensionless build-up pressure P_{DS} of zero since the well would return to the initial pressure. The build-up lines appear to move to the right with the increase of producing time prior to shut-in increases.

Horner-type build-up graphs are particularly useful in geothermal well measurements. Investigations on the *skin effect* or *skin factor* are useful in determining whether a well is damaged, stimulated or normal. On the Horner graph, pressure build-up with time exhibit semilog straight lines. A deviation is an indication of stimulated or damaged situation. Fig. 6.7C and Fig. 6.7D exhibit pressure build-up graphs for two steam wells at The Geysers (adapted from Ramey, 1976) exhibiting stimulated and damaged conditions respectively.

In conclusion, it may be mentioned that with passage of time, more realistic models for pressure-transient analyses are being made and they are very useful in estimating well conditions.

Reservoir Modelling

The geothermal reservoirs are usually complex hydrothermal systems involving single- or multiphase fluid flow. Mathematical and/or laboratory simulation of these reservoirs is helpful in estimating the recoverable energy, in determining the optimum management techniques and improving the understanding of the reservoir geometry, and in estimating boundary conditions and rock properties. As the production progresses, newer data permit refinement of initial crude models and a more purposeful and increasingly more accurate prediction of the reservoir performance.

Mathematical Modelling

Simple mathematical modelling basically consists of the formulation of continuity equations of mass, momentum and energy for each phase and the reduction of this system of equations to two non-linear partial differential equations. The final simplified equations, whose solution is sought, have two unknown dependent variables. These variables can be chosen from fluid pressure, temperature, enthalpy, water saturation, etc. As an example, let us discuss a case where fluid pressure and enthalpy have been chosen as the two unknown variables (Faust and Mercer, 1976). Consideration of the thermodynamic state of water under simplifying assumptions allows total density, saturated steam density, saturated water density, saturated steam enthalpy, saturated water enthalpy, water saturation and temperature to be expressed as functions of fluid pressure and enthalpy (Fig. 6.8). Next, the partial differential equations are transformed to approximate integral equations and solved simultaneously using a finite element technique. The dependent variables, pressure and enthalpy, are approximated using piecewise polynomials.

Steady state, vertical, one-dimensional flow. Donaldson (1968) has considered one-dimensional vertical flow of water and heat in a porous medium involving compressed water with the boundary conditions: at the top of the system ($Z = 0$), the temperature is θ_1 and the pressure is p_1; at the base ($Z = Z_2$), the temperature is θ_2 and a mass flow rate per unit area is specified as M. The analytical solution for

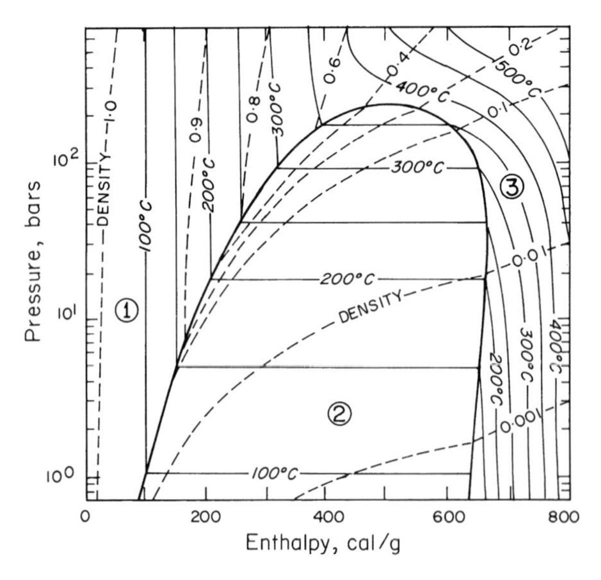

Fig. 6.8. Pressure–enthalpy diagram for pure water and vapor showing the three thermodynamic regions below the critical point (apex of parabola). *1* = compressed water, *2* = two-phase steam-water, and *3* = superheated steam (from White et al., 1971).

temperature assuming constant permeability, k, thermal conductivity, K, and heat capacity of water Q_w is given by

$$\theta = \theta_1 + (\theta_2 - \theta_1)\frac{1 - \exp\left(-\frac{MQ_wZ}{K}\right)}{1 - \exp\left(-\frac{MQ_wZ_2}{K}\right)} \tag{6.6}$$

The fraction MQ_w/K indicates the relative importance of convection in comparison with conduction. With convection becoming more dominant, the value of the fraction increases. In Fig. 6.9 results obtained for different values of MQ_w/K by Donaldson (1968) are presented. Additional data used in arriving at these results are: a thickness of 3 km; $\theta_1 = 116\,°C$; $\theta_2 = 224\,°C$; $p_1 = 3.0 \times 10^6\,Pa$; $M = 3.2 \times 10^3\,J\,kg^{-1}\,°C^{-1}$ and $K = 3.2\,W\,m^{-1}\,°C^{-1}$.

Multiphase horizontal flow

Faust and Mercer (1976) have mathematically investigated the behaviour of an initially hot-water geothermal reservoir that develops a two-phase (steam and water)

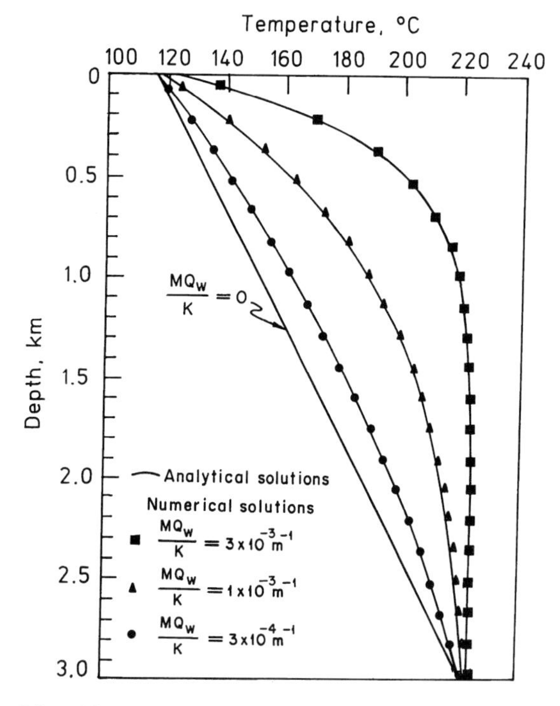

Fig. 6.9. Temperatures computed for one-dimensional steady-state vertical flow of compressed water (modified from Donaldson, 1968). For details, see text.

zone under the influence of production. Results obtained by them are shown in Fig. 6.10. In Fig. 6.10A, pressure values at various times are plotted against distance from the center of the reservoir and the times specified correspond to the duration of production. It is assumed that the production well bottoms in the center of the reservoir. During the initial stages of production, the pressure drops rapidly in the entire reservoir. After a month, the fluids split into hot water and steam at the well bottom and consequently the pressure is maintained there while it drops in the rest of the reservoir. After three years of production, the pressures in the entire reservoir have lowered to a value that is just above the pressure where two phases could exist. Further production reduces water saturation in the production zone. Fig. 6.10B indicates water saturation after five years of exploitation at various distances from the production well.

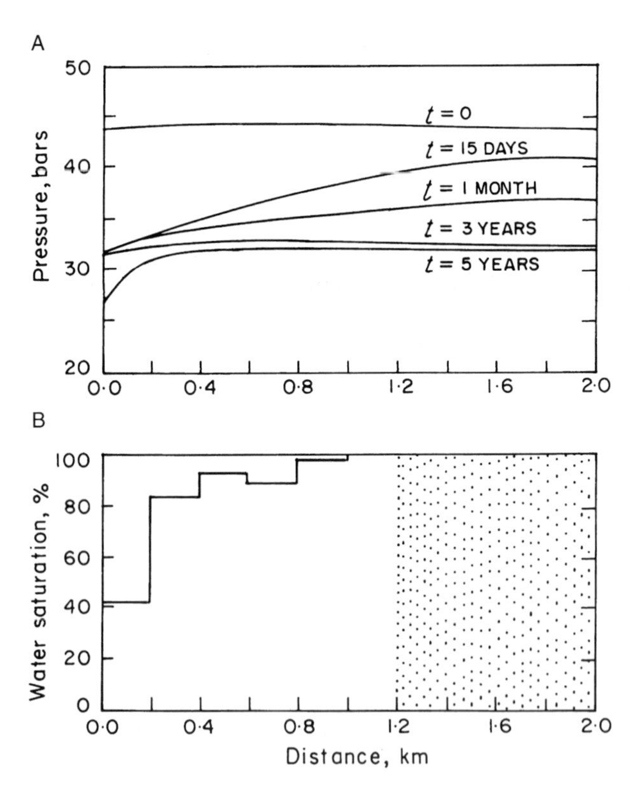

Fig. 6.10. Behavior of an initially hot-water geothermal reservoir that develops a two-phase zone under the influence of production (from Faust and Mercer, 1976). (A) Computed pressures versus distance from the center of the reservoir (left boundary of the model) at various times. (B) Computed water saturation versus distance from the center of the reservoir after five years of exploitation. Shading indicates water saturation.

In addition to investigating multiphase flow, convection pattern, heat and mass transfer and other problems concerning the dynamics of geothermal reservoirs, the effect of production rate on the ultimate heat produced and on the behavior of the ratio of steam to water production have been mathematically investigated. Robinson and Morse (1976) report that the total amount of heat that may be extracted from a geothermal reservoir is limited, and to obtain the maximum amount of heat, the reservoir pressure should be reduced to a point where the super pressured water will flash to steam within the pores of the rock.

Laboratory Modelling

Many parameters, important in estimating the production potential of geothermal fields, can be investigated through laboratory modelling. Results of an experimental study are briefly mentioned here.

In many known geothermal areas, porosity and permeability are too low to sustain adequate steam production rates. Also, as mentioned in Chapter 1, a large portion of the resource base consists of hot-rock formations with little water. The extent to which geothermal energy stored in a hot rock can be extracted under varying conditions is not known. To investigate the production potential of hot-dry rock under various circumstances, Hunsbedt et al. (1976) carried out experiments on a laboratory-size model simulating a fractured rock chimney. The model produces steam, either with or without simulated geofluid recharge, from initial temperature and pressure of up to 260 °C and 55×10^5 Pa. Transient behaviour resulting from steam production with two rock loads having 44% and 35% porosity was investigated. These experiments demonstrated that thermal energy stored in a fractured rock could be extracted effectively by reducing system pressure to allow boiling to take place in the rock formation. Under optimum operating conditions, 90% of the geothermal energy stored could be extracted. Maximum energy extraction was achieved when boiling took place near the top of rock mass and was controlled by recharge.

PRODUCTION TECHNOLOGY

Production technology deals with the movement of geothermal fluids from their sources to the sites where they are finally used either for the generation of electric power or for other industrial, agricultural or domestic purposes. Between these two ends, a number of practical problems dealing with the choice of proper equipment, designing of the transmission pipework and taking preventive measures against corrosion and scaling need to be tackled.

As mentioned earlier, the characteristics of the fluids produced vary considerably from one geothermal field to another. At places, only steam or high-temperature water with low impurities is produced, while at others a mixture of steam and water with a high percentage of incondensible gases like H_2S and CO_2 is produced. Sometimes, water dissolved with a variety of other chemicals, whose recovery may be profitable, is produced. During the initial stages of production, soon after drilling has been completed, rock fragments and debris are ejected. Gradually, the well blows itself clear of these drilling remains, but small particles from the formations may continue to be entrained by the fluid produced. The chemical properties of the fluid produced, the effect of changing wellhead pressure on the rate of production, the ratio of steam to water, as well as the maximum wellhead pressure in a closed well, are the characteristics which need to be known before designing an efficient collection and transmission system. Usually a geothermal field has a number of producing wells, and the fluids produced by wells at similar pressures may be mixed, forming separate transmission lines for each pressure group. The fluids may be transmitted as hot water, steam or a mixture of the two when separation is not done at the wellhead.

Wellhead Equipment

An arrangement for typical wellhead equipment is shown schematically in Fig. 6.11. Closed wells have much higher wellhead pressures compared to those prevailing under normal production conditions. To relieve very high pressures, a bleed valve is usually installed below the wellhead master valve. The shut-off valve sometimes experiences abnormally high forces due to the impact of geothermal fluid moving at high velocities at the time of closure. Using valves with large seating surfaces prevent damage. It is advisable to operate the master valve only when the flow is completely cut off and it regulates the flow with other downstream valves. The master valve should be used as the last resort when the other valves malfunction. When the fluid produced is a mixture of steam and water, a cyclone separator, like the one shown in Fig. 6.6, is commonly used to separate the two phases. The cyclone separator consists of a vertical cylindrical vessel in which the mixture of steam and water enters tangentially through a spiral inlet pipe. The water swirls around the wall and moves downward under gravitational force, while steam collects at the top and leaves through a pipe emerging from the bottom. The water level is controlled by an orifice, which is helpful in preventing water from entering the steam line. In the scheme shown in Fig. 6.11, the separated water is discharged to waste. However, it could be used for the purposes of district heating, greenhouses, etc., or steam at low pressures could be obtained from it by flashing and passing it through a second-stage cyclone separator. The ball float valve prevents possible carry-over water from entering the steam main. When the water in the valve reaches a critical level, the ball rises and closes the steam flow. Consequently, the pressure rises and the bursting disc ruptures, causing the excess of water to flow out. It is useful to provide a bypass line capable of discharging the total flow from the well. This provides a means of isolating and inspecting the wellhead equipment without closing the well, as well as stabilizing the flow before diverting it to the separator.

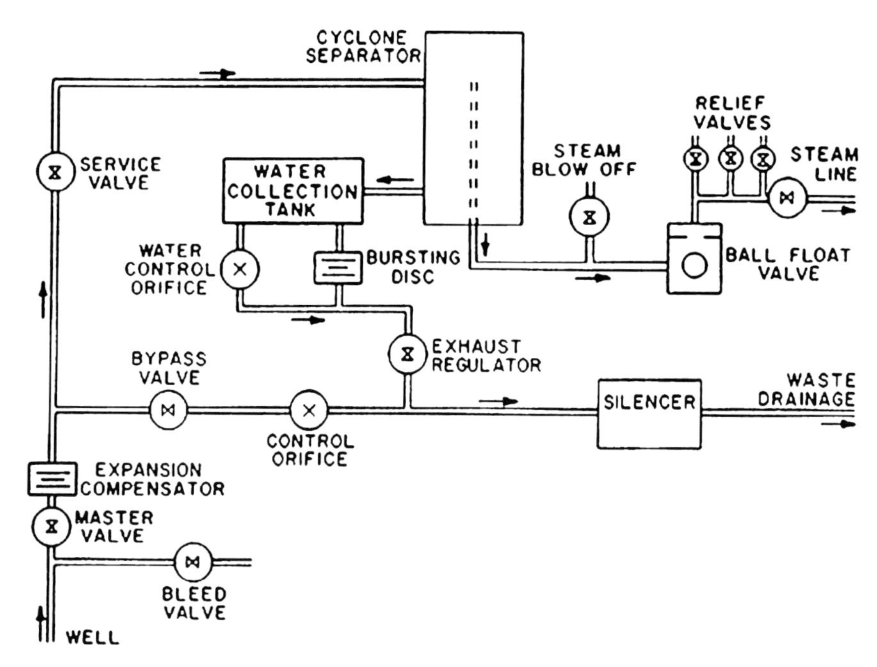

Fig. 6.11. Schematic representation of wellhead equipment.

Noise reduction

When waste fluids are allowed to escape to the atmosphere through an open-ended pipe, the noise level could be as high as 140 decibels (dB). Besides being a nuisance, this noise can affect the hearing of people at work. With the use of an efficient silencer the noise level can be reduced to 100 dB.It can be reduced most effectively by discharging the waste under water. However, for this process, the proximity of a lake or some other suitable body of water is necessary. At low noise levels, horizontal silencers, in which the pipe diameter increases gradually, are useful. Twin-cyclone silencers are very efficient at high noise levels. In these silencers, the entering fluid mixture is bifurcated by a flow splitter and made to flow in clockwise and anti-clockwise rotational swirls. Friction and turbulence destroy most of the kinetic energy of the fluid. The steam expands and emerges at low velocity from the top of the silencer, whereas swirls of water impinge and discharge from the bottom.

Waste disposal

The environmental effect of waste disposal from geothermal fields is dealt with briefly later in this chapter. However, it needs to be mentioned here that the problems arising from the chemistry of the waste hot water should be given due consideration during the planning of the project. Wherever acceptable, disposal of the hot water in rivers, lakes of the sea provides the least problem. Another alternative is to reinject the wastewater back into the subsurface formations. A suitable drainage system connecting each wellhead to the main drainage channel becomes necessary where many wells are under operation. The drainage channels are usually constructed of concrete. During drainage, the hot water gradually deposits considerable quantities of its mineral contents. Open channels make the cleaning operation easier.

Transmission of Hot Water

One of the most important considerations in hot-water transmission is to maintain adequate pressure so that the water does not boil and formation of steam is suppressed. The consequence of lowering and subsequently rising of the pressure could be disastrous. When the pressure is lowered, pockets of steam as well as air bubbles are formed. These steam pockets and bubbles collapse when the pressure is increased again. This causes a water hammer, which could burst the pipe. An adequate margin of pressure should be maintained by increasing the pressure through pumping, or by adding cold water, or through a combination of both. The pressure also drops gradually due to friction in the pipe. Topographic elevation changes of the pipeline also alter the pressure.

Surging is another important aspect of hot-water transmission, which deserves considerable attention. Surging occurs when the velocity of the moving water changes as a result of operating the control valve at the delivery end of the transmission pipeline. Opening or closing of this valve causes pressure oscillations. A rapid closure would suddenly decelerate the mass movement, consequently causing an increase in pressure. Similarly, a rapid opening would cause a pressure decline that would result in the formation of steam, which would later on collapse. Hence, it is important to investigate the surge effects systematically and to change the delivery end opening so that the resulting pressure variations are within the tolerance of the pipeline network.

Transmission of Steam

One of the chief considerations in designing a pipeline for the transmission of steam is the choosing of an appropriate diameter for the pipe so that there is not an excessive drop of pressure between the wellhead and the delivery end. In a typical geothermal field, the steam produced at several wellheads is transmitted through thinner pipes to the main transmission line. An optimum pressure, at which transmission should be performed, would mostly depend on the output characteristics of a group of wells feeding one particular transmission line. The density of steam affects the pressure drop during transmission. For the same pressure drop, low-density steam can flow at higher velocity than a high-density steam under other similar conditions. However, very high velocities can erode and damage the pipe. Mostly the velocities are kept in the 20–50 $m\,s^{-1}$ range for transmission of saturated steam and are not permitted to exceed 60 $m\,s^{-1}$, even in large-diameter pipes. Additionally, pressure drops are due to frictional resistance of flow and due to sudden enlargements or contractions of the diameter of pipes at the joints. For a given pressure and length of the pipe, the pressure drop is inversely proportional to the internal radius of the pipe and directly proportional to the square of the velocity. Uchiyama and Matsuura (1970) used the following empirical relation in deciding the inside pipe radius for steam transmission for the Matsukawa geothermal power plant in Japan:

$$r = \frac{1}{2}\left(\frac{0.785\, w\, v}{3600\, U}\right)^{1/2} \tag{6.7}$$

where r is the inside radius of the pipe, w the amount of steam being discharged, v the specific volume of 30% moist steam and $U =$ velocity.

Heat losses can be minimized to some extent by providing an insulated covering on the pipeline. Unless the steam is super heated, some of it condenses during its passage and it is useful to remove the water through *catch pots* installed at intervals. This also helps reduce the salt content which, when present in excess, is damaging for the turbine blades. As a precautionary measure against the malfunctioning of the equipment intended for preventing water from entering the steam mains, float-switches can be installed in the *catch pots* to detect the presence of excessive water and issue a warning signal.

Usually, the wellhead valves are kept fully open and the pressure is controlled from the delivery end of the pipeline. Thus the wellhead pressure would be higher than the delivery end pressure by an amount equal to the pressure loss during transit of steam and flow at each well would be automatically adjusted. Under extraordinary conditions, the wellhead pressure may be very high, necessitating its control through a throttle.

To ascertain that enough steam is available all the time in a central steam pipeline, usually more than the minimum number of wells required is connected. The excess steam is discharged through a vent valve, which also provides the necessary buffer to accommodate the fluctuations in steam requirements. A large reduction in demand causes excessive pressure increase, which is relieved through the safety valves provided at the wellhead.

Transmission of a Steam–Water Mixture

There are many advantages in transmitting a mixture of steam and water from wellhead to the powerhouse in comparison with separate single-phase transmission. This scheme is particularly very useful when the hot water produced is intended to be utilized. The main advantages are due to the elimination of much equipment used for separating the phases at the wellhead, no need to suppress boiling in the pipeline and generation of more power from the same amount of fluid produced.

As outlined by Smith (1973), depending upon pressure, temperature and ratio of water and steam, the flows can broadly be classified as:

Bubble flow
Bubbles of steam move along the upper part of the pipe at approximately the liquid velocity.

Plug flow
Alternate plugs of steam and water move along the upper part of the pipe above the water.

Stratified flow
Water flows in the bottom of the pipe with the steam above, over a smooth interface.

Wave flow
Due to higher velocity, the interface has become disturbed by waves traveling in the direction of flow.

Slug flow
A slug of water is picked up periodically by the more rapidly moving steam and continues at much greater velocity than the mean velocity of the water.

Annular flow
The water forms in a film around the inside of the pipe and the steam flows at high velocity in a central core.

Spray flow
Nearly all of the water is entrained in the steam.

It had been the general belief that concurrent transmission of steam and hot water, even in steady state, would cause excessive vibration, water hammer or surging because of instability. However, a series of tests conducted in New Zealand (James et al., 1970) and Japan (Takahashi et al., 1970) provided favorable results and no appreciable vibration, water hammer or surging in the steady state transmission was noted. The pressure drop of the steam–water mixture was moderate and acceptable from the point of view of practicality.

More encouraging results have been obtained from experiments conducted on two-phase transmission. Soda et al. (1976) conducted experiments at Hatchobaru, Japan with a 0.3-m-diameter, 200-m-long pipe arranged sloping downwards at an angle of 7° and provided with a shutoff valve at the delivery end. They report the pipeline to be sufficiently stable without any sounds or vibrations, even when the flow of mixture was shut off quickly. The stability was confirmed with visual as well as oscillographic testing of the pressure fluctuations. The sudden rise of the pressure at a point just before the shut-off valve was found to be lower than that predicted from theoretical considerations. At ordinary velocities, the flow was found to be of semi-annular mist type, whereas for lower velocities it was found to be stratified. The inside surface of the pipe was inspected after 74 days of experimentation and was found to have very little deposition and the surface was found to be free from erosion and corrosion at the moderate flow rates of the mixture.

These results indicate that in the near future, wherever possible, two-phase flow of geothermal fluids could be favored in designing the transmission lines.

Scale Deposition and Corrosion Control

Geothermal fluids are generally contaminated with chemical impurities of underground origin and may further be contaminated with impurities from the atmosphere. These cause scale deposition as well as erosion of the various equipment, piping and disposal channel surfaces that come in contact with the geothermal fluids. Knowledge of the amount of the chemical impurities present in the geothermal fluids and their corrosional and depositional effects on various surfaces under varying conditions of temperature, pressure and fluid-movement velocity are helpful in preventing the damage caused by deposition and erosion. Under extreme conditions, continuous scaling could choke the pipelines and continuous corrosion could make sections of pipelines and other equipments weak and incapable of handling the pressures they were originally designed to withstand.

Scale Deposition Prevention Research

Up to the present, geothermal brines are mostly discarded because of their scaling effect, in spite of their useful heat energy content. Experiments have been conducted to determine the relationship between the brine chemistry, process conditions and scale deposition. Wahl and Yen (1976) have reported the results of experiments conducted using brine from the East Mesa geothermal field, California. They found that deposition of calcite occurred at a rate that depended on the flashing of the brine and separation of the vapors but was independent of the heat flux. An examination of the chemical and the physical nature of the deposition indicated that it was quite sensitive to process conditions. The effect of electrical potential on scale formation has been investigated at the Lawrence Livermore Laboratory, California (Schock and Duba, 1975). The scale formed at the Salton Sea geothermal field, California was found to consist mainly of silicon, iron, copper, silver and aluminum. Significantly more scale was found to deposit on negative electrodes as compared to the positive or the neutral electrodes. The scale formed on the negative electrode contained up to 1,000 times more lead than the other electrodes. The amount of scale formed on the positive electrode appeared to be the same as on the negative electrode, however, the scale on the former was much richer in iron and zinc. The results of these experiments indicate that a system could be designed where, through the control of acidity and/or oxidation, the scale-forming constituents could either be kept suspended in the fluid or collected in a separator. In another investigation carried out at the Lawrence Livermore Laboratory (Jackson and Hill, 1976) on

Table 6.4. Summary of corrosion test results in nonaerated steam (150 days) (after Tolivia et al., 1976)

Materials	Corrosion rate (mm yr^{-1})	Corrosion Rate law*	Characteristics†	Maximum depth of pits or cavities (mm yr^{-1})	Stress corrosion
12Cr steel	0.010 (0.004–0.013)	L	SP-6	<0.024	Not sensitive
12Cr–1Mo–1W steel	0.006 (0.004–0.010)	P	SP-6	<0.024	Not sensitive
12Cr–0.2Al steel	0.021 (0.0005–0.023)	RP	GC-2	General corrosion	Not sensitive
15Cr–1.7Mo steel	0.0045 (0.0015–0.0075)	L	SP-1	0.05	Not sensitive
1Cr–1Mo–1/4 V steel	0.044 (0.022–0.059)	I	GC-2	General corrosion	Sensitive
3.5Ni-1 3/4Cr- 0.5Mo-0.1 V steel	0.020 (0.011–0.038)	L	SP-2	0.12	Sensitive
1.5Cr-1Al-1/4Mo steel (nitrided)	Weight gain	–	Nitrided layer wereremoved partly	0.97	Not tested
Carbon steel ASTM-A285	0.038 (0.038 ~0.13)	P	GC-2	General corrosion	Not tested
Aluminum	Weight gain	–	GC-1	General corrosion	Not sensitive

†SP-1 slight scattered pitting corrosion in wide range;
SP-2 scattered pitting corrosion in wide range;
SP-3 scattered pitting corrosion in remarkably wide range;
SP-4 remarkable scattered pitting corrosion in wide range;
SP-5 remarkable densely scattered pitting corrosion in wide range;
SP-6 scattered pitting corrosion in local range;
SP-7 densely scattered pitting corrosion in local range;
GC-1 slight general corrosion with uneven surface;
GC-2 general corrosion with uneven surface;
GC-3 remarkable general corrosion with uneven surface;
GC-4 general corrosion with wavelike uneven surface;
GC-5 remarkable general corrosion with wavelike uneven surface.
*L = linear, P = parabolic, RP = reverse parabolic, I = irregular, SA = slightly accelerate.

preventing scales formed by sulfides, it has been found that the problem could be solved through partial oxidation or combined oxidation-acidification processes.

In addition to the experiments conducted on chemical and electrical modification of the fluid to prevent scale formation in flowing geothermal brines, the other preventive procedures researched include fluid pretreatment to remove scale-forming constituents and hardware surface treatment to prevent adhesion and chemical or mechanical removal.

Corrosion Control

Marshall and Braithwaite (1973) have given a good account of the corrosion problem in the geothermal energy production industry and its control. Here we shall summarize their findings and the results of some other experiments carried out to control corrosion. Tables 6.4–6.6 are useful in selecting suitable material for constructing equipment and specific items to be used for handling geothermal fluids.

Surface corrosion

Geothermal fluids containing free sulfuric, hydrochloric or hydrofluoric acid cause extensive surface corrosion, thereby prohibiting their practical use. However, geothermal fluids discovered and used so far do not contain these acids in proportions, which may prohibit their use, and provided the fluids are not contaminated by atmospheric oxygen, their corrosive effect on common structural material is not high. Aeration of geothermal fluids causes depolarization and drastically increases their corrosional effect. Austenitic stainless steel, titanium and chromium (plating) are among a few exceptional metallic alloys that can be used in engineering construction. Adequate protection against corrosion by aerated geothermal fluids can also be provided by covering a steel surface with epoxy coating.

Erosion– corrosion

The combined action of erosion and corrosion on metals is significant, particularly in items like the blades of the turbine where wet steam at high velocity could be very damaging. Experiments show that a 13% chromium-stainless steel blading alloy provides adequate erosion–corrosion resistance. Corrosion-resistant alloys, such as austenitic stainless steels, provide good resistance against erosion–corrosion.

Stress corrosion

The presence of chloride solutions at high temperatures and the presence of tensile stresses cause stress corrosion. There does not seem to be a lower stress limit below which stress corrosion cracking would not occur. It appears that austenitic stainless steels are not susceptible to stress corrosion in air-free geothermal fluids. However, the presence of dissolved oxygen in geothermal fluids would cause stress corrosion of austenitic stainless steel.

Table 6.5. Summary of corrosion test results in aerated steam (150 days) (after Tolivia et al., 1976)

Materials	Corrosion rate, mm/yr	Ratio to corrosion rate in non- aerated steam	Corrosion Rate law*	Characteristics†	Maximum depth of pits or cavities ($mm\,yr^{-1}$)	Stress corrosion
12Cr steel	0.09 (0.085–0.14)	9.0	P	SP-5 microscopic cracks	1.7	Sensitive
12Cr–1Mo–1W steel	0.09 (0.04–0.23)	15	P	SP-4 microscopic cracks	1.6	Sensitive
12Cr–0.2Al steel	0.11 (0.06–0.16)	5.2	P	GC-3 microscopic deep pitting	General corrosion	Not sensitive
15Cr–1.7Mo steel	0.019 (0.005–0.023)	4.2	L	SP-6	1.2	Not sensitive
1Cr–1Mo–1/4 V steel	0.16 (0.13–0.50)	3.6	P	GC-3	General corrosion	Sensitive
3.5Ni–1 3/4Cr–0.5Mo–0.1 V steel	0.20 (0.12–0.52)	10	P	SP-2 inter-granular corrosion	0.70	–
1.5Cr–1Al–1/4Mo steel (nitrided)	Weight gain	–	–	SP-6	0.85	–

Carbon steel ASTM-A285	0.11 (0.065–0.44)	2.9	L	GC-3	General corrosion	–
Aluminum	0.065 (0.01–0.095)	–	SA	SP-3	2.9	–
Copper (deoxidized)	0.57 (0.51–1.11)	–	P	GC-5	General corrosion	Not sensitive

†SP-1 slight scattered pitting corrosion in wide range;
SP-2 scattered pitting corrosion in wide range;
SP-3 scattered pitting corrosion in remarkably wide range;
SP-4 remarkable scattered pitting corrosion in wide range;
SP-5 remarkable densely scattered pitting corrosion in wide range;
SP-6 scattered pitting corrosion in local range;
SP-7 densely scattered pitting corrosion in local range;
GC-1 slight general corrosion with uneven surface;
GC-2 general corrosion with uneven surface;
GC-3 remarkable general corrosion with uneven surface;
GC-4 general corrosion with wavelike uneven surface;
GC-5 remarkable general corrosion with wavelike uneven surface.
*L = linear, P = parabolic, RP = reverse parabolic, I = irregular, SA = slightly accelerate.

Table 6.6. Final evaluation of corrosion resistance in condensate (150 days) (after Tolivia et al., 1976)

Material	Velocity of condensate	Condensate rate law*	Average corrosion rate weight loss ($mm\,yr^{-1}$)	Maximum local penetration rate ($mm\,yr^{-1}$)	Corrosion characteristics†	Stress corrosion
Deoxidized copper	High	L	0.64	–	GC-5	–
	Low	L	0.24	–	GC-2	Not sensitive
Aluminum	High	RP	0.48	3.65	SP-4	–
	Low	P	0.01	1.16	SP-2	Not sensitive
Naval brass	High	RP	0.22	–	GC-2	–
	Low	P	0.07	–	GC-2	Not sensitive
18-8 stainless steel	High	–	0.0003	–	No corrosion	–
	Low	–	0.0008	–	No corrosion	Not sensitive
12Cr steel	High	L	0.090	0.85	SP-6	–
	Low	P	0.015	0.97	SP-6	–
Low-carbon steel	High	P	0.66	–	GC-5	–
	Low	P	0.29	–	GC-4	Sensitive

†SP-1 slight scattered pitting corrosion in wide range;
SP-2 scattered pitting corrosion in wide range;
SP-3 scattered pitting corrosion in remarkably wide range;
SP-4 remarkable scattered pitting corrosion in wide range;
SP-5 remarkable densely scattered pitting corrosion in wide range;
SP-6 scattered pitting corrosion in local range;
SP-7 densely scattered pitting corrosion in local range;
GC-1 slight general corrosion with uneven surface;
GC-2 general corrosion with uneven surface;
GC-3 remarkable general corrosion with uneven surface;
GC-4 general corrosion with wavelike uneven surface;
GC-5 remarkable general corrosion with wavelike uneven surface.
*L = linear, P = parabolic, RP = reverse parabolic, I = irregular, SA = slightly accelerate.

Sulfide stress cracking
Experimental investigations and observations in the field have shown that even high-strength steels are susceptible to cracking when simultaneously stressed and exposed to aqueous solutions of H_2S. Since the presence of H_2S is commonly observed in geothermal fluids, sulfide stress cracking is a factor of major importance in the design and selection of material for equipment. Medium- and high-strength carbon and alloy steels are not very suitable except when the environment is known to be free from a high concentration of H_2S.

Hydrogen infusion
It has been discovered that infusion of hydrogen into steel occurs when geothermal fluids containing aqueous solutions of H_2S corrode the steel surfaces. Under favorable circumstances, hydrogen fusion causes blistering, renders steel brittle, and is found to be associated with sulfide stress cracking and delayed fracture of stressed steels. The hydrogen permeation rate falls off rapidly with time at temperatures in excess of 100 °C. It is estimated that after 2–30 weeks, the corrosion product developed at the exposed surface prevents permeation of hydrogen. Surface coatings with paint or chemically formed magnetite coating and oxidation are helpful in reducing hydrogen infusion. The decrease in permeation activity with the increase of temperature is probably due to change in the corrosion mechanism at high temperatures.

Infusion-induced delayed fractures
Hydrogen infusion is found to be the cause of delayed fractures observed in high-strength steels subjected to tensile stress. Detailed field experimental data on delayed fracture of bore-casing steels exposed to geothermal fluids for a long time under closely controlled tensile loading have been provided by Marshall and Tombs (1969). In these tests, no delayed fracture occurred at high temperature (> 100 °C); however, when exposed to cold geothermal steam condensate, both low- and high-strength steels exhibited delayed fracturing. Results of experiments on delayed fracture under controlled stress distribution and stress corrosion suggest the use of steels with strengths in excess of 88,000 pounds per square inch (psi) in geothermal steam media devoid of condensate or water.

Field data on corrosion investigations
Tolivia et al. (1976) report on the corrosion resistance of different turbine materials in geothermal steam environment. These investigations were conducted at Cerro Prieto geothermal field in Mexico. Fig. 6.12 schematically explains the experimental set up. Steam was separated from the hot water by a separator and introduced into steam and aerated steam chambers (Figs. 6.12A,B). The separated steam was led through a cooler to obtain condensate, which was stored in a tank and from there introduced into low- and high-velocity test chambers (Fig. 6.12C). The test conditions and chemical compositions are listed in Tables 6.7 and 6.8 respectively.

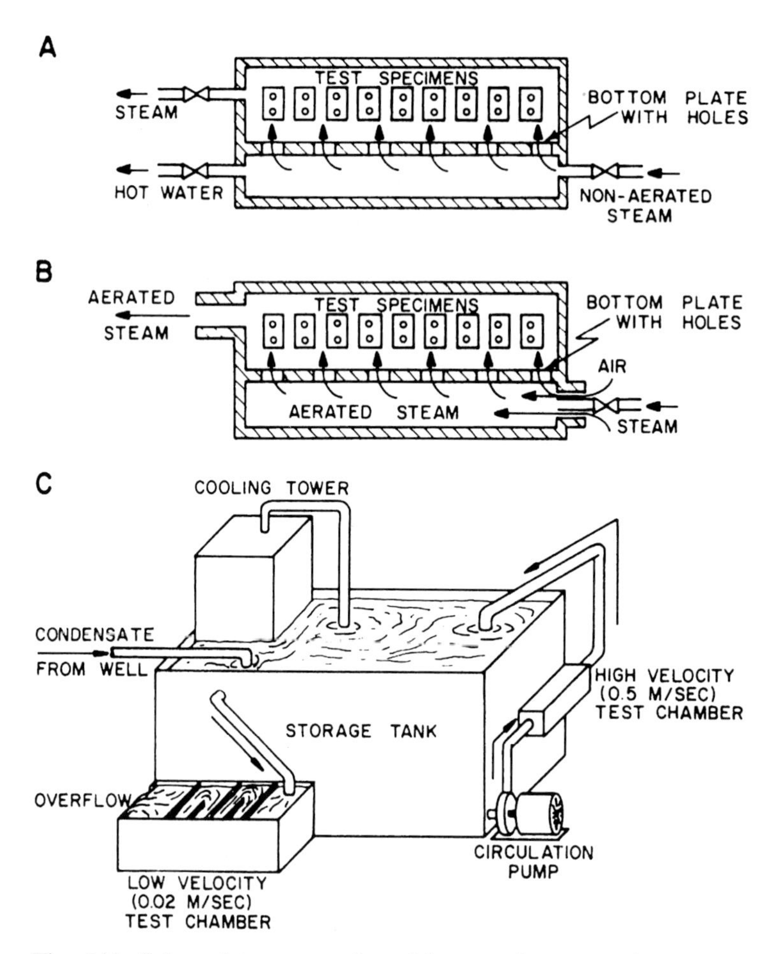

Fig. 6.12. Schematic representation of the corrosion test equipment (modified from Tolivia et al., 1976). (A) Test chamber for nonaerated steam. (B) Test chamber for aerated steam. Test apparatus for immersion test in condensate.

General corrosion tests were performed on plate pieces measuring 0.03 m × 0.06 m × 0.003 m. At intervals of 30 days exposure, for a total duration of up to 150 days, two test pieces of each material were removed from the test chamber and corrosion rates were measured. Stress corrosion tests were conducted on U-bend test pieces to which yield strength was applied. Visual and microscopic observations were made to check for the presence of cracks. Corrosion-fatigue tests were conducted on shank-type plane-bending specimens. Summary results of these experiments are presented in Tables 6.4–6.6.

It has been concluded from these experiments that the carbon steel used in turbine shells and piping has satisfactory corrosion resistance in non-aerated steam, and its

Table 6.7. Corrosion test conditions in separated steam and condensate (Tolivia et al., 1976)

	Velocity, ($m\,s^{-1}$)	Pressure (psi)	Temperature, (°C)
Separated steam			
Nonaerated	<10	70–194	152–194
Aerated	<10	Atmosphere	100
Condensate			
High velocity	0.5	–	38–58
Low velocity	0.02		

Table 6.8. Average chemical compositions of separated steam and condensate (Tolivia et al., 1976)

		Chemical composition (ppm by weight)
Separated steam	CO_2	19,500.00
	H_2S	2,580.00
	Cl^-	13.30
	SO_4	6.80
	Na	1.29
	K	0.58
	pH	8.35
Condensate	pH	7.10
	Cl^-	42.00
	Conductivity	648 μmho

corrosion rate increases in aerated steam and condensate. Hence adequate corrosion allowance should be made in design, and coating with epoxy resin would be required in low-temperature conditions. The deterioration of the fatigue endurance limit is more important than the general corrosion rate for the 12Cr steel used for the bucket. For making the rotor, low-alloy steels are almost as good as carbon steel in corrosion rate. Dioxized copper and aluminum are very poor in corrosion resistance in the condensate and as such cannot be used for constructing thin-walled heat exchanger tubes and the use of highly resistant material like titanium is desirable for their efficient functioning.

Corrosional effect on non-metallic material

Not much is yet known about the corrosional effect of geothermal fluids on non-metallic material. However, concrete and grout are widely and successfully used unless atmospheric oxidation of hydrogen sulfide occurs. Results reported by Lorensen et al. (1976) on experiments conducted to investigate the corrosive effect of geothermal brines at 300 °C on polymeric and other composite materials, show that polymers containing substantial aromaticity, fluorine, carbonized structure or aliphatic unsaturation have a high resistance against brines. The presence of water-sensitive groups such as imide, amide or ester affects even thermally very stable polymers, rendering them unsuitable. In field tests carried out with flowing brine, a fluorocarbon polymer is reported to be very suitable against erosion and scale deposition by Lorensen et al. (1976).

Pilot-scale equipment for corrosion and scaling studies

It is well understood that scaling and corrosion processes are correlated in geothermal systems. Stáhl et al. (2000) have developed pilot-scale equipment to model scaling and dissolution, and investigate corrosion. They point out that the corrosion and scaling characteristics of any given geothermal water are controlled by the composition of the geothermal fluid and the parameters under which a geothermal well is operating. Corrosion and scaling processes have to be treated together. Calcium carbonate, the least soluble component of the geothermal fluids, needs to be studied from the point of view of scaling. Several factors, such as the dissolved carbon dioxide, temperature, pH, dissolved salt content, and the rate of flow influence the solubility of calcium carbonate. The pilot scale experiment was designed to investigate solubility of calcium carbonate and corrosion and scaling under varying conditions of the influencing factors.

A suite of experiments was conducted. The major findings are:

(1) Scaling and corrosion are inseparable. Even in very simple experiments, scaling is accompanied by simultaneous corrosion.
(2) The dissolution process and equilibrium solubility of calcium carbonate is influenced by the type of structural material placed in the pipeline.
(3) The approach adopted of increasing the temperature from 35 °C to 55 °C at a constant carbon dioxide pressure in experiments extending to one day period, was suitable for modeling calcium carbonate scaling and corrosion at the same time. This approach provides adequate inputs to design inhibition experiments.
(4) It is important to avoid fast and intense corrosion at the start of scaling using appropriate inhibitors.

Operation and maintenance of geothermal wells

In an important overview, Thorhallsson (2003) has underlined the importance of proper maintenance in increasing the life of a geothermal well and thereby increasing the total power generation potential of a geothermal field. He underlines that geothermal wells cost about one-third to one-half of the total value of a geothermal electric generation plant. Earlier, a 20-year life span was considered to be adequate for a high-temperature well. However, with the improvement of maintenance and operating procedures, these wells are now expected to operate for twice as long. In the case of the low temperature wells, the expected life is much longer.

Thorhallsson (2003) has discussed several important factors that are helpful in extending the life of a geothermal well. These include geothermal well design, the main parameters (such as wellhead pressure, total flow rate, temperature, pressure, possible change in diameter of casing due to scaling or corrosion, and steam purity) that need to be monitored to constantly assess the health of the entire system, indicators showing that a well is not operating normally (such as sudden changes in pressure or flow rate), precautions to be taken while starting or closing the operation of a well, well cleaning, well head maintenance, and handling of casing damage or leak. The oldest high-temperature well in Iceland has been producing for about 50 years. There are several wells, which are more than 25-years old.

In the concluding remarks, Thorhallsson (2003) observes that the best way to increase the life of a geothermal well and minimize costly repair and maintenance is to ensure continuous flow and avoid thermal cycling. In many cases, the best protection against corrosion of the steel casing is to keep the geothermal well hot by steam bleeding. It is also important to assess the reservoir behavior and well performance to ensure adequate supply of geothermal fluids and optimum operation of the well.

ENVIRONMENTAL ASPECTS

When compared with nuclear and fossil fuels, geothermal energy is an environment-friendly energy resource primarily because no combustion of fuel takes place during its production. On the other hand, use of geothermal energy for power production and other direct uses, while replacing few tens of million tons of precious oil fuel per year, effectively contributes to a reduction in CO_2 emissions by several tens of million tons every year. Geothermal fields produce only about one-sixth amount of CO_2 that a natural gas-fueled electrical generating power plant produces and none of the nitrous oxide or sulfur-bearing gases. Binary-cycle power plants, which are the most used ones for power production worldwide, neither emit polluting gases in the atmosphere nor liquid discharges at the surface. Geothermal fluids, which contain certain minerals and salts, are often re-injected into the reservoir at depths well below the groundwater aquifers. This process not only prevents atmospheric and groundwater pollution, but also replenishes the water table in the reservoir and minimizes hazards such as land subsidence due to continuous, large-scale, withdrawal of fluids. Nevertheless, as with any other source of energy, production of geothermal energy also causes environmental pollution, although on a much smaller scale. Therefore, the governments in geothermal energy-producing countries should encourage the sustainable development of geothermal energy resources in an economic and environmentally responsible manner.

The environmental aspects of geothermal power production have been discussed at length by several authors (Armstead, 1976a; Crittenden, Jr., 1981; Barbier, 2002; Kristmannsdóttir and Ármannsson, 2003; Clauser, 2006). In the following, we shall briefly touch upon some of the major concerns in geothermal development, such as (i) physical effects of fluid withdrawal, (ii) thermal effects, (iii) chemical pollution, (iv) biological effects, and (v) noise. We shall also briefly discuss the measures for their mitigation.

Physical Effects of Fluid Withdrawal

The withdrawal of geothermal fluids from a reservoir at rates higher than that of natural recharge by surface water, as is often the case in production of electrical power, causes several physical effects such as *lowering of groundwater table, land subsidence*, and *triggered seismicity*. Sometimes, changes in surface-geothermal manifestations occur such as disappearance of hot springs and fumaroles due to sustained production of geothermal fluids and steam. Withdrawal-caused depletion in groundwater levels can result in mixing of fluids between aquifers and an inflow of corrosive water. The problem of anthropogenic-induced subsidence due to withdrawal of large quantities of fluids has been dealt with extensively by Narasimhan and Goyal (1984). Subsidence at variable rates, depending on the nature of the rock formation and the type of the geothermal field, has been reported from Wairakei ($\sim$400 mm yr^{-1}); Larderello ($\sim$250 mm yr^{-1}); Svartsengi, Iceland ($\sim$10 mm yr^{-1}) and other geothermal fields (Hunt, 2001; Allis, 2000; Eysteinsson, 2000). In Wairakei, in the area of maximum subsidence, a pond up to 6 m deep has formed. Flooding caused destruction of several trees. The subsidence has also caused damage to well casings and pipelines, and a section of a nearby highway had to be rebuilt and resurfaced. In general, subsidence is less in the case of hard-rock formations and water-dominated fields when compared to sedimentary rock formations or vapor-dominated fields. The effects of lowering of groundwater table and subsidence can be minimized by re-injection of the used fluids (i.e., waste water after extraction of energy) into the reservoir. However, prolonged geothermal exploitation, subsidence, and re-injection could trigger earthquakes in critically stressed areas. Experiments carried out at the Rangely oil field, Colorado, have demonstrated that injection of fluid can change subsurface pore fluid pressure and thereby trigger earthquakes (Raleigh et al., 1972). The problem of triggered earthquakes has been dealt with in detail, particularly in the case of non-geothermal areas, by several authors (see for example, Gupta and Rastogi, 1976; Gupta, 1992 and references therein). The treatment remains more or less similar for geothermal areas also.

Thermal Effects

Thermal effects refer to the substantial loss of thermal energy from spent waters to the environment primarily due to the low efficiency of geothermal power plants. Excess heat emitted in the form of steam may affect cloud formation and change the weather locally. Wastewater disposed into streams, rivers, lakes, or local groundwater may seriously affect the biology and ecological system. Cooling in evaporation ponds is an option for minimizing the pollution, but it is generally not considered as a long-term solution as the ponds tend to increase in size and may cause chemical pollution of the environment. Therefore, subjecting the geothermal fluids to multiple uses such as power production, and subsequently, other low-temperature direct uses should be considered to utilize the maximum possible heat before re-injecting them into deep reservoirs.

Chemical Pollution

Chemical pollution is caused by the discharge of chemicals into the atmosphere through steam and wastewater. As mentioned by Kristmannsdóttir and Ármannsson (2003), the major pollutant chemicals in the liquid fraction are hydrogen sulfide (H_2S), arsenic (As), boron (B), mercury (Hg) and other heavy metals such as lead (Pb), cadmium (Cd), iron (Fe), zinc (Zn) and manganese (Mn). Other harmful constituents, although present in smaller quantities, are lithium (Li), ammonia (NH_3) and aluminum (Al). High salt concentrations in certain geothermal brines cause additional problems. Disposal of water of this type is a risky endeavor, as As and Hg, in particular, may accumulate in sediments and organisms. Re-injection of the wastewater into the reservoir appears to be the best solution at the present time. To prevent mixing of wastewater with potable ground waters in shallow aquifers, both the production and injection wells are lined with steel casing pipe and cemented to the surrounding rocks. Care must be taken to prevent, and if necessary, detect and repair any leakage due to damage of casing or the cement surrounding the casing pipe.

Air pollution is caused primarily due to the presence of non-condensable gases in the steam. These gases constitute predominantly CO_2 (up to 90% by weight of steam) and to some extent, H_2S, besides smaller amounts of NH_3, CH_4, N_2, H_2, Hg, B and Rn, with the total ranging from a few grams to about 60 grams per kilogram of steam (about 5% by weight of steam, on an average). The total emissions of these gases is only a small fraction of that emitted by power plants driven by burning fossil fuels such as coal, oil, and gas. As an example, the amount of CO_2 produced per MWh of power is, typically, 0.5 kg from geothermal flashed steam, whereas it is about 990 kg from coal, 839 kg from petroleum, and 540 kg from natural gas (Reed and Renner, 1995). Nevertheless, it is important to take these small emissions into account because of impending severe restrictions on greenhouse emissions as follow-up of the Kyoto protocol recommendations.

Hydrogen sulfide can reach moderate concentrations in the steam produced from some geothermal fields. Some systems contain up to 2% of H_2S by weight in the separated steam phase. This gas presents a pollution problem because it is easily detected by humans at concentrations of less than 1 ppm in air. Ammonia occurs in small quantities in many geothermal systems. In conventional flashed-steam geothermal power plants, the ammonia is oxidized to nitrogen and water as it separates from the steam phase and passes into the atmosphere. Because the high pressures of combustion are avoided, geothermal power plants have none of the nitrogen oxides emissions that are characteristic of fossil fuel plants. Boron, mercury, and ammonia are leached from the atmosphere by rain, leading to contamination in soil and vegetation. Radon is present as ^{222}Rn-isotope in geothermal steam, typically in the range 3,700–78,000Bq kWh^{-1} (Layton et al., 1981), and is not considered a serious pollutant in the present scenario.

Binary geothermal power plants do not allow a steam phase to separate, so carbon dioxide and the other gases remain in solution and are re-injected into the reservoir, without discharging to the atmosphere. Therefore, this type of plants does not contribute to air pollution.

Silica is one of the most troublesome products of wet geothermal fields. It often occurs in saturated solution with geothermal water at depth. Consequent to reduction of pressure and flashing, silica either precipitates immediately or remains in solution for a limited time. A considerable amount of money is spent annually at the Wairakei geothermal field in cleaning the silica deposits from open-bore water discharge channels from the field to the river. In district heating installations, where bore-water remains contained for a long time within the pipes, silica scaling in distribution pipelines becomes a serious problem. The fear of subterrancan silica precipitation has often acted as a deterrent for re-injection experiments.

Biological Effects

The impact of gas and water discharges from geothermal wells on terrestrial and aquatic life will depend on contaminant dilution, dispersion, chemical speciation, and heat. These effects should be carefully monitored so that appropriate steps for their minimization can be devised in specific cases.

Noise

Noise pollution during the construction and operation stages of a geothermal power plant includes those by drilling and maintenance (90–120 dB) and discharge of fluids (~120 dB). Brown (1995, 2000) has compared these noise levels with other common sounds that occur in our day-to-day life. The only way to mitigate the sound pollution is to ensure softening of the noise using appropriate silencers, to levels lower than the pain threshold of human ear (120 dB in the frequency range 2,000–4,000 Hz).

Chapter 7

THE CERRO PRIETO GEOTHERMAL FIELD, MEXICO

INTRODUCTION

The Cerro Prieto geothermal field, located in the northern part of Baja California, Mexico ranks among the three largest geothermal power-producing fields in the world. With a total installed capacity of 720 MWe, the Cerro Prieto stands at par with the Tongonan field in Leyte, Phillipines, and is exceeded only by the Geysers, which has an installed capacity of about 1,420 MWe. The electrical power generation from this field alone accounts for 75% of Mexico's geothermal capacity of 953 MWe and meets more than 50% of Baja California's total power requirements. The geothermal field has been producing power successfully for the past 30 years. Researches carried out by Lawrence Berkley Laboratory (LBL), Comisión Federal de Electricidad (CFE) and other groups from both within and outside of North America, over the past several decades, have generated a wealth of information on Cerro Prieto. It is one of the most well-studied geothermal reservoirs in the world today. Data and information obtained from research experiments carried out here have been gainfully utilized in exploration and development of other geothermal fields worldwide, particularly in the area of reservoir engineering studies.

A historical perspective of the geothermal exploration at Cerro Prieto is given in Mañón et al. (1978). Exploration efforts started way back during 1958–1959 with geological reconnaissance surveys and aerial photo interpretation studies. The first well was drilled in 1960. This was followed by extensive and deeper drilling during the middle and late 1960s. The Cerro Prieto volcano was considered as a probable heat source in the initial discovery period. During 1965, isotherm maps were drawn from measurements in 50 shallow wells. The first geochemical survey was performed in 1965 and 1966 (Mercado, 1968). Between 1966 and 1968, 15 production wells were drilled, from which the field productivity was estimated to be 75 MWe. Hydrological and stable isotope studies were subsequently carried out, followed by production testing. Production of electric power from geothermal resources at Cerro Prieto commenced in April 1973, when the first 37.5 MWe unit was commissioned. From 1972 to 1977, 18 more wells were drilled, 9 of which were to supply fluids to another 75 MWe plant (Units 3 and 4). Additional geochemical studies were performed by Molina and Banwell (1970), Mercado (1976, 1977), and Crecelius et al. (1976). During 1977 other geoelectric and geophysical studies were performed, including seismic, magnetic and gravity surveys, which covered a total area of 620 km^2 (CICESE, 1977; Razo and Arellano, 1978; Razo and Fonseca, 1978).

By the year 2003, there were 149 production wells and 9 injection wells in operation at Cerro Prieto, producing about 5,855 metric tonnes of steam per hour and 8,060 metric tonnes of geothermal brine per hour (Gutiérrez-Negrín and Quijano-León, 2005). The geothermal field is wholly managed and operated by the CFE. In this chapter, we shall review the geological and tectonic setting of the Cerro Prieto geothermal field, strategies adopted for exploration and assessment of power potential, and commercial production of electric power. The material presented in this review is based on previous exploration studies and information obtained from numerous wells drilled in the Cerro Prieto geothermal field during the past five decades, which have been published by several authors.

GEOLOGICAL AND TECTONIC SETTING

The Cerro Prieto geothermal field is the largest among several high-temperature, water-dominated geothermal fields located in the Salton Trough of southern California, U.S.A. and northern Baja California, Mexico (Fig. 7.1). The Salton Trough is a tectonically active, structural depression, forming the landward extension of the Gulf of California. The tectonic regime is transitional between the East Pacific Rise to the southeast and the San Andreas Fault System to the northwest (Elders, 1979). The depression is partially filled up by continental sediments of the delta of the Colorado River. The other fields in the Salton Trough are located at Salton Sea, Westmorland, Brawley, Glamis, East Brawley, East Mesa, Dunes, Border, Heber in California and Tulecheck, Panga de Abajo, Mesa de San Luis and Mesa de Andrade in Baja California. These are much smaller compared to Cerro Prieto, and many are yet to be fully explored.

The major geothermal fields in Mexican territory can be broadly divided into two classes: (a) those located within or in close proximity of the Mexican volcanic belt of central Mexico, such as Los Azufres, Los Humeros, Ixtlan de los Hervores, Los Negritos and La Primavera and (b) outside the Mexican volcanic belt in the northern parts of Mexico, such as Cerro Prieto, Las Tres Virgenes and Piedras de Lumbre. The locations are shown in Fig. 7.2. The geological setting, surface manifestations and temperatures of surface discharge observed in these fields are summarized in Table 7.1. All of these fields are characterized by surface features such as hot springs, fumaroles and mud pools, indicating a volcanic source of heat for the hot water and steam fields. The surface geothermal manifestations had been exploited in a rudimentary way by the local inhabitants for centuries. In the 1940s and 1950s the power-generating potential of these surface manifestations was realized and intensive geological, geophysical and geochemical investigations were mounted with the objective of fully exploring the geothermal fields associated with them and assessing their energy potential. The Cerro Prieto geothermal field is the most promising among them. The discussion in this chapter is therefore restricted to Cerro Preito only.

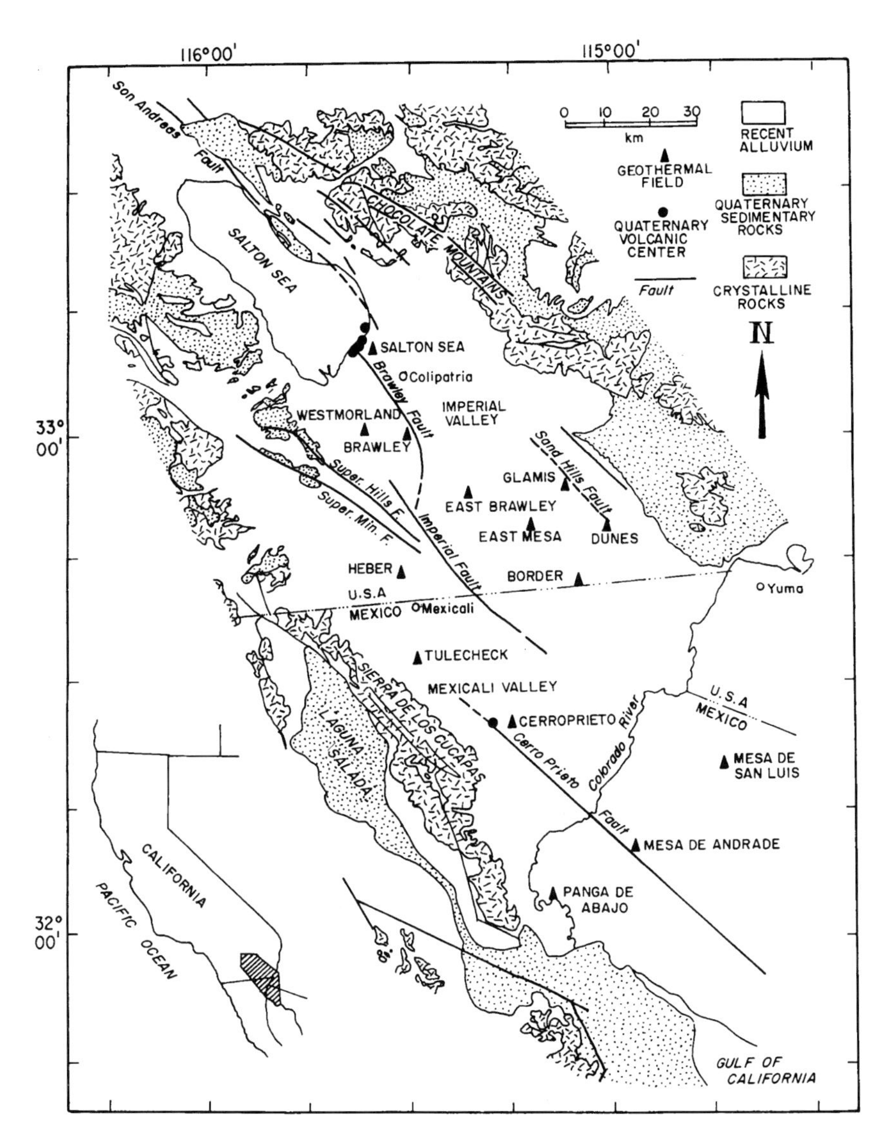

Fig. 7.1. Geological map of the Cerro Prieto geothermal field located in the Salton Trough of southern California and northern Mexico (modified from Elders et al., 1984). The distribution of geothermal fields (filled triangles) is shown with respect to the Quaternary volcanic centers (filled circles) and major faults identified in the region.

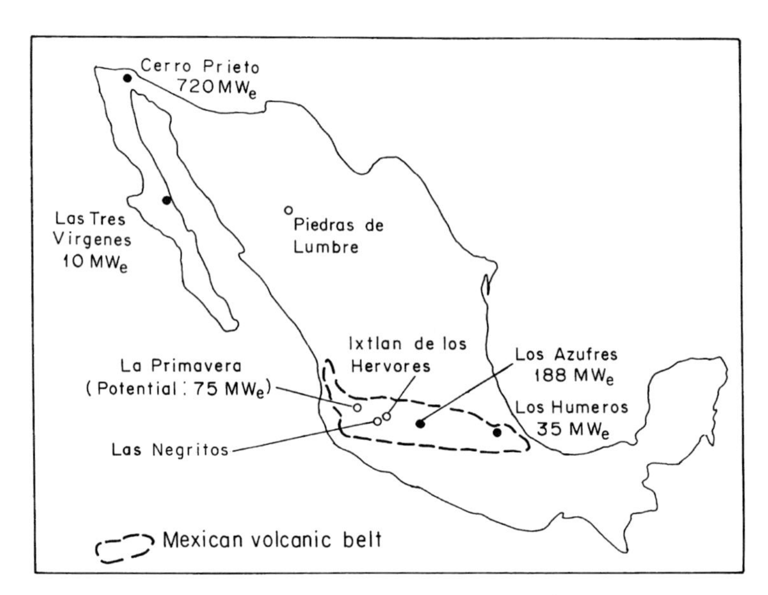

Fig. 7.2. A simplified map showing the locations of both producing (filled circles) and non-producing fields (open circles) in Mexico. The Mexican volcanic belt is indicated. (Modified from Gutiérrez-Negrin and Quijano-León, 2005).

The geology of the Cerro Prieto geothermal fields has been briefly described by Banwell and Gomez Valle (1970). The Mexicali Valley, in which the field is located, is a prolongation of the Imperial Valley, California and it retains the physiographic features of the latter. Thick deltaic sedimentary deposits consisting of successive horizons of sands, muds and clays with various degrees of compaction are the main constituents. The geothermal field covers an area of approximately 18 km^2. On the west, it is terminated by the Sierra de los Cucapa (Fig. 7.1). On the east, it extends to form the bed of the Colorado River. These deltaic sediments rest unconformably over intrusive granitic rocks, affected by the San Andreas—San Jacinto fault system. The granitic intrusion has been severely fractured and dislocated, resulting in numerous faults and fissures providing hydraulic continuity to depth, with a steam–water mixture escaping from the surface at a number of places.

Tectonically, the Cerro Prieto field lies in a "pull-apart" basin of the San Andreas Fault system, bounded by two major right-lateral strike-slip faults, the Imperial and the Cerro Prieto (Fig. 7.1). These NE–SW oriented faults are interlaid by several NE–SW faults that act as collectors of geothermal fluids. The heat source is a regional thermal anomaly resulting from the thinning of continental crust at the bottom of the basin (Quijano-León and Gutiérrez-Negrín, 2003). The heat, along with the hydrothermal fluids, is transferred through the Late Cretaceous granitic basement rocks to deep aquifers within sandstones and shales.

Table 7.1. Characteristics of some Mexican geothermal fields (modified after Banwell and Gomez Valle, 1970; Quijano-León and Gutiérrez-Negrín, 2003)

Name of geothermal field	Physiographic expression	Types of surface activity	Surface temperature (°C)
Cerro Prieto (~7 m a.s.l.)	Delta valley, with surrounding farmlands	Hot springs, fumaroles, mud volcanoes, pools and salt deposits	45–100, with higher values observed in the wells
Las Tres Virgenes (~720 m a.s.l.)	Volcanic complex composed of three volcanoes—La Virgen, El Azufre and El Viejo; Pliocene to Quaternary extensional tectonic regime	Hot springs, fumaroles, mud volcanoes	35–85
Piedras de Lumbre	In the high country of the Sierra Tarahumara within the Basin & Range tectonic province	Fumaroles, hot springs of sodium chloride composition and wide alteration zones (kaolin)	40–90
Los Negritos (~1,533 m a.s.l.)	Caldera-like structural valley, open toward NW; flows of Tertiary and Quaternary lavas, surrounded by farmlands	Mud volcanoes and bubbling lakes aligned NE–SW	36–90
Ixtlan de Los Hervores (~135 m a.s.l.)	E–W faulted structural valley, with flows of basalt and andesite; acid soils unfit for cultivation in the hydrothermal area	Hot springs, boiling mud lakes, steam discharges and low-pressure wells; deposits of geyserite with ceramic residues	45–100
Los Humeros (1,800 m a.s.l.)	Centrally faulted caldera; explosion craters, deposits of basalts, sands and stratified tuffs	Low-pressure fumaroles, hot ground and alteration zones	85–90
Los Azufres (~2,800 m a.s.l.)	Intensely fractured double caldera; altered rhyolites and volcanic ash	Superheated steam (fumaroles), hot ground and alteration zones	45–110

(*Continued*)

Table 7.1 (*Continued*)

Name of geothermal field	Physiographic expression	Types of surface activity	Surface temperature (°C)
La Primavera (~1,570 m a.s.l.)	Faulted caldera; the geothermal area is located on two N–S faults which bound a graben	Superheated steam (fumaroles), alteration zones and hot ground	55–100

The granitic basement is exposed to the west of the Cerro Prieto field and is a part of the California batholith. The basement deepens to 4,000 m or more in the central part of the field, overlain by about 2,400 m—thick sequence of gray shales and sandstones that hosts the geothermal fluids (Quijano-León and Gutiérrez-Negrín, 2003). The geothermal fluids rise from depth in the east of the geothermal field and discharge at the surface to the west of the field. Two major producing aquifers have been identified within the sandstone formations, aquifer A and aquifer B (often referred in literature as Alpha and Beta reservoirs). The A-reservoir is located at depths lower than about 1,200 m in the western part of the field and it supplied geothermal fluids to the CP-I power plant during the 1970s and early 1980s. The B-reservoir is deeper ($>1,600$ m) and hotter (310–330 °C) relative to the A reservoir, and it extends over the entire field (Lippmann et al., 1991; Truesdell et al., 1997). From 1986 to the present time, most wells produce from the deeper B-reservoir. Hydrothermal alteration of the sandstone formation has given rise to deposition of secondary minerals such as chlorites, calcite, silica, epidote and amphiboles.

The geological evolution of the Cerro Preito geothermal field has been traced from the Miocene times to the present by Lyons and van de Kamp (1980). The evolutionary history is summarized in Table 7.2.

SURFACE MANIFESTATIONS

The surface manifestations of the Cerro Prieto geothermal field consist of numerous hot and cold springs, fumaroles and pools. Majority of the features occur in the areas to the northwest, west and southwest of the present production area. A list of surface features is provided in Table 7.3 and their locations are plotted in a map shown in Fig. 7.3. Gas samples were collected from these features and subjected to chemical analysis in the laboratory (Nehring and D'Amore, 1984). Results of analyses for major gases are given in Table 7.3. In general, nitrogen content is found to be higher in the surface manifestations located to the south of the production zone. When compared with the gas compositions from the exploration and production wells, the surface features indicate (i) depleted levels of hydrogen sulfide, probably

Table 7.2. Geological history of the Cerro Prieto geothermal field (after Lyons and van de Kamp, 1980)

1.	Basin formation	Miocene
2.	Early stage basin fill including volcanics	Miocene
3.	Onset of Colorado delta progradation filling trough from northeast	Mid-Pliocene
4.	Termination of lithofacies deposition west of field and shoaling of marine connection to Imperial Valley area	Middle Pleistocene
5.	Onset of NE–SW and NNE–SSW faulting. Onset of thermal activity: (i) volcanism northwest of field and (ii) induration of sediments by hydrothermal alteration	Late Pleistocene (1,10,000 year or older)
6.	Episodic faulting and thermal activity: (a) repeated faulting, fracturing and fracture mineralization, (b) secondary solution porosity, and (c) volcanism	Late Pleistocene to present

due to its reaction with iron to form iron sulfides and sulfates and (ii) elevated levels of nitrogen, interpreted to have been introduced with fresh groundwater entering the field from the eastern side and mixing with thermal waters during its ascent to the surface. Studies carried out by Valette-Silver and Esquer-Patino (1979) brought out similarity in isotopic composition and ratios of major elements between the hot springs and the production wells, suggesting that the geothermal reservoir and the hot springs are closely related.

Large-scale hydrothermal alteration of surface rocks is associated with the surface manifestations (Valette-Silver and Esquer-Patino, 1979). The alterations are caused by one or more of the following processes locally active in the area: boiling, steam condensation, evaporation, mixing with meteoric water and surface oxidation. Alterations of surface sediments within the hot spring zones exhibit increase in some minerals such as quartz, felspar and illite, and destruction of other minerals such as kaolinite, montmorrilonite, calcite and dolomite (Valette-Silver et al., 1981).

GEOCHEMICAL SURVEYS

Geochemical characterization of fluids and gases emanating from surface geothermal manifestations and those produced by the different wells in the area represent one of the early studies in the exploration of Cerro Prieto geothermal field. Results of analyses of gas samples collected from surface manifestations were summarized in the previous section. Several authors have described the composition of the reservoir fluids, the source of the fluids, their spatial and temporal variation, and computation of reservoir temperatures on the basis of geothermometers (Molina and Banwell, 1970; Mercado and Samaneigo, 1978; Mañón et al., 1979; Fausto et al., 1979, 1981; Truesdell et al., 1984).

Table 7.3. List of surface features (as shown in Fig. 7.3) and chemical composition of their gases in mole percent (from Nehring and D'Amore, 1984)

Location	Description	CO_2	H_2S	NH_3	He	H_2	Ar	O_2	N_2	CH_4
N-3	Cold spring in canal	97.2	0.84	0.0066	0.00012	0.0028	0.025	0	0.80	1.07
N-5	20 °C pool	94.3	2.02	0.023	0.00022	0.0018	0.041	0	1.39	2.19
N-7	30 °C pool	97.8	1.03	0.0078	0.000037	0.0042	0.0059	0	0.17	1.01
N-18	Large cold pool	88.4	0.22	0.034	0.0019	0.00071	0.056	0	2.18	8.82
N-20	60–80 °C pool	79.9	0.99	0.065	0.00064	1.87	ND	0	2.42	14.8
N-31	100 °C fumarole along canal	80.7	1.25	12.5	0	3.61	0.046	0	1.97	0.23
N-36	Small 100 °C pool	83.0	3.54	1.83	0.00036	0.66	0.13	0	5.81	4.73
N-43	90 °C spring in marshy pool	88.1	1.40	0.083	0	1.88	0.094	0	2.78	5.52
Near N-46	Spring near edge of Laguna Volcano lake	77.1	2.78	0.090	0.00061	1.73	0.15	0.53	8.88	8.11
N-49	Large cold pool	63.6	1.06	0.075	0	0.00039	(as O_2)	3.98	22.6	6.51
N-50	Large cold pool	80.8	2.21	0.058	0.0015	0.028	0.12	0	4.64	11.7
N-61	Cold spring in canal	91.8	0	0.085	0	0.012	(as O_2)	0.80	2.58	4.25

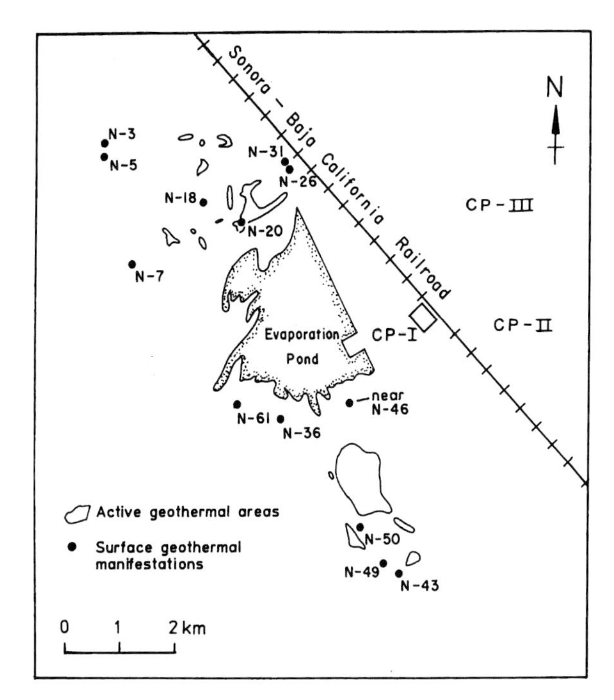

Fig. 7.3. Map showing the locations of surface geothermal manifestations in the Cerro Prieto area (modified after Nehring and D'Amore, 1984). Most features occur to the west of the production area.

The Cerro Prieto area is very arid, with an average annual rainfall of about 13 cm, and the elevation is only 7 m above the mean sea level. Consequently, the water balance is controlled mainly by evaporation. The observable surface runoff is very small, mainly through groups of large mud volcanoes, muddy bubbling lagoons and occasional phreatic eruptions. Molina and Banwell (1970) compared the atomic ratios, $Cl : HCO_3 : B$, $Cl : HCO_3 : SO_4$ and $Na : SiO_2 : K$, for Mexican geothermal fields. On the $Cl : HCO_3 : B$ and $Cl : HCO_3 : SO_4$ plots, the Cerro Prieto field separates very sharply from the rest (Fig. 7.4A, B). Cerro Prieto samples are also distinguished by a relatively higher proportion of sodium and potassium relative to silica (Fig. 7.4C). Temperatures inferred on the basis of chemical analyses of samples from drillholes in the Cerro Prieto region show large variations, which are consistent with the wide range of temperatures and chemical conditions observed over the field.

These initial studies were followed by several geochemical and isotopic studies, results from which are summarized by Lippmann et al. (1983). Some of the salient information derived from these surveys are:

(i) On the basis of computed deuterium and chloride concentrations in the reservoir brines and measured Cl/Br ratios, Truesdell et al. (1981) conclude that the geothermal fluids are a mixture of Colorado River water and a saline brine of

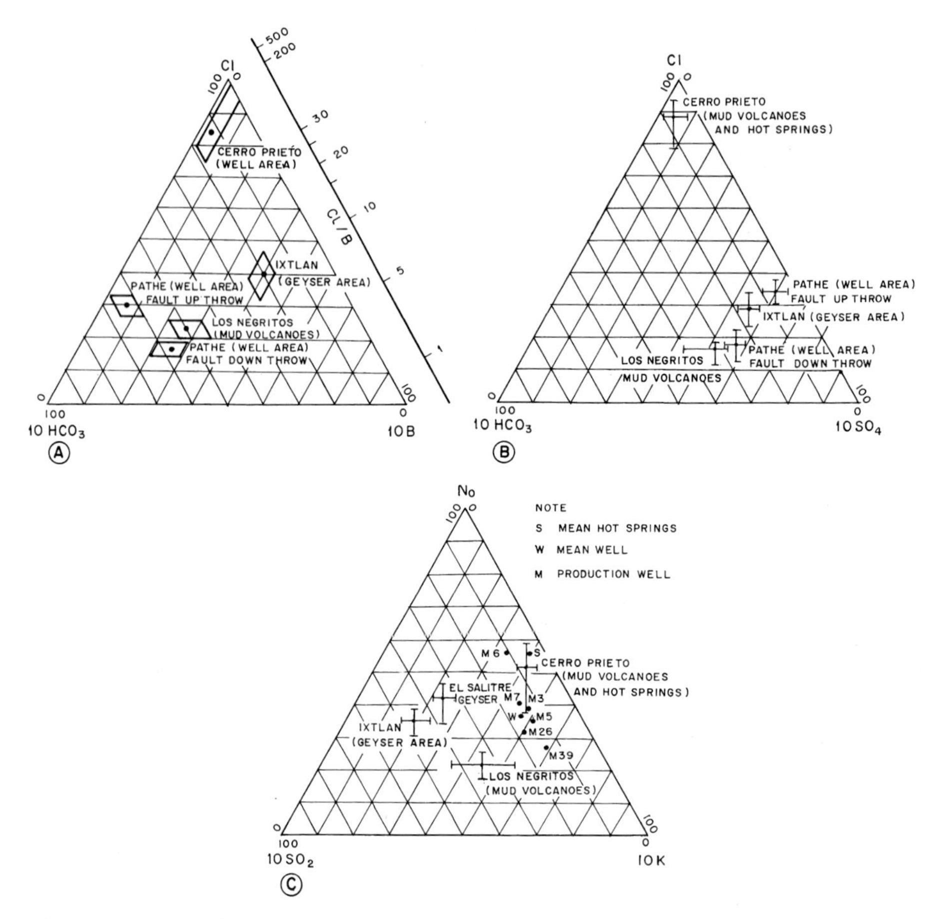

Fig. 7.4. Geochemistry of geothermal fluids in Mexico (modified from Molina and Banwell, 1970). Plots of atomic ratios for the geothermal fluids shown are: (A) Cl : HCO_3 : B, (B) Cl : HCO_3 : SO_4, and (C) Na : SO_2 : K are shown. For details, see text.

seawater origin. Deep circulation of these fluids and extensive alteration of reservoir rocks by high-temperature reactions have been inferred.

(ii) Similarities in isotopic composition and ratios of major elements between the hot springs (surface manifestations) and producing wells, as discussed earlier, indicate a close relationship between the two (Valette and Esquer-Patino, 1979). Chemical changes observed in the hot springs after initiation of fluid production from the field tend to support the relationship. Further, with respect to the wells, the surface manifestations show a loss of H_2S and increase of nitrogen. The former has been interpreted to be due to reaction with metal ions or air, and the latter due to mixing of geothermal waters with ground water during their upward movement along the faults.

(iii) Tracer studies by Mazor and Truesdell (1984) using radiogenic (He and ^{40}Ar) and atmospheric noble gases (Ne, Ar and Kr) show that the geothermal fluids are dominantly of meteoric origin. Their studies further suggest that the meteoric waters penetrate deeper than 2,500 m (i.e., below the level of first boiling) and get mixed with radiogenic He and ^{40}Ar formed in the reservoir rocks. Afterwards, small amounts of steam are lost by continuous removal as it forms, and mixing occurs with shallower cold water.

(iv) A model developed from the analyses of chemical, production and reservoir-engineering datasets by Grant et al. (1984) suggests that the relatively shallow, western part of the Cerro Prieto reservoir has low-permeability rocks at the bottom. At the top and at the sides, it has an interface with cooler water. There is no continuous permeability barrier around or immediately above the reservoir. Permeability within the reservoir is dominantly intergranular. The dominant cooling process in the natural state is through mixing with cold water rather than by boiling. Production causes displacement of hot water by cold water, and not by vapor. Local boiling occurs near most wells in response to pressure decrease, but no general vapor zone has formed. The model is shown schematically in Fig. 7.5.

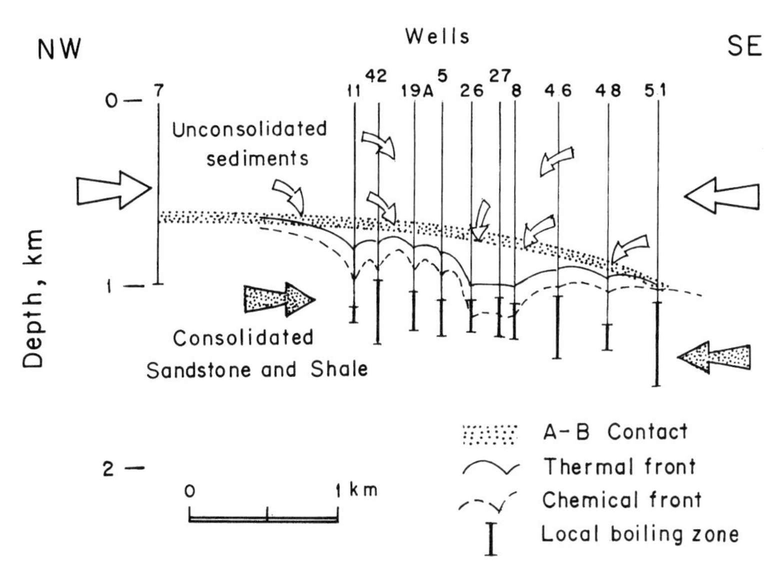

Fig. 7.5. A schematic subsurface model along a NW–SE section of the western Cerro Prieto reservoir (modified after Grant et al., 1984). The model, derived from analyses of chemical, production and reservoir-engineering datasets, shows the (i) relatively shallow, western part of the Cerro Prieto reservoir bounded below by low permeability rocks and above and at the sides, by an interface with cooler water, (ii) flow of hot (shaded arrow) and cold waters (unshaded arrow) toward the producing wells, and (iii) local zones of boiling, inferred for each well. The A/B contact, as discussed in the text, is indicated by the shaded zone.

GEOPHYSICAL MEASUREMENTS

Geophysical investigations by the CFE in the Mexicali Valley near the Cerro Prieto volcano began in the 1960s. Initially, gravity and seismic refraction methods were used for structural information related to faults and basement configuration. Emphasis shifted to electrical resistivity surveys in the early 1970s after it was determined that a large area of low-resistivity coincided with the thermal manifestations south of the volcano and the known high temperature zone at depth. Supplemented by ground magnetic and gravity measurements, the resistivity data have been interpreted to yield a detailed picture of the structure concealed by valley fill and to identify promising exploration targets. The primary objective of geophysical surveys was to unravel the subsurface structure of the reservoir and identify the locales of geothermal fluids within it, which would constrain the locations for drilling production and injection wells.

Detection of Geothermal Anomaly

The preliminary subsurface temperature distribution in the Cerro Prieto area was delineated from temperature measurements in a number of wells (Mercado, 1976; Díaz, 1978). The distribution of wells along two perpendicular profiles passing through Cerro Prieto and the subsurface temperature distributions revealed by them are shown in Figs. 7.6 and 7.7. Subsequently, Rivera et al. (1982) carried out further studies and produced contoured maps of temperatures at depths of 1,500 and 2,500 m. Large variations in the subsurface temperatures can be gleaned from the datasets. Moreover, the temperatures measured in the boreholes under such conditions are commonly lower than the true

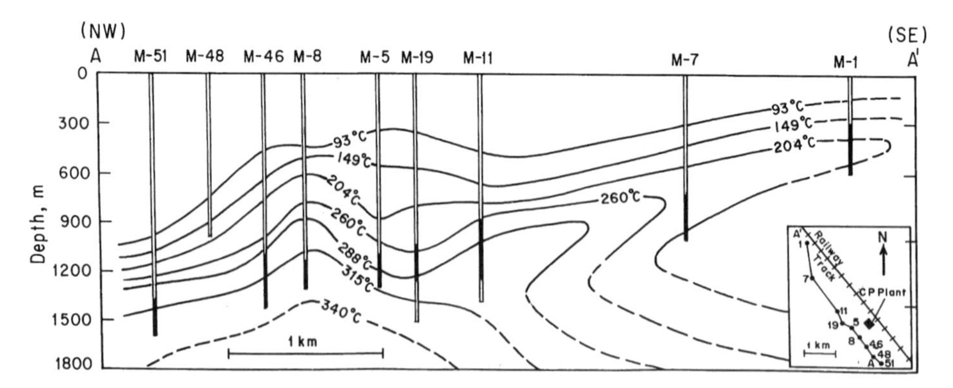

Fig. 7.6. Subsurface temperature distribution in the Cerro Prieto area as revealed from temperature measurements in borewells located along a profile parallel to the Sonora–Baja California railway line. The filled segment toward the bottom of each borehole indicates the producing zone for geothermal fluids (Mercado, 1976).

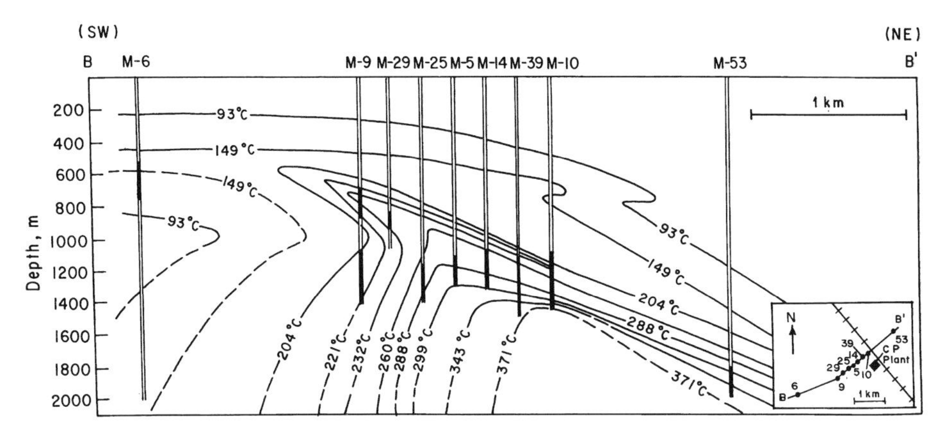

Fig. 7.7. Subsurface temperature distribution in the Cerro Prieto area as revealed from temperature measurements in borewells located along a profile perpendicular to that shown in Fig. 7.6. The filled segment towards the bottom of each borehole indicates the producing zone for geothermal fluids (Mercado, 1976).

formation temperatures because of cooling by circulation of drilling fluids. From these measurements it was concluded that (a) thermal fluid is ascending in the eastern and central portions of the field and moving in a westerly direction, (b) the gradual temperature increase in the eastern area suggests the location of the reservoir heat source, (c) a recharge zone occurs in the western part, in the area of the alluvial fans of the Sierra de los Cucapas, and (d) another recharge zone is located in the eastern part of the field, where meteoric water from the Colorado River feeds the reservoir.

A number of geophysical techniques have been used in the geothermal exploration program, but with varying degrees of success. These include direct current resistivity, self-potential, magnetotellurics, precision gravity, magnetics, seismicity studies and seismic reflection. Over the past four decades, several workers have carried out detailed geophysical surveys in the area for various purposes. In the following, we shall mention briefly the significant achievements of a few methods in detection and delineation of the geothermal anomaly at Cerro Prieto.

Gravity and Magnetic Surveys

Bouger gravity and magnetic anomaly maps of Cerro Prieto area, as given by Razo and Fonseca (1978), are shown in Figs. 7.8 and 7.9, respectively. Positive magnetic and coincident gravity anomalies located about 9 km northeast of Cerro Prieto volcano are possibly caused by a shallow magnetic basement in the area. Low-magnetic and gravity anomalies to the south of the volcano could be attributed to the presence of a thick sedimentary sequence in this area. This interpretation is supported by seismic reflection studies. Another feature, a "ridge"-like gravity high extending southeastward from the well M-96, could be caused due to densification of shales

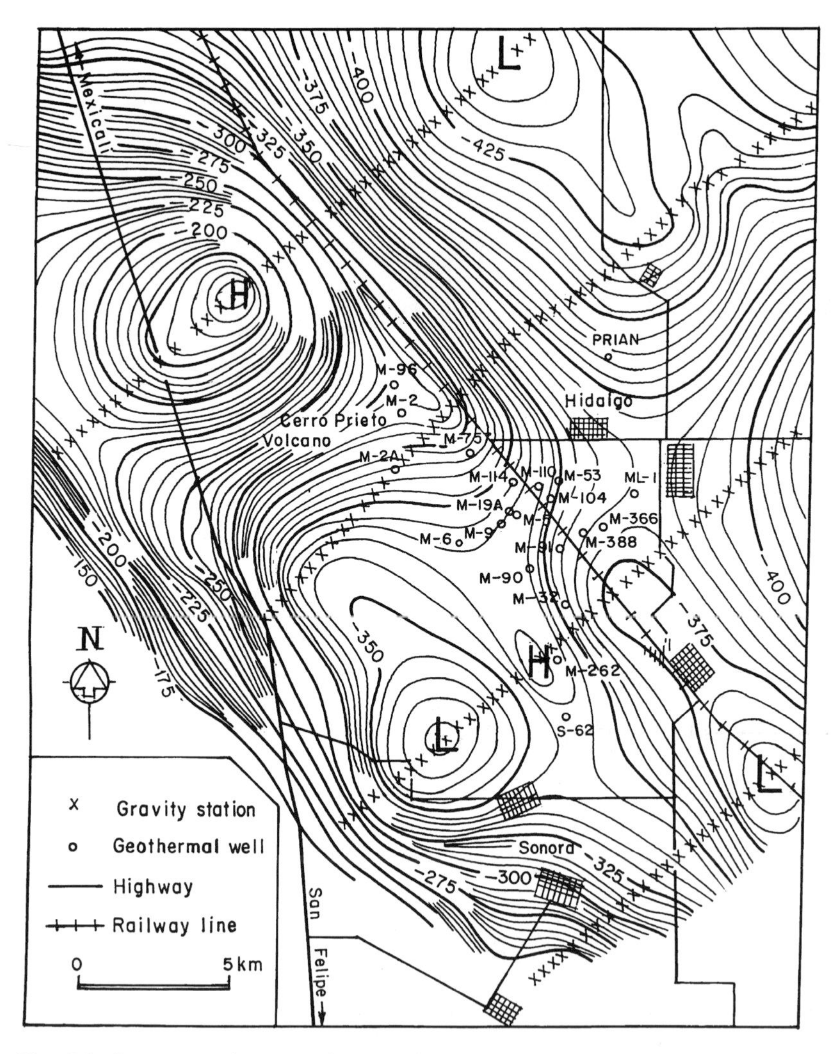

Fig. 7.8. Bouger gravity anomaly map of Cerro Prieto area (modified from Razo and Fonseca, 1978). The contour interval is 1 mgal. The localized gravity highs (H) and lows (L) are indicated.

underlying the sedimentary sequence. The "gravity highs" in this area are commonly associated with subsurface high-temperature anomalies.

Electrical Surveys

Electrical resistivity studies have been carried out in Cerro Prieto area employing mainly Schlumberger and dipole–dipole configurations (e.g., Garcia, 1976; Mañón

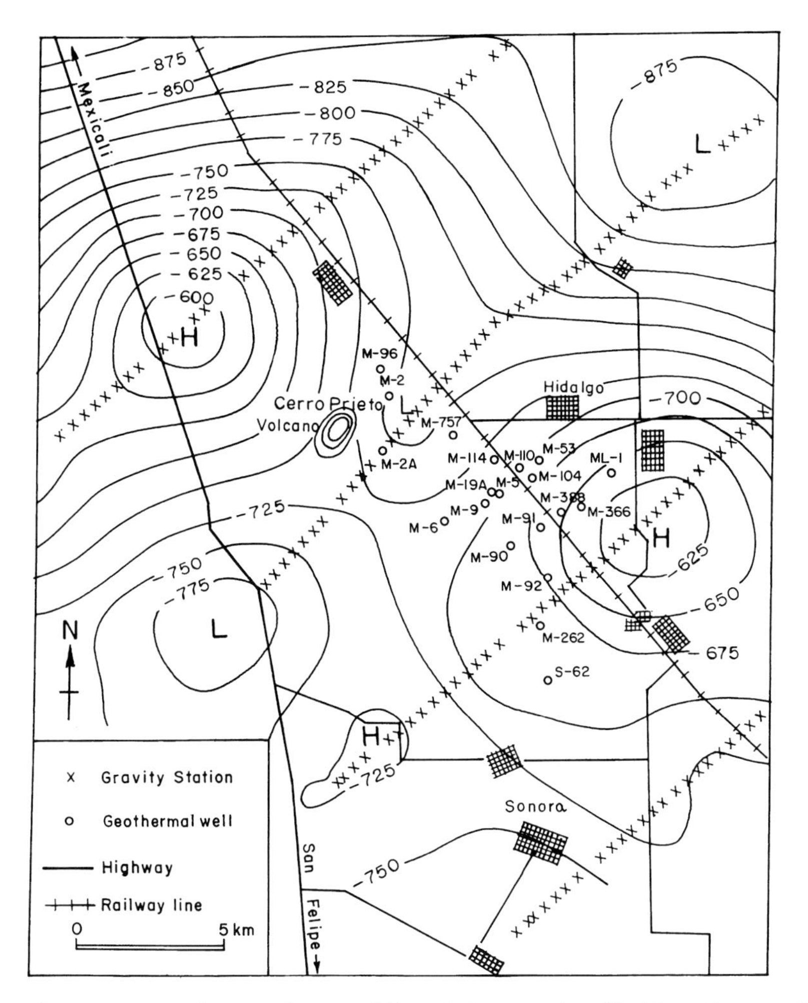

Fig. 7.9. Magnetic anomaly map of Cerro Prieto area (modified from Razo and Fonseca, 1978). The contour interval is 25 gamma. The localized magnetic highs (H) and lows (L) are indicated.

et al., 1978; Wilt et al., 1980; Wilt and Goldstein, 1981, 1984; and others). Garcia (1976) reported one of the earliest studies including apparent resistivity measurements at constant depth through profiling. An area of about 70 km^2 having an apparent resistivity $<2\,\Omega$m within an overall low-resistivity ($<10\,\Omega$m) region surrounding the Cerro Prieto volcano was delineated. In the second phase of survey carried out within a 40-km^2 zone around the previously producing area, actual resistivity values were obtained employing Schlumberger depth soundings with current electrode spacing ranging between 20 and 4,000 m. From the resistivity contours at the deepest interpreted depth of 1,200 m, a NW–SE trend for the geothermal field, parallel to the regional fault system, could be gleaned.

Mañón et al. (1978) reported results of additional Schlumberger soundings using AB/2 spacing of 500 m. The resistivity map of the Cerro Prieto area is shown in Fig. 7.10. Although quite simple, this map provides significant clues for further

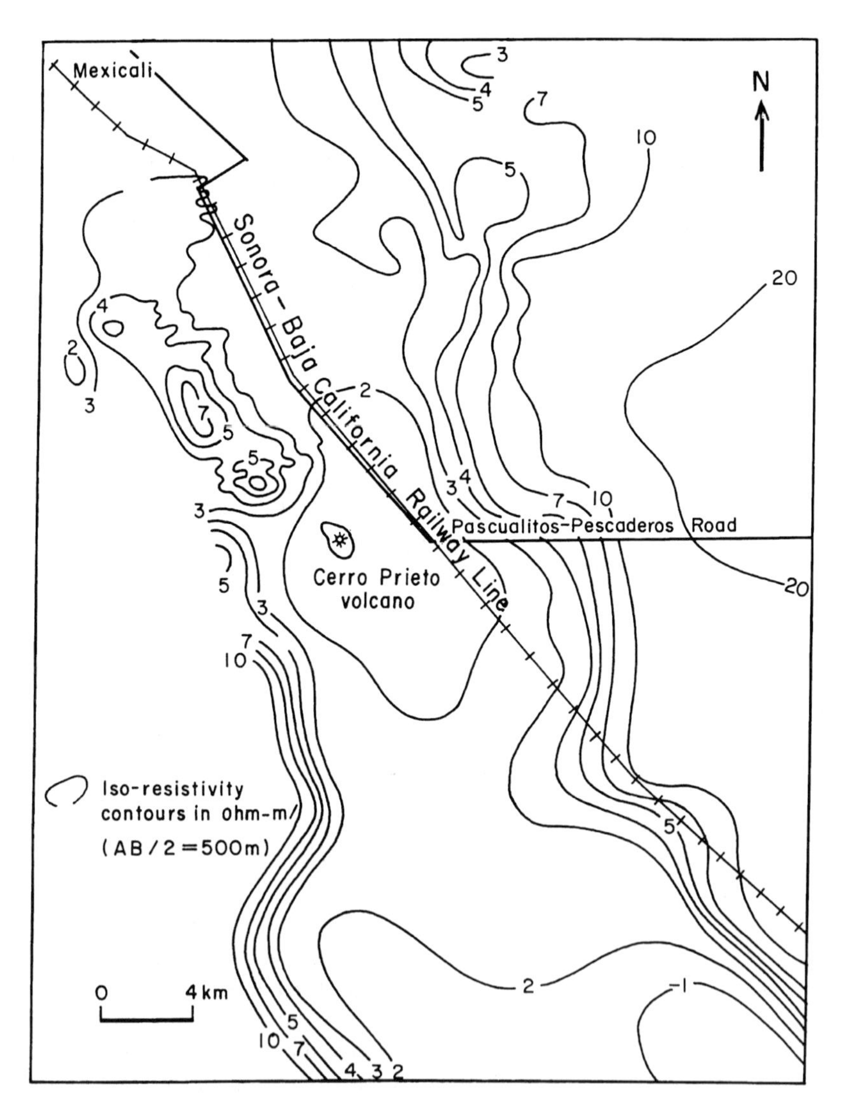

Fig. 7.10. Preliminary resistivity map of the Cerro Prieto area prepared on the basis of Schlumberger soundings with AB/2 spacing of 500 m (modified from Mañón et al., 1978). Contour interval is 1 Ω m.

exploration. Unlike other volcanic terrains where low resistivity anomalies in geothermal areas are often associated with hot water producing zones, the Cerro Prieto is best identified with high resistivity. Low resistivity ($<2\,\Omega$ m) appears to delineate the extent of the geothermal area, but the production zone is characterized by higher resistivity (5–10 Ω m) coinciding with the "ridge-like" gravity high mentioned earlier. Resistivity surveys carried out by Wilt et al. (1980) also show a high-resistivity body below about 800 m, which approximately coincides with the productive zone in the older part of the geothermal field. High resistivity associated with production zones has been attributed to porosity loss by hydrothermal alteration in shales. The correspondence between the top of the high-resistivity shales observed on the well logs and the high-resistivity body interpreted from resistivity surveys supports this inference. Dipole–dipole surveys carried out by Wilt et al. (1980) again showed that the production region is characterized by high resistivity (4 Ω m) relative to the surrounding region (1–2 Ω m). Wilt and Goldstein (1981) further showed that the 4 Ω m body coinciding with the producing zone has an eastward dip of 30°–50°, up to a depth of at least 2 km.

From the studies carried out so far, it appears unlikely that geothermal anomalies in the Cerro Prieto area can be detected on the basis of resistivity data alone.

Self-Potential Surveys

The production area is characterized by a very prominent and large amplitude dipolar self-potential anomaly, as shown in Fig. 7.11 (Corwin et al., 1978, 1980). The

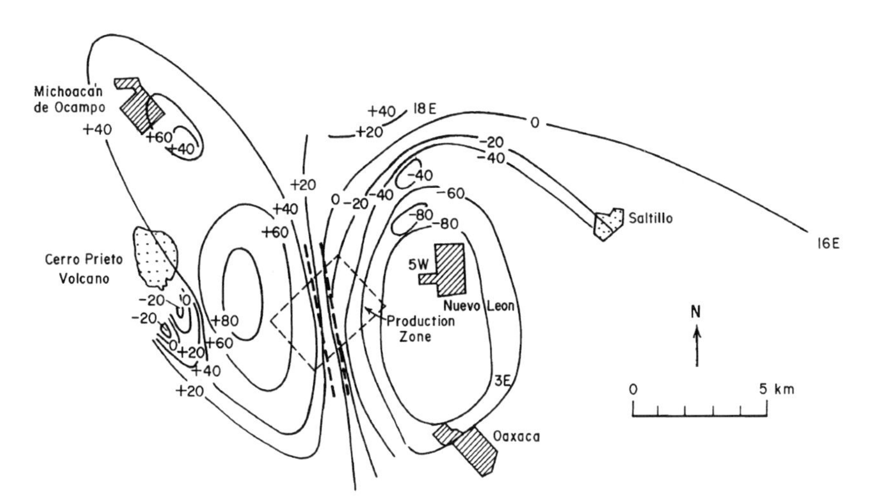

Fig. 7.11. Large amplitude, long wavelength dipolar self-potential anomaly in the Cerro Prieto area (modified from Corwin et al., 1978). The contour interval is 20 mV.

inflection points of the two anomalies are centered over two major faults. The anomaly has been interpreted to have been caused by a combination of fluid flow and temperature gradient, indicating that the faults of the volcano system provide the main conduits for ascending thermal fluids.

Magnetotelluric Soundings

Magnetotelluric surveys were conducted for the first time in 1978. Preliminary results showed a marked similarity with the results obtained from the electrical resistivity surveys of the dipole–dipole and Schlumberger arrays. The results of the magnetotelluric survey gave better information on the deeper strata, which, in the electrical method are masked by shallow saline layers of low resistivity. The magnetotelluric method was, therefore, used to verify the consistency of the resistivity models developed from other methods.

Seismic Reflection Surveys

Seismic reflection studies in the Cerro Prieto area have brought out fairly close correspondence between the producing intervals for geothermal fluids and reflection-poor zones in the reflection profiles (Lyons and van de Kamp, 1980). The lack of reflections in the producing zones probably results from reduction in porosity due to hydrothermal alteration, which may have obiliterated the original acoustic impedence stratification in the producing zones. A more-or-less similar situation occurs in the case of East Mesa geothermal field, where poor reflection zones are found associated with producing zones. However, the lack of reflections in this case has been explained due to reduction in porosity due to hydrothermal alteration and extensive fracturing of rocks in the high-temperature zone resulting in dispersion of seismic energy.

The seismic reflection method does not provide direct evidence of possible production zones below the surface. However, the occurrence of a reflection-poor zone coincident with the hydrothermal alteration zone is a useful indicator in identifying geothermal targets. It may be noted that before inferring a possible producing area, it is important to preclude other potential sources of reflection-poor zones such as structural complexities, single rock type sequences, reduced reflectivity of densified shales due to presence of vapor phase, and shallow basement highs. Seismic reflection data can also be gainfully used to refine the interpretation of basement configuration and faults that have been delineated on the basis of other geophysical methods.

HYDROGEOLOGICAL MODEL

The hydrogeological model of Cerro Prieto, developed by Halfman et al. (1984, 1986) using results of geological studies, electrical well logs and subsurface temperature measurements in numerous wells, explains the movement of geothermal fluids under natural state conditions. The model is shown in Fig. 7.12. The locations of the main faults that control the subsurface flow of geothermal fluids in the Cerro Prieto geothermal field are shown in Fig. 7.13. The salient features of the model have been summarized by Lippmann et al. (1989). The model clearly shows that the circulation of geothermal fluids takes place mainly from east to west. The hot fluids, with temperatures of about 350 °C, enter the system from the east and southeast directions, travel westward, and are manifested at the surface in the form of hot springs, fumaroles and mud pools in the western and southwestern parts of the geothermal field. The lithology and fault geometry play a dominant role in governing fluid circulation. The geothermal fluids come up from deeper levels through the SE-dipping, normal fault H, and move laterally into the more permeable sandy layers, and recharge the B-reservoir located at depths in excess of 1,600 m.

At certain locations, due to absence of a shale caprock above the sandstone layers, the fluids come up closer to the surface until they are stopped from further ascent by the presence of low permeability sandy materials sealed by mineral precipitation. The fluids therefore continue to flow westward through a shallower permeable layer located between about 1,000 and 1,500 m depth. This storage forms the A reservoir, and it is restricted to the western part of the field only, i.e., west of the railroad tracks. The normal fault L guides the fluids up toward a shallower aquifer where mixing with cold groundwater occurs.

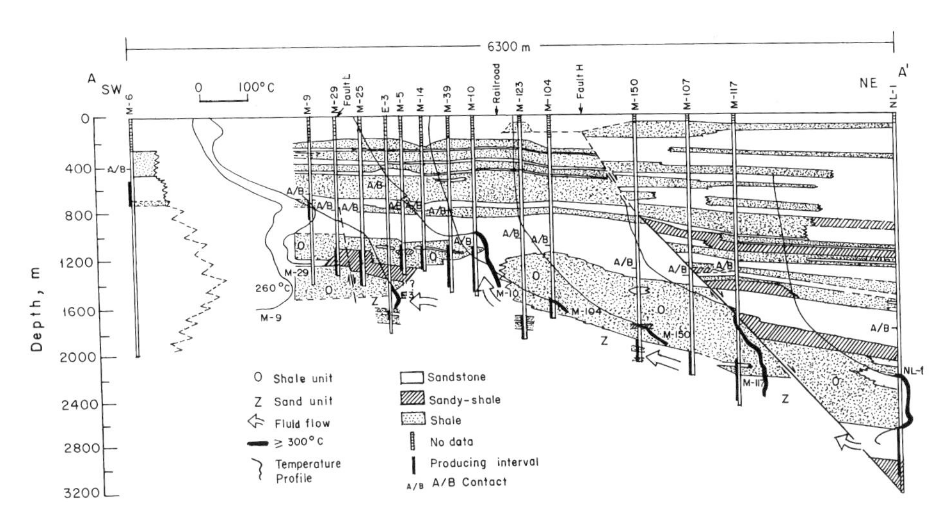

Fig. 7.12. Schematic hydrogeological model of Cerro Prieto along a SW–NE geologic cross-section AA′ (shown in Fig. 7.13), developed using results of geological studies, electrical well logs, and subsurface temperature measurements in numerous wells (modified from Halfman et al., 1984). The model explains the movement of geothermal fluids vis-à-vis the lithology and fault configuration in the area.

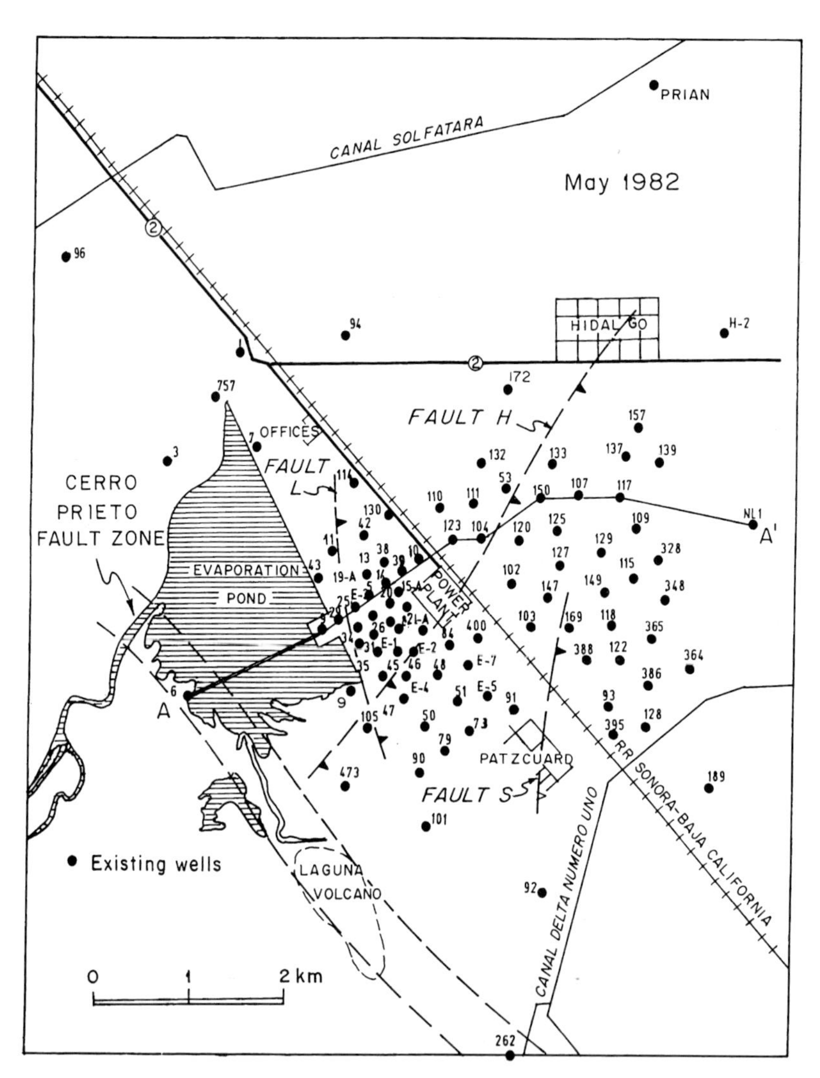

Fig. 7.13. Distribution of main faults controlling the movement of geothermal fluids in the Cerro Prieto geothermal field (modified from Halfman et al., 1984). In addition to the Cerro Prieto fault zone in the western part, three other faults (L, H and S) have been identified. The approximate location of the power plant and the distribution of wells as in May 1982 are also shown. The geological cross-section along AA′ is shown in Fig. 7.12.

A simple model for the natural circulation of geothermal fluids in Cerro Prieto, integrating available data has been given by Lippmann (1983). The salient features of this model are:

(i) Waters of Colorado River and seawater origin penetrate deep into the sedimentary fill of the Mexicali Valley. These waters/brines are heated by

1,000–10,000-year old diabase dikes and gabbro intrusions. A melt zone and magmatic injection at 9–10 km depth is likely.

(ii) The heated fluids move, as a result of the pressure gradients and buoyancy, toward the western side through a sandstone unit (reservoir B) overlain by a thick shale layer.

(iii) The geothermal fluids during their passage through the sedimentary rock formations cause hydrothermal alteration in them. Several secondary minerals have replaced the original cement of the rocks with depositions of calcites, chlorites, silica, epidote, amphibole and prehnite. The fluids themselves are often extensively altered due to their reactions with the reservoir rocks at high temperatures.

(iv) During their westward movement toward the present production area, the geothermal fluids tend to flow upward though faults and permeable zones in the overlying shale horizons. The upward movement of the heated fluids causes boiling near the surface due to rapid reduction in pressure. Mineral deposition also takes place during this process.

(v) The geothermal fluids mix with shallow, cold groundwater along the western edge of the field causing secondary mineral precipitation and consequent porosity reduction. Some of the hot fluids that ascend to the surface through fault zones appear as numerous hot springs, fumaroles and mud pools along the western edges of the field.

High $^3He/^4He$ ratios, up to 6.3 times the atmospheric ratio, confirms the magmatic constituent of the fluids (Welhan et al., 1979). Such high ^3He contents are usually associated with volcanic gases, young volcanic rocks and high-temperature geothermal fluids heated by magmatic heat sources at depth (Welhan et al., 1988).

DRILLING

As discussed in Chapter 6, drilling of geothermal wells differs considerably from drilling for oil. At Cerro Prieto, after having chosen a drilling site on the basis of geophysical, geological and geochemical studies, the ground around it is compacted to ensure its stability. Next, a trench is dug, to a depth of 2–3 m, for the drilling cellar, and the rotary drilling equipment is installed. A typical drilling rig and accessories, used at Cerro Prieto, are shown in Fig. 7.14.

The drill hole diameter decreases with depth. At Cerro Prieto, drilling typically starts with a 22″-diameter hole. As the depth increases, it is reduced to 15″, and at the depth of 1,000 m, it is further reduced to 10.6″. This diameter is normally retained to the bottom of the hole, which typically averages about 2,500 m for production wells and about 1,500 m for injection wells at the present time. The drilling fluid consists of bentonite slurry with several additives to withstand the high temperatures encountered. A cooling tower is installed to cool the circulating slurry.

The drilling fluid is often lost through fissures and fractures encountered during the course of drilling. In the event that these fissures and fractures are unproductive,

Fig. 7.14. A rotary rig at Cerro Prieto (photograph by HG).

they are sealed with vegetable fibers, mica flakes, cellophane chips or in some cases by cement. If these fissures lie in an adequately productive zone, drilling is terminated and the hole is considered to have been completed. With the progress of drilling, casing is provided to hold the formations from caving. At Cerro Prieto, seamless steel pipes of 16″ diameter are used for anchoring, 11.75″diameter pipes for lining and 7.6″ diameter pipes for production. The production casing is slotted. Portland cement, with additives to delay setting, is used for binding the casing with the borehole. Fig. 7.15 schematically shows a typical borehole.

Geothermal fluid-producing horizons are detected by means of electrical, thermal and sonic logging. During drilling, the necessary safety equipment is installed at the anchor pipe to keep the well under control. This essentially consists of a blowout preventor, which automatically seals the well when pressures become dangerously high. After termination of drilling, the safety equipment is withdrawn and a valve

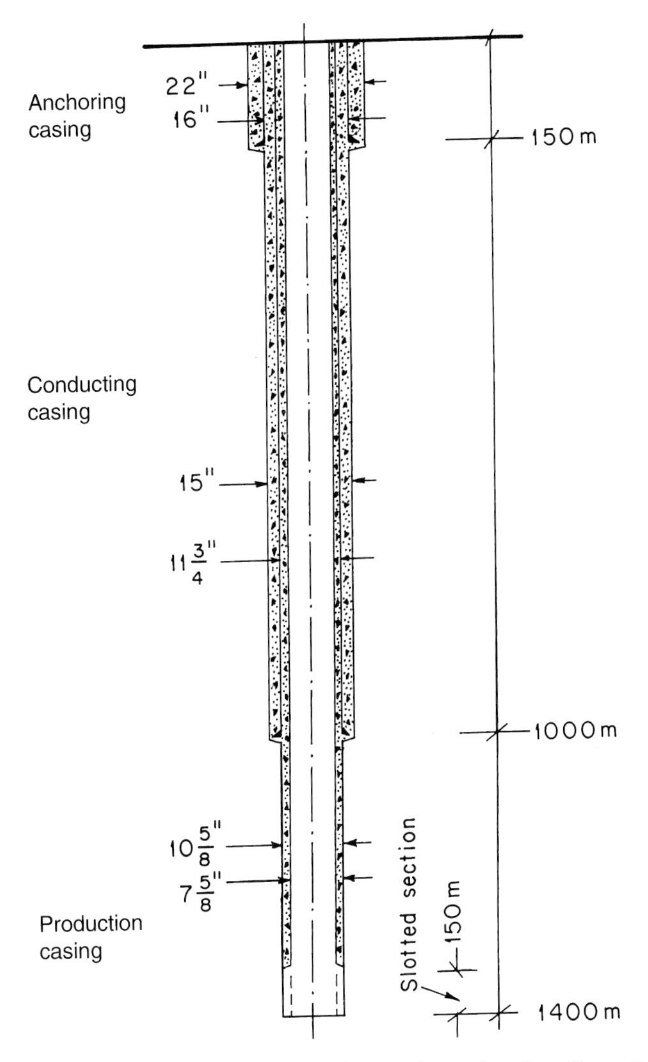

Fig. 7.15. Schematic section of a geothermal well at Cerro Prieto. For details, see text.

tree (Fig. 7.16) is installed on the pipe head. Geothermal fluids are taken out through different ducts from the valve tree.

Up to the end of 2003, a total of 315 wells had been drilled in the Cerro Prieto geothermal field (Gutiérrez-Negrín and Quijano-León, 2005). The combined depth of all the wells is 7,25,500 m. In general, the wells at Cerro Prieto are deeper than those at other Mexican fields. The productive life of each production well in this field is about 10 years, with a work over around the fourth or fifth year.

Fig. 7.16. A typical valve tree at Cerro Prieto (photograph by HG).

PRODUCTION

Cerro Prieto geothermal wells produce a mixture of hot water and steam. It is necessary to separate them and send the dry steam to the turbine of the power plant. This is achieved by passing the well output through a centrifugal separator (Fig. 7.17). The centrifugal force makes the water adhere to the walls of the separator, and it finally descends to the bottom, while the steam rises in the separator and is removed through a pipe. The separated water is discharged into the silencers and later into the drainage channels.

Fig. 7.17. A centrifugal separator at Cerro Prieto (photograph by HG).

The silencers are useful in reducing the noise produced by discharging the geothermal fluids at a high pressure into the atmosphere. At Cerro Prieto, horizontal as well as twin-barrel vertical silencers are used. The horizontal silencer uses a series of pipes with increasing diameters. As the fluid passes through it, the flow speed gradually decreases and the fluid is discharged at a much diminished noise level. The twin-barrel vertical silencer consists of two vertical wooden pipes resting on a concrete structure. The discharge pipe from the separator releases fluids at the concrete structure. With the release of pressure, steam is formed, and the water and steam become separated due to centrifugal action, similar to that in the separator. Eventually, steam is released at a low pressure, and consequently with reduced noise, to the atmosphere and the water to the drainage channel.

TRANSPORTATION OF STEAM

Hundreds of meters of pipeline are used to bring the steam from the separator to the electric power plant turbine. Several steam traps are installed to remove water formed by condensation of steam, since the steam must be dry when it enters the turbines. As a measure of additional precaution, the steam is passed through a moisture absorber before entering the turbines. All production wells are connected through four branches of pipelines. Pipeline joints are so constructed that they are capable of absorbing the mechanical stresses generated by thermal expansion of the pipes. At the power plant, the four branches of pipelines are connected to two steam collectors. The total length of pipeline, with a pipe diameter of 16–22 in., is in excess of 6 km. The secondary pipelines of small diameters add up to several more kilometers. A schematic representation of the wells and the pipelines for the plant CP-I at Cerro Prieto is shown in Fig. 7.18. The layout, however, has been upgraded

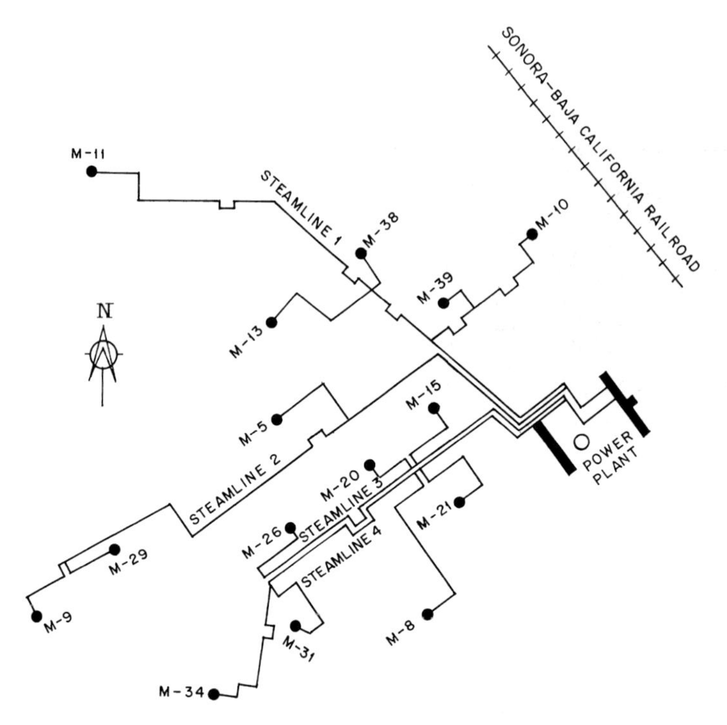

Fig. 7.18. A sketch of the location of wells and steam transportation pipes (modified from "Cerro Prieto—Underground Power", a brochure issued by CFE). This network has been expanded by adding a few hundred wells during the past three decades leading to about 10-fold increase in power production.

significantly over the past 3 decades due to addition of a large number of wells and distribution pipelines.

THE POWER PLANT

The Cerro Prieto geothermal power plant is a prestigious project of CFE, Mexico. The infrastructure of the plant is quite conventional. The structure, made of galvanized steel, covers an area of $60 \times 14.5\,m^2$ and has a height of 30 m. Figs. 7.19 and 7.20 give general views of the powerhouse. A schematic layout of the powerhouse and the accessories is given in Fig. 7.21. The turbogenerator and auxiliary equipment are installed inside, while the condensers, because of their size, are installed outside the engine house. The powerhouse is a two-storied structure with the turbogenerators, oil and hydrogen systems and several laboratories, operational and administrative offices. All turbogenerators are built with special metal alloy to reduce corrosion and erosion. Fig. 7.22 shows a turbine.

Fig. 7.19. A general view of the powerhouse at Cerro Prieto (photograph by HG).

Fig. 7.20. Another view of the powerhouse at Cerro Prieto (photograph by HG).

The Cerro Prieto geothermal field is located in desert and the scarcely available water does not meet the requirement of the power plant. It is therefore necessary to recover water formed by condensation of steam and use it for cooling purposes. This is achieved through a cooling tower (Fig. 7.23) built for the purpose. The cooling tower is huge ($146 \times 25 \times 18\,m^3$) and is capable of cooling $27{,}000\,m^3$ of water per hour to 16 °C.

At the present time, the Cerro Prieto produces power from four plants, CP-I, CP-II, CP-III and CP-IV, employing both single and double flash units with a total

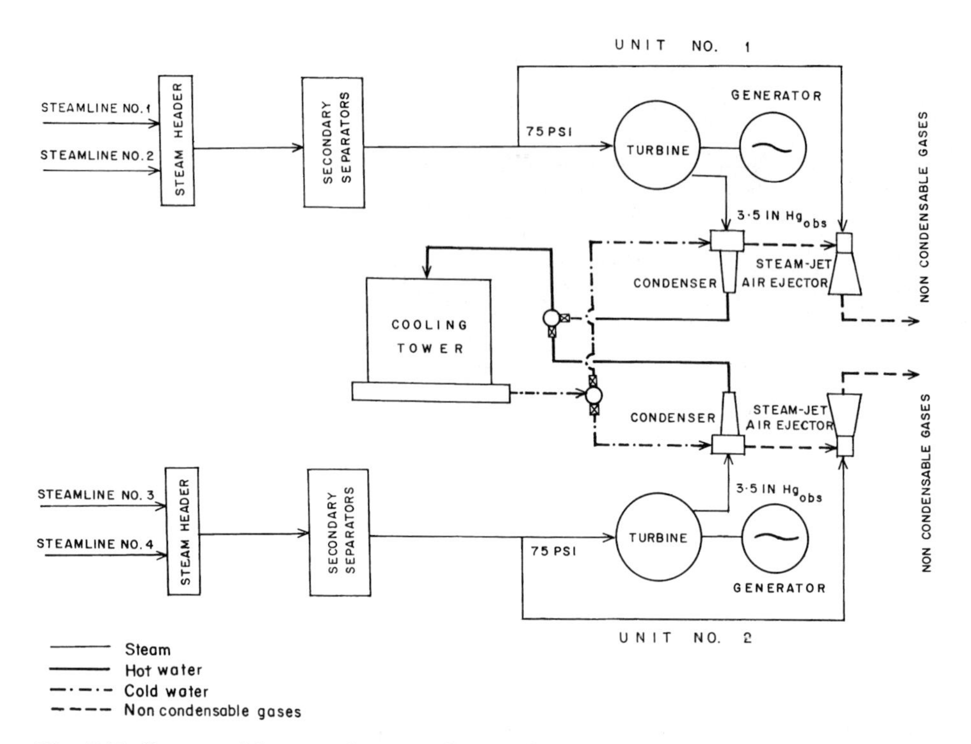

Fig. 7.21. Layout of the powerhouse at Cerro Prieto (modified from "Cerro Prieto—Underground Power", a brochure issued by CFE).

installed capacity of 720 MWe. The details of power production from each plant are listed in Table 7.4.

COST OF GEOTHERMAL POWER

Generation costs for geothermal power at Cerro Prieto are quite comparable to the power produced from fossil fuels (Hiriart and Andaluz, 2000). In the case of the first three power plants, CP-I, CP-II and CP-III, the cost turns out to be 3.46 U.S. cents per kWh. The cost is made up of four major components (Table 7.5):

In the case of the new plant, CP-IV, commissioned in 2000, the costs are even lower. The cost of installed capacity has been projected to be USD 797 per kW, and that for power generation turns out to be as low as 2.81 U.S. cents per kWh. It must be mentioned here that the costs for generation of geothermal power vary widely for different locations. Factors influencing the cost include the pressure of steam supply, production and transmission of steam and hot water from the production site to the power plant, corrosive nature of the geothermal fluids (that affects the overall maintenance costs) and several others. The power plants are operated and maintained

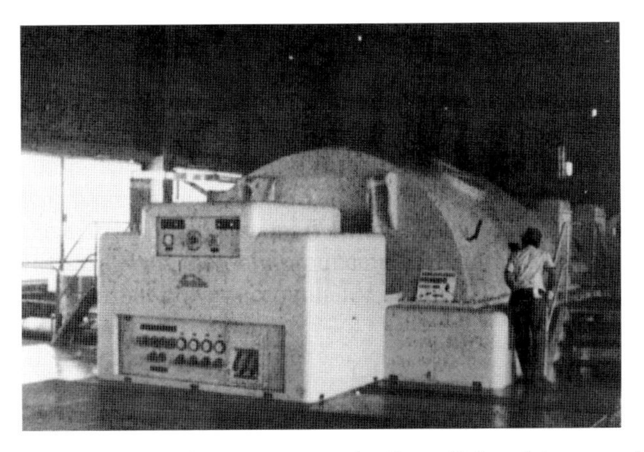

Fig. 7.22. A turbogenerator at the Cerro Prieto (photograph by HG).

Fig. 7.23. A view of the cooling tower at Cerro Prieto (photograph by HG).

wholly by the CFE, a Mexican government entity, which enjoys certain cost exemptions such as geothermal royalties and development fees when compared to the private power companies. This scenario results in lowering the total cost of generating geothermal power at Cerro Prieto and other Mexican fields.

PRESENT STATUS OF THE GEOTHERMAL RESERVOIR

Electric power generation at Cerro Prieto geothermal field has progressed steadily over the last 35 years since its inception in 1973 (Fig. 7.24, Table 7.4). During the last

Table 7.4. Summary of power production from the Cerro Prieto geothermal plants (Source: Comisión Federal de Electricidad; Gutiérrez-Negrín and Quijano-León, 2005)

Power plant	Unit no.	Type of unit	Year of commissioning	Installed capacity, MWe	Total installed capacity, MWe
CP-I	Unit-1	Single flash	1973	37.5	180
	Unit-2	Single flash	1973	37.5	
	Unit-3	Single flash	1979	37.5	
	Unit-4	Single flash	1979	37.5	
	Unit-5	Double flash	1982	30.0	
CP-II	Unit-1	Double flash	1986	110.0	220
	Unit-2	Double flash	1987	110.0	
CP-III	Unit-1	Double flash	1986	110.0	220
	Unit-2	Double flash	1987	110.0	
CP-IV	Unit-1	Single flash	2000	25.0	100
	Unit-2	Single flash	2000	25.0	
	Unit-3	Single flash	2000	25.0	
	Unit-4	Single flash	2000	25.0	

Table 7.5. Break-up of power generation costs in U.S. cents per kWh at Cerro Prieto plants CP-I, CP-II and CP-III (from Hiriart and Andaluz, 2000)

Component of generating cost	Cost (U.S. cents per kWh)
Power plant investment	1.63
Operation and maintenance of plant	0.36
Steam supply	1.17
Operation and maintenance of field	0.30
Total	3.46

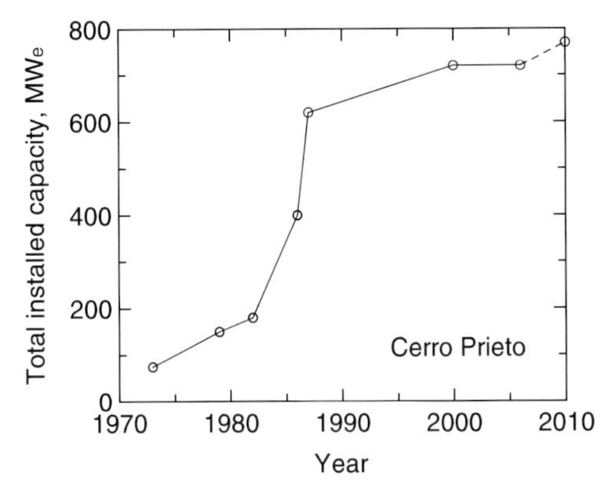

Fig. 7.24. Plot showing the growth in power-electric power production from Cerro Prieto geothermal field with time. The dashed segment of the curve represents the projected addition of 50 MWe by the CFE (Hiriart, 2001).

five years, 44 production wells and three injection wells were drilled (Gutiérrez-Negrín and Quijano-León, 2005). The production wells (average depth of 2,835 m) were drilled mainly to replace exhausted, old wells as no new areas were explored. At the present time, most of the production wells extract geothermal fluids from the deeper, B-reservoir. Succinct summaries for the current status of the geothermal reservoir, on the basis of geochemical, geological and temperature data are given by Truesdell and Lippmann (1998) and Lippmann et al. (2004).

A simplified, present-day, layout of the geothermal field, showing the distribution of production wells and the location of power plants (CP-I, CP-II, CP-III and CP-IV) vis-à-vis the controlling fault network is shown in Fig. 7.25. The normal fault, Fault H, strikes NE–SW and dips about 45° to the SE. The production wells in the proximity of CP-III and CP-IV are mostly located in the upthrown block of the Fault H. The wells in the proximity of CP-II are mostly in the SE downthrown block, with a few in the fault zone or in the adjacent upthrown block. The average

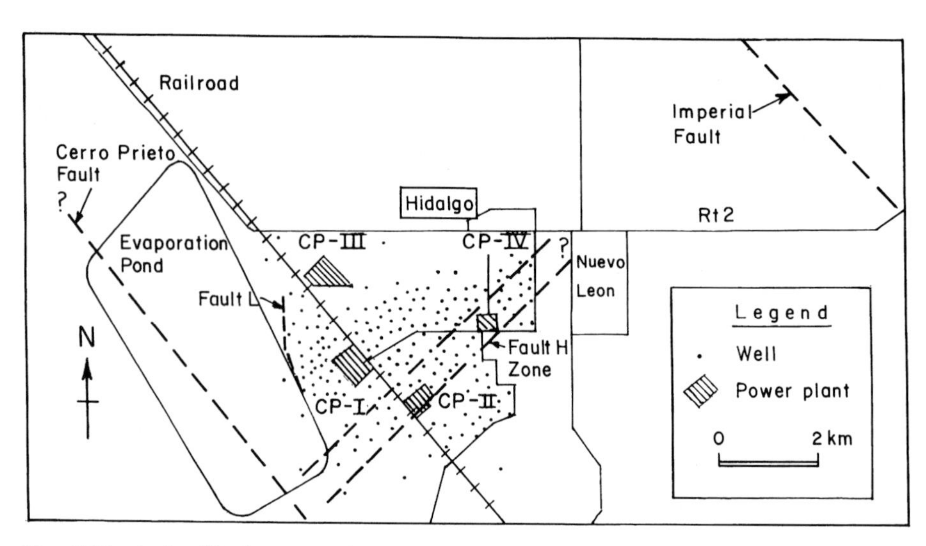

Fig. 7.25. A simplified, present-day, layout of the geothermal field, showing the distribution of production wells and the location of power plants (CP-I, CP-II, CP-III and CP-IV) vis-à-vis the controlling fault network (modifed from Lippmann et al., 2004).

depth to the top of the reservoir in the downthrown block is 2,500 m, whereas that in the upthrown block is 2,000 m.

A comparison of the present-day, much exploited, state of the reservoir as in the year 2000 vis-à-vis the relatively less-exploited state as in 1986 has been made by Lippmann et al. (2004). The comparisons are primarily based on four chemical and physical parameters including reservoir fluid temperature, the well inlet vapor pressure (reflecting excess enthalpy), reservoir chloride concentrations, and total discharge oxygen isotope ($\delta^{18}O$) data. Chloride concentration is particularly useful for evaluating reservoir processes because it is not leached from or taken up by the reservoir rock minerals, or partitioned into steam, and is not affected by processes other than boiling and mixing. As discussed previously in Chapter 5, reservoir temperature derived using Na–K–Ca geothermometer is usually unaffected by near-well boiling or mixing with condensate or low-salinity groundwater.

The comparison clearly indicates that the geothermal reservoir has responded to the progressively increasing production demands with a reduction in pressure and enthalpy over the past few decades. This results in localized boiling, inflow of cold groundwater into the reservoir from the surroundings, condensation and phase segregation. With the commissioning of the fourth power plant, CP-IV, further declines in pressure are anticipated, particularly in the eastern parts of the B-reservoir. Reservoir temperatures have dropped by more than 10 °C except in the SE sector near CP-II. Detailed analysis of data showed that the temperature drops in different parts of the reservoir can be attributed to different processes: (i) reservoir boiling in the NW sector of the Fault H, as brought out by the high chloride concentrations, possibly of residual brines, (ii) vertical groundwater recharge down the fault in the NE region, as brought out by the low chloride and low $\delta^{18}O$ values, and (iii) entry of cold fluids both from the surroundings and due to injection through wells in the SW sector. Significant changes have been observed in the chemistry of extracted fluids also, as a result of increased exploitation. The geothermal field has undergone subsidence as a result of the reduced pressures. The subsidence rate during 1994–1997 was greater than 8 cm per year with a peak subsidence rate of 11 cm $year^{-1}$ in the production area (Glowacka et al., 1999, 2000). Of this, the peak tectonic subsidence is 0.45 cm $year^{-1}$, or about 4% of the observed total subsidence, and the remaining 96% is attributed to man-induced subsidence due to withdrawal of geothermal fluids for power production (Glowacka et al., 2005). Therefore, it is important to assess the stress changes due to deformation caused by fluid extraction vis-à-vis the stress change caused by tectonic motion, and their role in triggering earthquakes in the geothermal area. Ongoing detailed monitoring studies will throw more light on the behavior of the Cerro Prieto geothermal reservoir.

Chapter 8

WORLDWIDE STATUS OF GEOTHERMAL RESOURCE UTILIZATION

INTRODUCTION

Geothermal resources vary widely from one location to another, depending on the temperature and depth of the resource, the rock chemistry and the abundance of groundwater. Utilization of geothermal resources can broadly be classified into electric power generation and non-electric use. The type of the geothermal resource determines the method of its utilization. High-enthalpy resources, such as dry steam and hot fluids that are found in volcanic regions and island chains, can be gainfully utilized to generate electric power. Modern technologies allow generation of electric power even from medium-enthalpy resources using binary cycle plants. On the other hand, moderate-to-low temperature resources such as warm-to-hot waters that are found extensively in most continental areas are best suited for direct heating and cooling (non-electrical) purposes. Exhaustive summaries have been provided by several authors, both for geothermal power plant technologies (Wood, 1973; Armstead, 1976b; Hudson, 1995; Barbier, 2002; Clauser, 2006) and non-electrical uses (Dickson and Fanelli, 1995; Lienau, 1995; Lund et al., 1998, 2005a). A historical perspective of geothermal power production over the past about 100 years has been succinctly summarized by Lund (2004). In the following, we shall briefly describe the various uses of geothermal resources, the methods of utilization and their underlying principles, and the status of worldwide utilization of geothermal energy.

ELECTRIC POWER GENERATION

Geothermal power plants in operation at the present time are essentially of three types: "dry" steam, flash and binary. A "dry" steam reservoir (such as the Geysers in United States of America and the Larderello in Italy) produces steam but very little water. The steam is piped directly into a "dry" *steam power plant* to provide power to run the turbines (Fig. 8.1). The spent steam (condensed water) can be used in the plant's cooling system and injected back into the reservoir to replenish water and pressure levels. A hot water reservoir (such as the Wairakei in New Zealand) is used in a *flash power plant*, in which hot fluids with temperatures usually in excess of

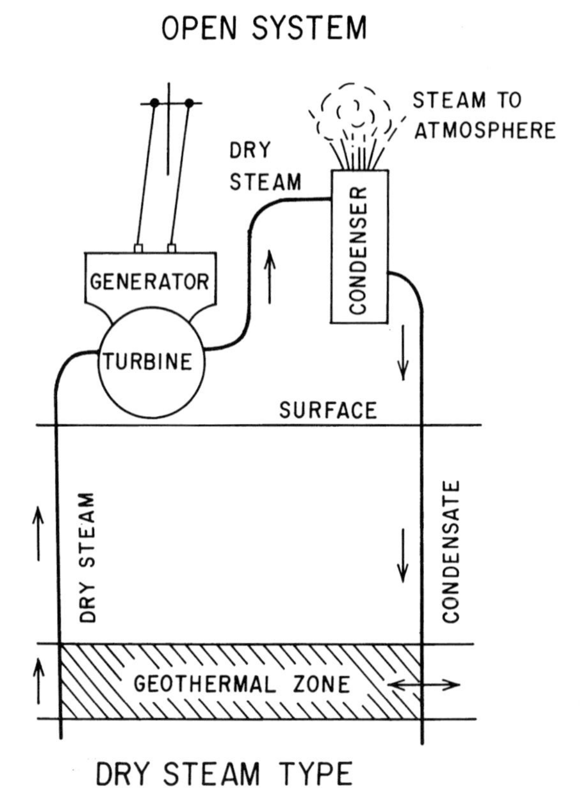

Fig. 8.1. Working principle of a dry-steam-type geothermal plant.

~180 °C are brought up to the surface through a production well where, upon being released from the pressure of the deep reservoir, some of the water flashes into steam in a "separator". The steam then powers the turbines (Fig. 8.2). The steam is cooled and condensed and either used in the plant's cooling system or injected back into the geothermal reservoir. Flash-steam plants are the most commonly used for electric power generation mainly because most geothermal reservoirs are formed by liquid-dominated hydrothermal systems.

Low-to-medium enthalpy reservoirs with temperatures between about 85 and 150 °C are not hot enough to flash enough steam but can still be used to produce electricity in a binary *power plant* (also, referred to as *Organic Rankine Cycle* (*ORC*) power plant). In a binary system, the geothermal fluids are passed through a heat exchanger, where their heat is transferred into a low-boiling point binary (secondary) liquid such as propane, isobutane, isopentane and ammonia. When heated, the binary liquid flashes into vapor, which, like steam, expands and powers the turbines (Fig. 8.3). The vapor is then re-condensed to a liquid and is used repeatedly.

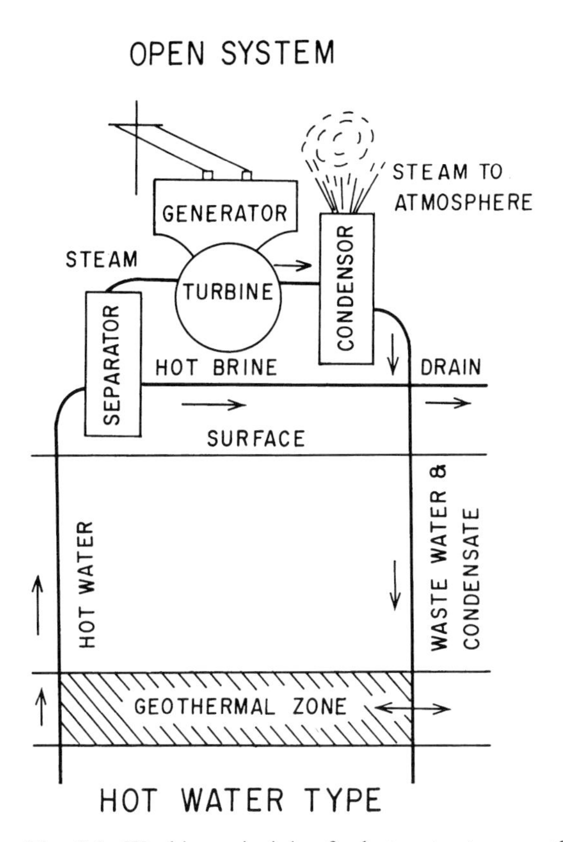

Fig. 8.2. Working principle of a hot-water-type geothermal plant.

Unlike the dry-steam and flash-steam plants, the water or steam from the geothermal reservoir never comes in contact with the turbine and generator units. This relatively recent technology has resulted in utilization of geothermal resource from smaller reservoirs worldwide using low-capacity binary units and proper selection of a stable binary fluid. The other advantage with such small units is that they can be gainfully cascaded for optimum utilization of resources wherever feasible and depending on energy demands. A variant of the binary cycle technology, known as *Kalina thermodynamic cycle* (Kalina, 1984), potentially yields higher thermal efficiency as compared to the ORC. The Kalina cycle uses a two-component vapor containing typically 70% ammonia and 30% water as the working fluid. The improvement over the ORC is that the boiling of a mixture of ammonia and water occurs over a range of temperatures, unlike steam and pure fluids that evaporate at a specific boiling temperature. The Kalina cycle technology is presently undergoing active testing in Iceland (DiPippo, 2004), and it will take some more time to demonstrate the actual improvement in efficiency resulting from its use.

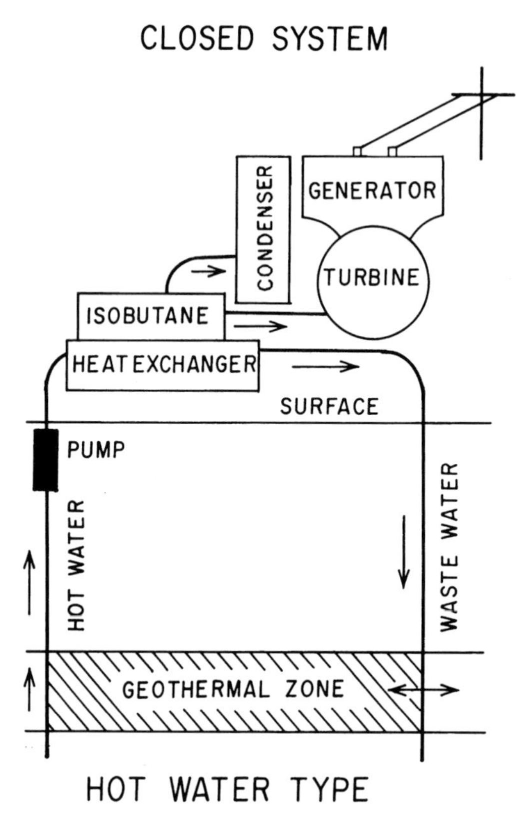

Fig. 8.3. Working principle of a binary power plant, using isobutane as a secondary working liquid.

Efficiency of Electricity Generating Cycles

In general, geothermal power plants have lower efficiency when compared with fossil-fueled and hydropower plants primarily due to (i) low temperature of the steam, which is usually much below 250 °C and (ii) presence of non-condensable gases such as CO_2, H_2S, NH_3, CH_4, N_2 and H_2 in the steam, which have to be removed from the condensers of power plants (Barbier, 2002). The efficiency of geothermal power plants depends upon the type of the power generating cycle used, which in turn is determined on the basis of the nature (dry steam or hot water) and temperature of the geothermal resource. An up-to-date summary of efficiencies generally achieved with commonly used geothermal power generating cycles is given by Barbier (2002).

Steam-dominated generating cycles have the highest efficiencies ranging from 10% to 17%, due to the high enthalpy of steam (typically, 2,800 kJ kg^{-1}). When geothermal steam contains substantial amounts (greater than 15% in weight) of non-condensable gases, it is preferable to use *non-condensing cycles* where steam from the production well is passed through a turbine unit and exhausted to the atmosphere. Such cycles consume about 15–25 kg of steam per kW h of power generation. On the other hand, when the gas content of steam is less than about 15% in weight, *condensing cycles* are employed to condense the used steam for re-injection into the reservoir. The steam consumption in this case is lower than that for non-condensing cycles, typically 6–10 kg of steam per kW h power generated, with lower consumption for higher turbine inlet pressures and vice versa.

Hot-water-based geothermal plants using the *binary cycle* have much lower efficiency, typically 2.8–5.5%, when compared with steam-dominated generating cycles. However, the latter can be installed only at very few places worldwide because vapor-dominated geothermal resources are rare, whereas low-enthalpy hot water resources are abundant and therefore can be effectively used to generate electric power in smaller quantities but at a large number of sites. Modern technologies have made small-capacity binary power-generating units readily available on a commercial basis. When appropriate, a number of small binary cycle units can be cascaded for obtaining larger production rates.

Geothermal Power Plants vis-à-vis Fossil-Fueled and Hydropower Plants

Geothermal power plants offer several distinctive features compared to the conventional fossil-fuel power plants and hydropower plants (Energy Alternatives: A Comparative Analysis, 1975; Dickson and Fanelli, 1995). Some of these are listed as follows:

(1) No combustion of fuel is involved in a geothermal plant, which in addition to saving infrastructure costs, also result in minimization of environmental pollution.
(2) Low temperature and pressure of steam result in lower plant efficiencies (typically ~15%), contrasting with comparatively higher (35–38%) for fossil-fueled plants.
(3) Long and complicated start-up procedures render geothermal power units more suitable as base-load units than as peak-load units, although availability of pre-assembled and factory tested modular units over the past few years have reduced the start-up times significantly. Base-load power is the power that is delivered at all times throughout the day, as against peak-load power that is used only at peak-consumption hours of the day.
(4) Because transportation of hot water and steam over long distances through pipelines results in substantial loss of heat, geothermal plants need to be installed close to the site of steam or hot water production.
(5) In general, a 100 MWe capacity geothermal power plant typically requires 80 tons of steam every hour. This rate is usually achieved from multiple production wells tapping the same reservoir.

(6) Steam and cooling water mix directly in the contact condensers.
(7) Usually, no external make-up water is required for cooling. The cooling tower evaporation rate is exceeded by the steam flow to the turbines. Thus, condensed exhaust is used as cooling tower make-up water.
(8) Non-condensable gases are released into the atmosphere from the condenser and from the cooling tower. At the Geysers, California, 75–80% of the condensed steam evaporates in the cooling tower and 20–25% is re-injected into the steam-bearing formations.
(9) The water or brine in the hydrothermal system is passed through a separator, which draws off steam to run the turbine, and then routed to re-injection wells. At times, additional water from cooling towers also needs to be disposed off. This is done either through re-injection or diverting it to a river or some other suitable water body.
(10) The steam has a fair amount of minerals, which cause erosion and corrosion of the turbine. This requires continuous and extensive maintenance.
(11) Initial expenditure of geothermal power plants are higher because the geothermal wells and power plants must be constructed at the same time. However, over a period of time, it decreases substantially because the costs and availability of steam and hot water are stable and predictable.
(12) Unlike hydropower plants, geothermal plants are usually set up as small units generating electric power as little as 1 MW and up to a few tens of megawatts at the most. Thus, geothermal power production can be controlled as per energy requirements: when the demand is low, extraction of hot fluids can be restricted by using fewer production wells; when the demand is high, extraction of fluids can be upscaled by using greater number of production wells and cascading multiple geothermal power plant units.

NON-ELECTRICAL USES

There is a large potential for development of direct use of geothermal energy in many parts of the world. Direct use of geothermal energy refers to the use of the heat energy of low-to-moderate temperature geothermal waters without conversion to some other form of energy such as electrical energy. Rising prices of oil over the past few years, the rapid increase in atmospheric CO_2 concentration resulting from burning of fossil fuels, and the extensive fallout of both the factors on the world economy has generated renewed interest in efficient utilization of low-enthalpy geothermal resources as an energy alternative for several nonelectrical applications such as space heating, greenhouse and aquaculture facilities, heat pumps and many industrial applications. The long-term economics of using geothermal resources for nonelectrical uses work out to be more attractive when compared with the requirements of conventional fuel resources in well-endowed geothermal regions, especially in the cold-climatic regimes of middle- and high-latitude belts. According to recent world estimates, the equivalent savings in fuel oil add up to about 25 million tonnes per year and about 24 million tonnes in carbon emissions to the atmosphere per year (Lund et al., 2005a). Therefore, although the use of high-enthalpy geothermal resources for generation of electric power continues to be more popular, the economic as well as environmental benefits of using moderate-to-low enthalpy fluids to meet domestic heating, agricultural and several industrial energy needs should not be ignored today.

Direct use of geothermal waters turns out to be a more efficient process relative to generation of electric power from geothermal resources because the losses incurred in the former are not imposed by the laws of thermodynamics. However, one difficulty with direct use is the fact that low-temperature geothermal waters cannot be transported to large distances without substantial heat loss. In these usages, heat is lost primarily due to inadequate insulation, low flow rates, and terminal temperature differences in heat exchangers and drains.

The basic components of a low-enthalpy thermal distribution system are (i) assembly of transmission and distribution piping with pipelines of varying diameters depending upon flow rates, (ii) downhole and circulation pumps, along with regulators, valves, expansion joints, etc., and (iii) suitable heat exchangers for extracting heat from the warm geothermal waters. An up-to-date review of engineering and design practices is given in Lund et al. (1998). A critical requirement in the system is adequate thermal insulation to prevent excessive heat loss and temperature drop in the fluid. There is a continuous decline in fluid pressure along the pipelines due to viscous friction losses. Changes in elevation along the pipeline affect the hydrostatic head. Therefore, to prevent pressure from dropping below a value that would result in local boiling, and to maintain sufficient flow throughout the entire network, pumps are essential. Surge tanks are required to prevent sudden and potentially damaging changes in the pressure due to sudden changes or surges in the flow.

Impurities, such as hydrogen sulfides, carbon dioxide, chlorides, silica, bicarbonates and entrained sand particles cause corrosion, erosion and scaling in the distribution pipelines. Chemical corrosion and/or mechanical erosion can cause pipeline failure, whereas scaling increases pumping requirements as well as disrupting the operation of heat exchangers. Depending upon the severity of impurities, it may be necessary to provide special lining material for the pipeline.

Necessary allowance also needs to be made to accommodate axial expansion of metallic pipes due to high temperatures of the fluid. Excessive stresses caused by axial expansion could cause failure in pipeline, supports, joints or anchors. Conventional expansion devices; such as slip joints, bellows, U- or Z-shaped expansion joints; placed at regular intervals along the pipeline provide adequate protection. At the present time, the most commonly used piping materials are carbon steel (which can withstand fluid temperatures exceeding 100 °C), fiberglass-reinforced plastic, polyvinyl chloride and asbestos cement.

Various methods are in use to provide insulation for the pipeline system. These range from placing the prefabricated pipe within another pipe with insulation in between, to burying the pipe in trenches under the Earth. The insulation material must be waterproof and watertight. Carbon steel pipes are insulated with polyurethane foam, rock wool or fiberglass. Conventional steel pipes are usually wrapped with polyvinyl chloride before burying in the ground. The nature and thickness of insulation depends on many factors such as the temperature of the fluids, flow rates, transmission length, type of soil and local water table.

Many direct-use applications require geothermal fluids to be transmitted several kilometers from the site of its production. In such cases, it is necessary to estimate the temperature drop as a function of distance of transmission, flow rate, fluid temperature, etc. In general, with flow rates varying between 5 and 15 lps in a pipe of diameter 0.15 m, temperature drops in the range 0.1–1.0 °C km^{-1} in insulated pipelines and 2–5 °C km^{-1} in uninsulated pipelines are common (Ryan, 1981). Temperature losses are lower for larger diameter pipes or higher flow rates. As an example, plot of the temperature drop as a function of distance in a 0.45 m diameter pipe covered with 0.5 m urethane insulation is shown in Fig. 8.4 (Howard, 1975). For a soil temperature of 16 °C and flow rate of 1.5 m s^{-1}, the fluid temperature drops from 100 to 80 °C after moving a distance of 100 km.

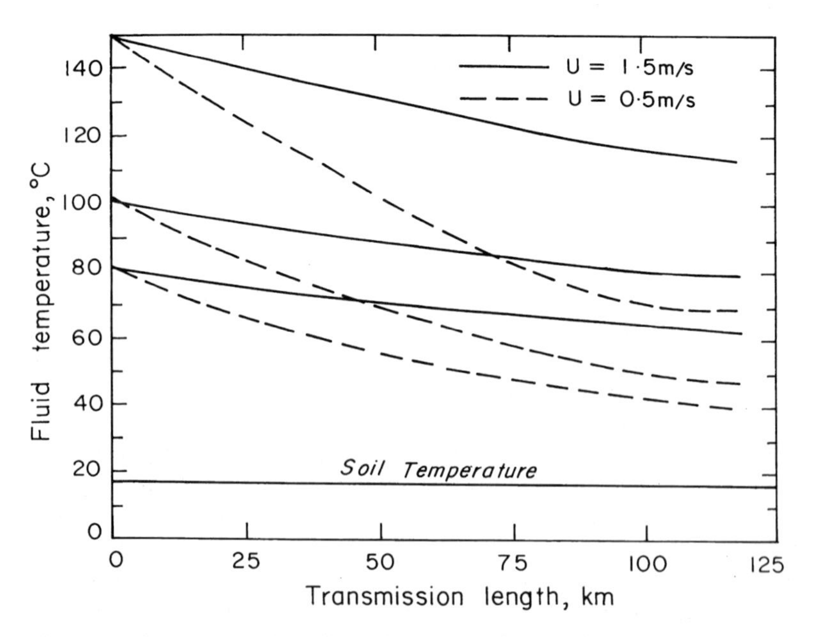

Fig. 8.4. Diagram showing effect of distance of transmission on fluid temperature in an 18″ diameter pipe insulated with 2″ thickness of urethane. The soil temperature is 16 °C. Two flow rates, 1.5 and 0.5 m s^{-1}, have been considered (modified after Howard, 1975).

Applications

Direct uses of geothermal resources span a wide variety of applications, depending upon the temperature of the resources, socio-economic conditions of the individual areas possessing such resources, and the energy needs of local people living in these areas. Direct utilization of the heat extracted from the geothermal waters can be broadly classified into three categories. These are as follows:

(1) Residential and commercial use: (a) space and district heating, space cooling, heat pumps, (b) water (potable, hot or cold utility, etc.), (c) bathing, swimming and balneology, (d) refrigeration, (e) de-icing, and (f) waste disposal and bio-conversion (i.e., extraction of methane, ethanol and other chemicals by anaerobic digestion and fermentation of municipal solid waste, thereby reducing the volume requiring disposal).

(2) Agriculture and related use: (a) animal husbandry, (b) aquatic farming, (c) greenhouse heating, (d) agricultural product processing such as drying, fermentation, waste disposal and conversion and canning.

(3) Industrial use: (a) pulp, paper and wood processing, (b) heap leaching for recovery of gold, silver and other minerals, (c) waste water treatment, (d) production of diatomaceous Earth (a naturally occurring substance comprised of fossilized remains of microscopic hard-shelled creatures found in marine as well as fresh waters; used as insect deterrents).

Lindal (1973) suggested a broad classification of the various direct uses of geothermal fluids on the basis of their temperature requirements (Fig. 8.5). The Lindal diagram, as it has been widely referred to in literature, is more or less valid even to this day and serves as a rough illustration of the temperature requirements. It must be noted that many of the applications are practiced over a range of temperatures rather than a single temperature as suggested in the diagram. It can be seen from the Lindal diagram that, in general, the agricultural and aquacultural applications require the lowest temperatures, followed by space heating and industrial applications. Few additional industrial applications with specific temperature requirements have been described by Lienau (1995).

One very common and extensive application in recent times has been in *geothermal (ground source) heat pumps*. A geothermal heat pump makes use of the relatively stable temperature at a depth of a few meters in the ground. During winter, the subsurface temperature is warmer than the room temperature inside a house, whereas during summer the subsurface temperature may be cooler. Therefore, heat pumps can be used both for heating as well as cooling. The advantage of ground source heat pumps relative to air source heat pumps is that the difference of room temperature from the ground temperature is always smaller than the difference of the room temperature from the outside air temperature. Therefore, ground source heat pumps need to do less work than air source heat pumps. Heat pump systems use groundwater aquifers and soil temperatures in the range 5–30 °C. A comprehensive review of the working of ground source heat pumps, efficiency and

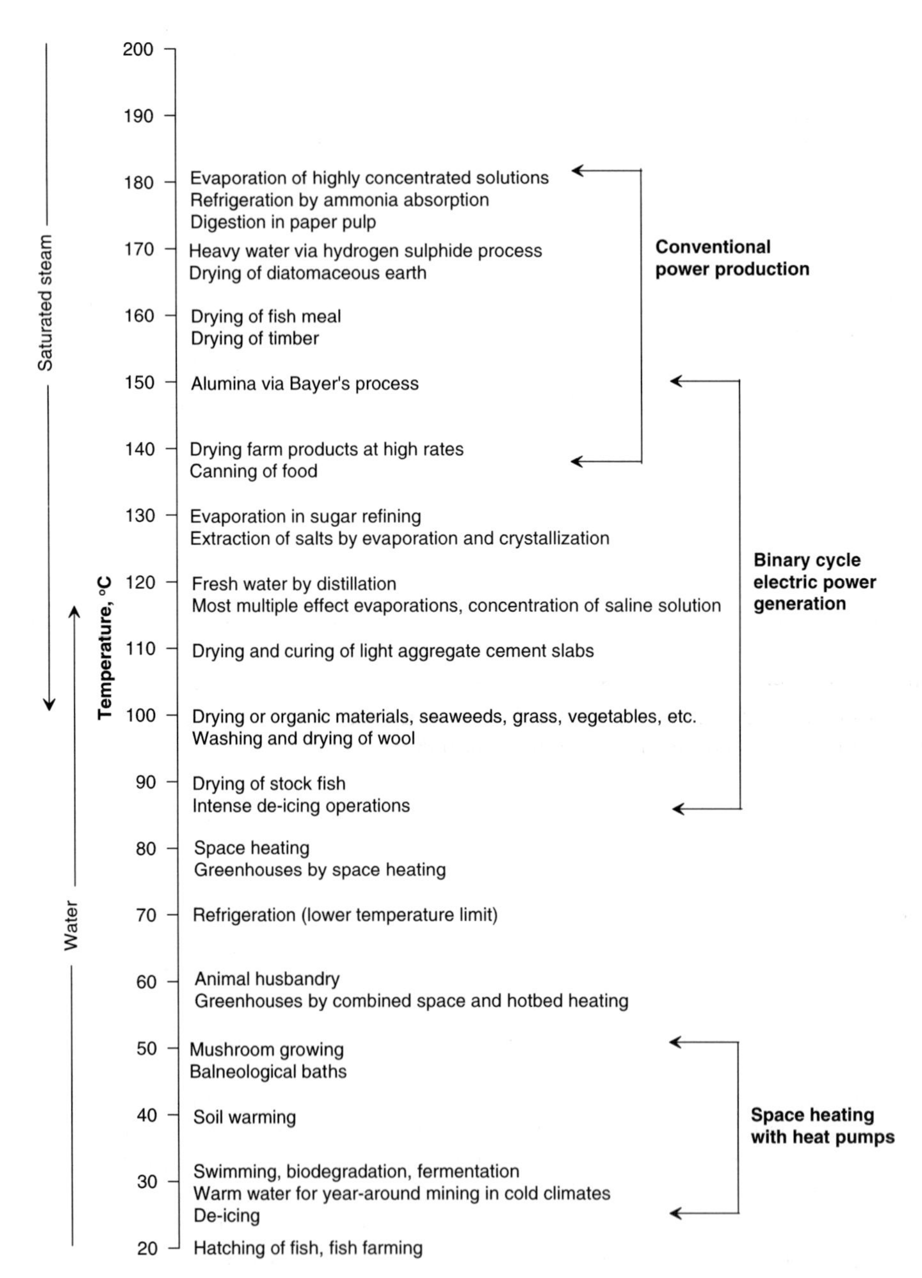

Fig. 8.5. Illustration showing general temperature requirements of a spectrum of direct-use applications of geothermal resources. Majority of these applications are practiced within a range of temperatures; the ranges for individual applications are not shown here to maintain clarity. The broad temperature ranges for electric power generation from geothermal resources are also indicated (modified after Lindal, 1973).

cost considerations is given by Clauser (2006). Heat pumps utilizing very low-temperature fluids have extended geothermal developments to countries that have not been using geothermal energy extensively such as France, Switzerland, Sweden and areas of the mid-westem and eastem United States of America.

GLOBAL SCENARIO

Utilization of geothermal energy resources for both electric power generation as well as direct use has become imperative in view of the rapidly rising energy demands and costs of energy in modern society. High-enthalpy resources, which are restricted to certain specific tectonic provinces only, are being exploited for generating electric power. Even relatively low-enthalpy resources are being developed for power generation using binary cycles and for other direct uses. Use of combined heat and power plants (CHP) is being promoted at several low-enthalpy geothermal locations. In these plants, hot waters with temperatures up to 100 °C are first utilized for power generation using binary cycle plants and then cascaded for low-temperature, direct-use applications such as space heating, greenhouse heating, aquaculture, bathing and swimming. Geothermal heat pumps are now extensively used for heating and cooling houses in the United States of America, northern Europe and China. Although the accumulated energy actually generated by using heat pumps appears to be miniscule, the impact of their use on the demands of energy for heating and cooling is substantial. Enhanced geothermal systems, although not yet commercially viable, holds considerable promise in catering to small-scale energy needs of the people. A welcome trend has been the active participation by industries, particularly from the private sector, in developing geothermal fields. From the environmental angle, usage of the relatively clean geothermal resource earns substantial "carbon-credits" when compared with other fossil-fueled power plants. The earned carbon-credits can be traded commercially, i.e., operators of high-CO_2-emission power plants in countries, that have signed the Kyoto Protocol, can purchase from low-CO_2-emission power plants a "carbon-credit" proportional to their respective CO_2 emissions, thereby escaping legal and financial penalties. It is therefore not surprising that researches focusing on new exploration and exploitation techniques toward optimum utilization of available geothermal potential are being promoted in several countries.

Today, 24 countries are engaged in producing electric power from geothermal energy resources. The total installed capacity is about 8,900 MWe from 468 power-generating units (Bertani, 2005). Power generation from geothermal resources has witnessed a steady, albeit modest, growth over the past three decades (Fig. 8.6). The present-day annual electrical energy produced from geothermal resources is about 57,000 GW h yr^{-1}, which works out to be less than 0.4% of all electricity generation worldwide. Therefore, the contribution of geothermal power to the overall world electricity generation is very modest. At a local level, however, geothermal power contributes quite significantly in a few developed as well as developing countries

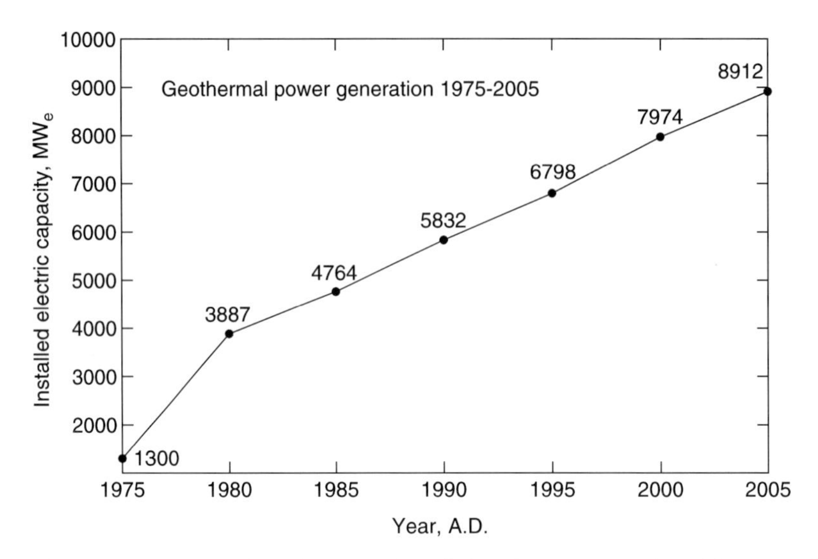

Fig. 8.6. Growth of installed capacity of electric power-produced from geothermal energy resources worldwide during the period 1975–2005 (Data source: Bertani, 2005).

(Fig. 8.7). In contrast, direct use of geothermal energy resources play an important role in several countries. According to available reports, 72 countries are utilizing this energy resource for a wide variety of applications with a crude estimate of about 28,000 MWt installed capacity. This amounts to an annual energy-saving equivalent to about 25 million tonnes of oil and a carbon emissions reduction of 24 million tonnes in to the atmosphere (Lund et al., 2005a). Fig. 8.8 shows the approximate utilization of geothermal resources for a variety of direct uses.

We will now briefly discuss the present-day scenarios for utilization of geothermal energy resources in a few countries in which installed geothermal electric capacity is about 100 MWe or more. The data sets used in making the summaries have been obtained primarily from the proceedings of the World Geothermal Congress held in Turkey during the year 2005, which contain detailed accounts of utilization of geothermal energy resources in individual countries.

United States of America

Geothermal fields in the United States of America (USA) are mainly concentrated in its western parts comprising the states of California, Nevada, Wyoming, Utah, Oregon, Idaho and New Mexico, and in the states of Alaska and islands of Hawaii (Fig. 8.9). A large part of the western United States is associated with high heat flow ranging from 60 to more than 100 $mW\,m^{-2}$ (Lachenbruch and Sass, 1977; Blackwell and Steele, 1992; Sass et al., 2005). The total installed capacity for power generation is about 2,534 MWe, the largest in the world (Lund et al., 2005b). Of this, the fields in

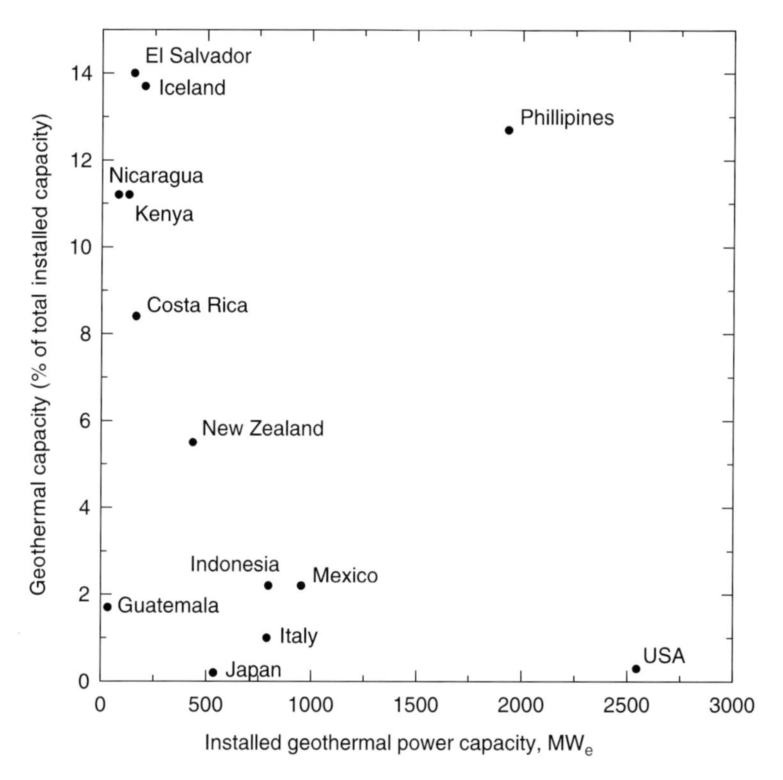

Fig. 8.7. Contribution of installed geothermal electric capacity to overall national electric power capacity (all sources) for major geothermal power-producing countries (Data source: Bertani, 2005).

California account for a whopping 88.5%, Nevada 9.5%, Utah 1% and Hawaii 1% (Table 8.1). An additional capacity of 632 MWe has been planned, with California accounting for 460 MWe. Extensive research and development activities have been carried out in the country as a part of federal sponsored exploration and exploitation programs over the past five decades (Muffler, 1979; Duffield and Sass, 2003; Hill, 2004; Lund et al., 2005b). A relatively recent trend has been the participation by the private power companies in a big way: from development of identified geothermal fields through power generation and distribution.

The Geysers geothermal field in California is the largest producing field in the world with an installed capacity of 1,421 MWe. The other fields in California are the Imperial Valley (comprising Salton Sea, East Mesa and Heber), Coso, Long Valley (Casa Diablo), Honey Lake Valley and Glass Mountain. At the Geysers, declining steam pressures and plant shutdowns since the late 1990s, has resulted in a 400 MWe reduction in installed capacity. This has been partially overcome through a re-injection experiment that started in 1998. Injection of recycled wastewater obtained

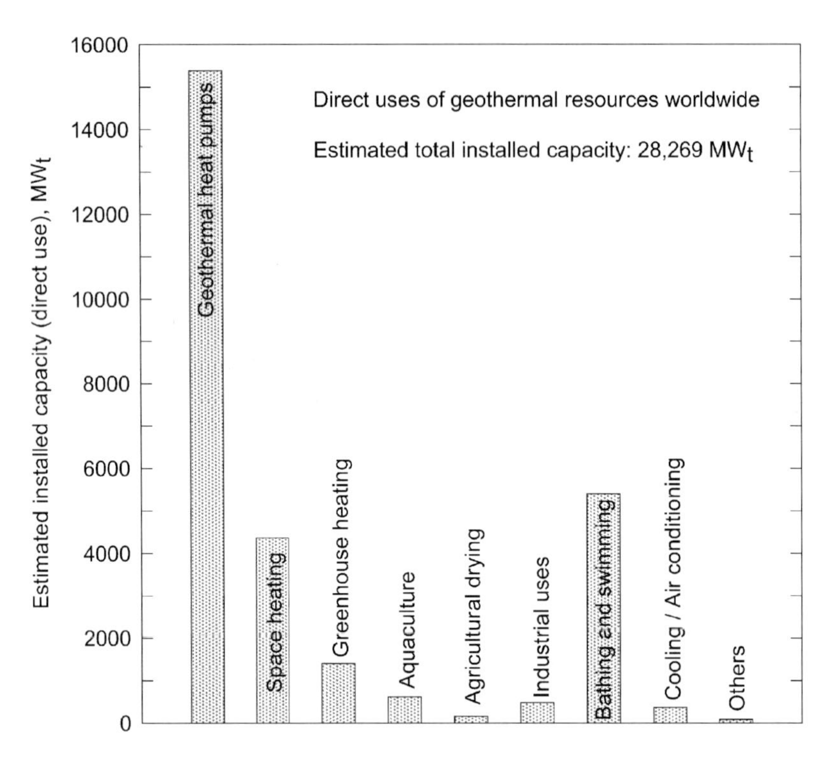

Fig. 8.8. Estimated installed capacities for utilization of geothermal energy resources for direct uses (Data source: Lund et al., 2005a).

from two nearby communities into the reservoir has made it possible to recover about 100 MWe of power-generating capacity (Lund et al., 2005b). With another pipeline ready for bringing in wastewater from Santa Rosa, the capacity of The Geysers is expected to improve further. Overall, geothermal power meets nearly 6% of the all energy requirements in California, which is a very significant contribution considering high energy requirements due to a large population and heavy industry in the state. In Big Island, Hawaii, geothermal power meets about 25% of its energy needs. On a national level, however, geothermal power accounts for less than 1% of the USA energy requirements. Even with such a small contribution, the total energy savings for the nation is estimated to be around 4.5 million tonnes of fuel oil per year and a corresponding reduction in air pollution to the tune of 4 million tonnes per year (Lund et al., 2005b).

The USA leads the world in direct-use applications of geothermal resources with an estimated installed capacity of 7,817 MWt. Owing to a lack of adequate records for the various uses throughout the country, this estimate should be taken as an indicator only. Ground source heat pumps, which have been employed extensively in the mid-west and eastern parts of the country for space heating and cooling, account

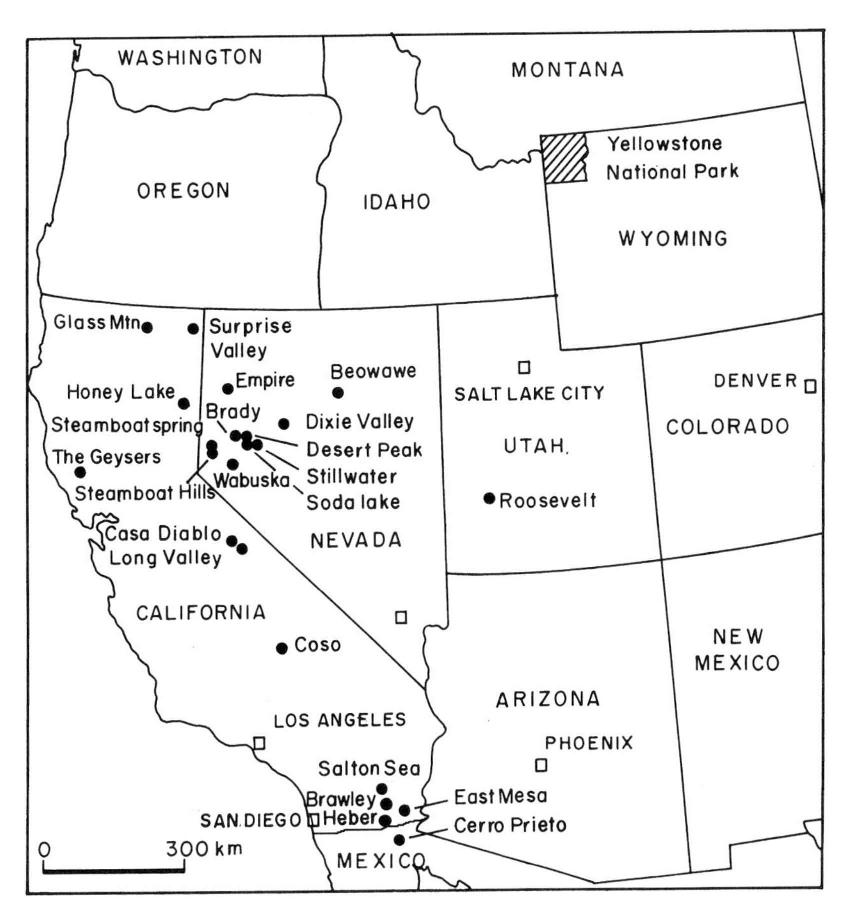

Fig. 8.9. Map showing the locations of geothermal power-producing plants (filled circles) in the United States of America and northern Mexico (modified after Bertini, 2005; Lund et al., 2005b).

for about 70% of direct-use applications. District heating and agriculture drying constitute other major forms of utilization. This is in addition to greenhouse heating, heating of pools and spas, aquaculture, snow melting and a variety of industrial applications (Lund et al., 2005b). Direct utilization of geothermal energy that are practiced in the USA clearly demonstrate the immense potential of further savings in terms of fossil fuel-driven energy sources as well as substantial reduction in carbon dioxide levels in the atmosphere. It also indicates how valuable these savings can be for smaller developing economies that have adequate geothermal resources, such as Guatemala, Honduras, El Salvador, Nicaragua and Costa Rica, located in and around the "Pacific Ring of Fire".

Table 8.1. Electric power generation from geothermal resources in United States of America (after Lund et al., 2005b)

State	Geothermal field	Installed capacity (MWe)
California	The Geysers	1,421.0
	East Mesa, Imperial Valley	79.0
	Heber, Imperial Valley	85.0
	Salton Sea, Imperial Valley	336.0
	Honey Lake Valley	3.8
	Coso	274.0
	Casa Diablo (Mammoth/Pacific)	40.0
Nevada	Beowawe	16.6
	Brady	21.1
	Desert Peak	12.5
	Dixie Valley	62.0
	Empire	4.8
	Soda Lake	26.1
	Steamboat	58.6
	Stillwater	21.0
	Wabuska	2.2
	Steamboat Hills	14.4
Utah	Roosevelt	26.0
Hawaii	Puna	30.0
	Total	2534.1

Phillipines

High-temperature (>250 °C) geothermal systems in the Phillipines are associated with Quaternary to Recent volcanism. Detailed information on development of geothermal resources in Phillipines is given in Sussman et al. (1993), and a recent update is provided by Benito et al. (2005). The assemblage of island arcs comprising the Phillipine archipelago have been accreting between the westward subducting Pacific plate (Phillipine trench) and the eastward subducting South China Sea plate (Manila trench). The opposing subduction zones have been mostly responsible for creating a discontinuous belt of Pliocene to Quaternary volcanoes all along the length of the archipelago. The Phillipine fault, a long and active strike–slip fault, occurs between the two active subduction zones. The very high helium isotopic ratios ($^3He/^4He$) found in gas samples suggest an upper mantle origin for helium and its association with magmatic heat sources (Giggenbach and Poreda, 1993). The unique geological setting of Phillipines has contributed immensely to the occurrence of abundant, high-enthalpy geothermal energy resources in the country. Locations of major geothermal fields along with the geological and tectonic settings are shown in Fig. 8.10.

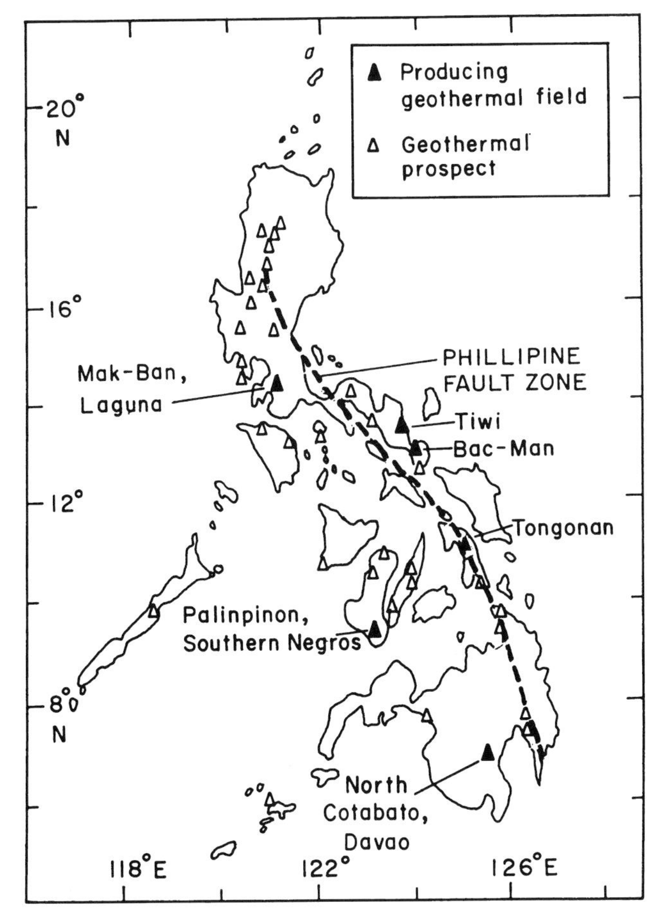

Fig. 8.10. Map showing the locations of geothermal power-producing plants (filled triangles) and major prospects (unfilled triangles) in the Phillipines (Source of data: Benito et al., 2005).

Generation of electric power from geothermal fluids started some 40 years ago. Today, the total installed electric power generating capacity of about 1,931 MWe from geothermal resources accounts for about 13% of the total installed electric capacity and almost 19% of the total energy mix of the country. Phillipines is next only to United States of America in installed geothermal capacity. With a total estimated geothermal potential of around 4,335 MWe and aggressive geothermal exploration programs planned in the next 10 years, Phillipines may soon become the largest producer of geothermal power in the world.

Tongonan field is the largest producer of geothermal power, followed by Mak-Ban, Tiwi, Palinpinon, Bac-Man and North Cotabato (Table 8.2). There are a large number of other potential fields; these are being taken up for further exploration and

Table 8.2. Electric power generation from geothermal resources in Phillipines (after Benito et al., 2005)

#	Geothermal field	Installed capacity (MWe)
1	Tiwi, Albay	330.0
2	Makiling Banahaw (Mak-Ban), Laguna	425.7
3	Tongonan, Leyte	722.7
4	Palinpinon, Southern Negros	192.5
5	Bacon-Manito (Bac-Man), Albay, Sorsogon	151.5
6	North Cotabato, Davao	108.5
	Total	1930.9

production. Direct uses of geothermal energy are limited mostly to agricultural drying; however their potential to replace the usage of expensive and polluting fossil fuels are being widely recognized.

Mexico

Geothermal energy resources in Mexico are predominantly used for generation of electric power. Four producing fields, Cerro Prieto, Los Azufres, Los Humeros and Las Tres Virgenes, with a total installed capacity of 953 MWe contribute to making Mexico the third largest geothermal power producer after the United States of America and the Phillipines (Gutiérrez-Negrin and Quijano-León, 2005). The geothermal power accounts for about 3% of the electric energy produced in Mexico. The four fields together produce 7,700 metric tonnes of steam per hour from 197 wells, which is then fed to as many as 36 power plants. The steam is mixed with brine, which is either injected back into the reservoirs or treated in a solar evaporation pond at Cerro Prieto. A fifth field, La Primavera, with an estimated potential of 75 MWe is expected to start production in the next few years. A detailed update of geothermal energy utilization in Mexico is given by Gutiérrez-Negrin and Quijano-León (2005).

The locations of the geothermal fields are shown in Fig. 8.11, and the installed capacities for power generation plants are summarized in Table 8.3. The Cerro Prieto, located close to the northern border of Mexico with the United States, is the largest producer of electricity with a generating capacity of 720 MWe. The remaining three fields, Los Azufres, Los Humeros and La Primavera are located within the Pliocene–Quaternary volcanic belt.

Indonesia

Indonesia is richly endowed with both dry-steam and hot-water geothermal resources associated with numerous active volcanoes all along its 7,000 km long plate

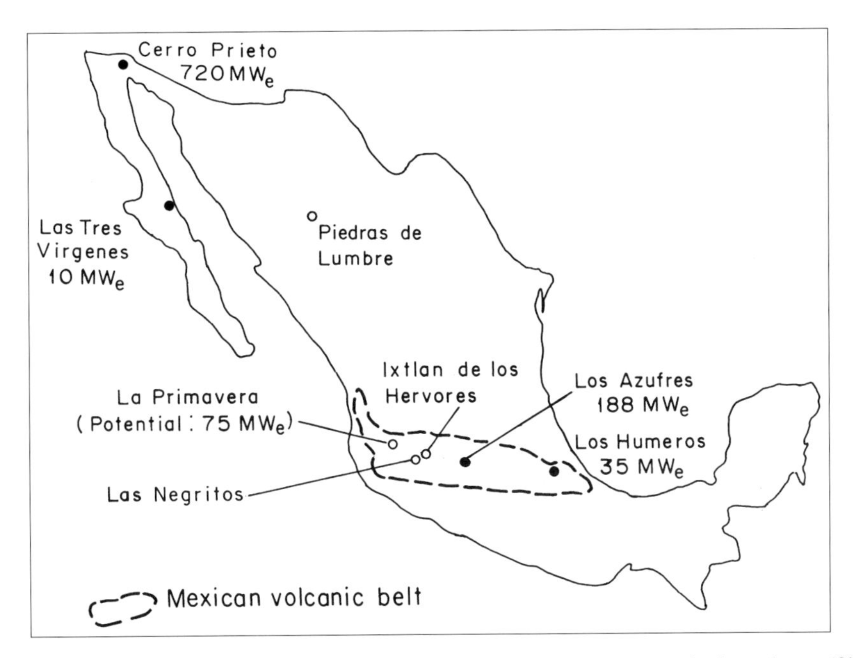

Fig. 8.11. Map showing the locations of geothermal power-producing plants (filled circles) and geothermal prospects (open circles) in Mexico (modified after Gutiérrez-Negrin and Quijano-León, 2005).

Table 8.3. Electric power generation from geothermal resources in Mexico (after Gutiérrez-Negrin and Quijano-León, 2005)

#	Geothermal field	Installed capacity (MWe)
1	Cerro Prieto	720
2	Los Azufres	188
3	Los Humeros	35
4	Las Tres Virgenes	10
	Total	953

boundary with the Indo-Australian plate to its south. Both convergent as well as strike–slip movement along the plate boundary is believed to have caused large concentration of high-temperature geothermal systems close to the plate margins in Sumatra, Java, Nusa Tenggara, Sulawesi and Halmahera (Fig. 8.12). At least 70 high-enthalpy geothermal fields have already been identified, of which seven are under production (Table 8.4). The producing fields in Java islands are predominantly steam fields whereas those in Sumatra and Sulawesi are hot water fields. The total installed capacity is estimated at about 807 MWe, with additional capacities planned in the near future. However, further exploration and development programs would

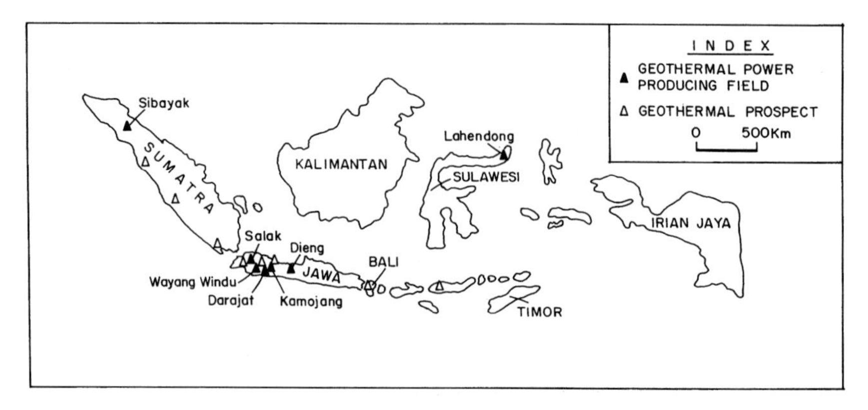

Fig. 8.12. Map showing the locations of geothermal power-producing plants (filled triangles) and major prospects (unfilled triangles) in Indonesia (modified after Sudarman et al., 2000).

Table 8.4. Electric power generation from geothermal resources in Indonesia (after Ibrahim et al., 2005)

Area	Geothermal field	Installed capacity (MWe)
Java	Kamojang	140
	Gunung Salak	330
	Darajat	145
	Dieng	60
	Wayang Windu	110
Sumatra	Sibayak	2
North Sulawesi	Lahendong	20
	Total	807

be required to utilize the vast geothermal energy potential available in the country, and contribute to its rapidly growing energy needs. Very few direct uses of geothermal energy are practiced in Indonesia.

Italy

Electric power generation from superheated steam at Larderello, the oldest producing geothermal field in the world, dominates the geothermal energy utilization scenario in Italy. Electricity was first produced from geothermal steam in 1904. A detailed update on utilization of geothermal energy in Italy is given by Cappetti and Ceppatelli (2005). The Larderello, with a total explored area of about 250 km^2 and containing 180 wells, along with the geothermal fields of Travale-Radicondoli and Mt. Amiata, amount to an installed capacity of about 790 MWe in Italy, making it the fourth largest geothermal power producer in the world. All the three fields are located in Tuscany area, where about 25% of the electricity requirements are met from geothermal power. However, geothermal power accounts for only about 1% of the energy needs of the country. Declining steam pressures at Larderello during the 1970s and 1980s have been largely controlled by re-injection of the steam condensate back into the reservoir and partly by deeper exploration programs. A further 60 MWe of power generation has been planned.

The major geothermal fields and the electric generating capacities of the installed power plants are listed in Table 8.5. The locations of the fields are shown in Fig. 8.13. In addition to power plants, active projects for direct use of geothermal resources for a wide variety of applications are in place. District heating, greenhouse heating, health spas and balneology, aquaculture, geothermal heat pumps and several

Table 8.5. Electric power generation from geothermal resources in Italy (after Cappetti and Ceppatelli, 2005)

#	Geothermal field	Installed capacity (MWe)
1	Larderello	542.5
2	Travale-Radicondoli	160.0
3	Mt. Amiata (Bagnore, Piancastagnaio)	88.0
	Total	790.5

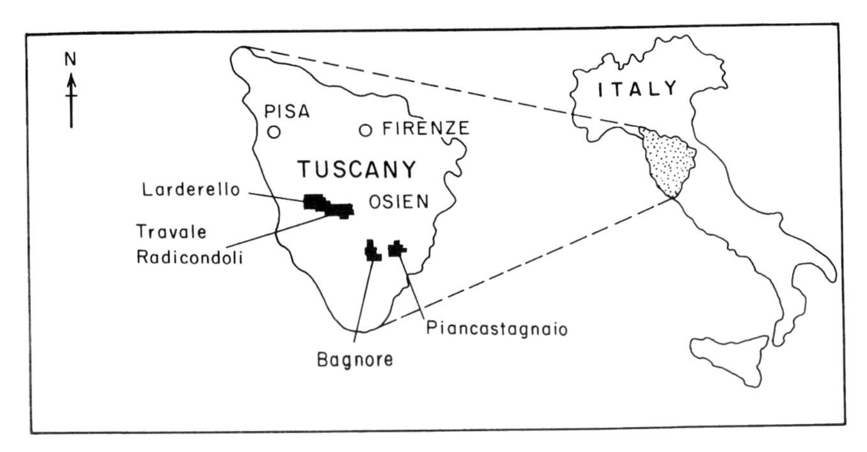

Fig. 8.13. Map showing the locations of geothermal power-producing plants (filled areas) in Italy (modified after Cappetti and Ceppatelli, 2005).

industrial applications utilize the heat available from spent steam as well as from low-enthalpy geothermal liquids in the Tuscany area (Lund et al., 2005a).

Japan

Japan has considerable geothermal resources which, while consisting mainly of wet steam, include an important dry-steam field at Matsukawa. It is tectonically one of the most active countries, with about 200 volcanoes of Quaternary age and abundant geothermal resources associated with them. Most of the geothermal fields are located along the volcanic front in eastern Japan, from Hokkaido through eastern Honshu island up to Izu islands. A few other fields are located in Kyushu district, which is a part of another volcanic front. A recent account of utilization of geothermal energy is provided in Kawazoe and Shirakura (2005). Locations of geothermal fields are shown in Fig. 8.14.

The total installed capacity for power production is about 535 MWe, which makes Japan the sixth largest producer of geothermal power. Hatchobaru, Kakkonda,

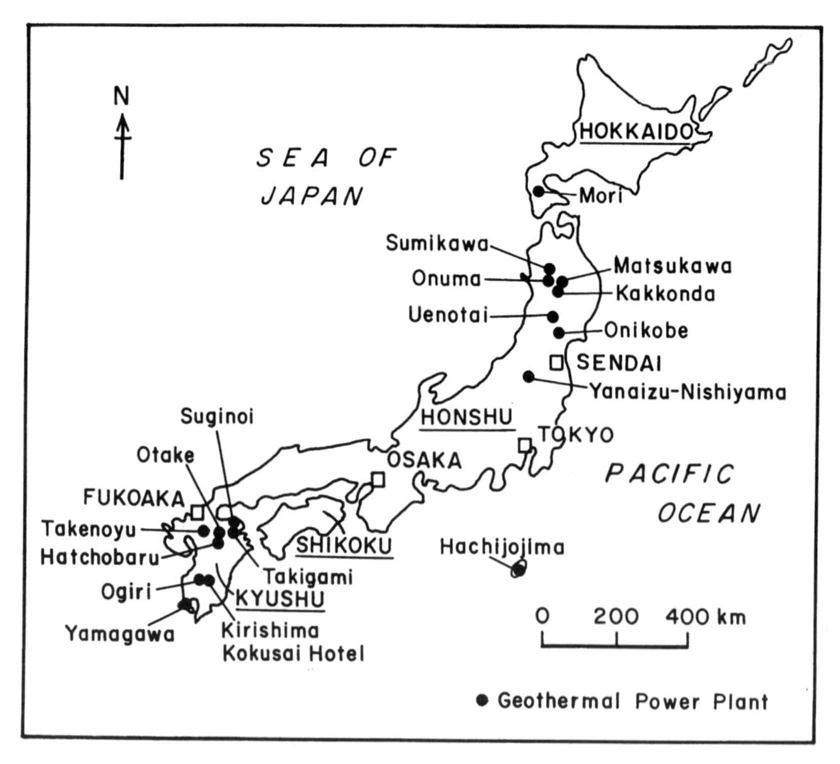

Fig. 8.14. Map showing the locations of geothermal power-producing plants (filled circles) in Japan (modified after Kawazoe and Shirakura, 2005).

Table 8.6. Electric power generation from geothermal resources in Japan (after Kawazoe and Shirakura, 2005)

Area	Geothermal field	Installed capacity (MWe)
Hokkaido	Mori	50.0
Akita	Sumikawa	50.0
	Uenotai	28.0
	Onuma	9.5
Iwate	Matsukawa	23.5
	Kakkonda	80.0
Fukushima	Yanaizu-Nishiyama	65.0
Miyagi	Onikobe	12.5
Oita	Otake	12.5
	Hatchobaru	110.0
	Suginoi	3.0
	Takigami	25.0
	Kujyukannko Hotel	2.0
Kumamoto	Takenoyu	0.05
	Kirishima Kokusai Hotel	0.1
Kagoshima	Yamagawa	30.0
	Ogiri	30.0
Tokyo	Hachijojima	3.3
	Total	534.45

Yanaizu-Nishiyama, Sumikawa and Mori are the largest power plants with individual capacities in excess of 50 MWe (Table 8.6). Even though geothermal power accounts for only a very small component, less than 0.5%, of the total electric power generation in Japan, it is considered very seriously as an alternative and environmentally benign source of energy. As part of a new policy initiated by the Japanese government in 2003, every electric power supplier is required to supply a certain quantity of power generated from renewable energy sources in proportion to their total sales.

Another recent trend in Japan has been in setting up of small capacity (a few megawatts each) power plants close to the existing large plants. These plants cater to the domestic energy needs of the local people. The economics of installing and operating small-scale plants have been found feasible, mainly because of the low-exploration risk and low-initial costs.

Geothermal energy also finds use in recreational, agricultural, animal husbandry and space heating activities in Japan. Ground source heat pumps have been employed abundantly for space-heating and cooling in a large number of localities, thereby reducing the dependence on fossil fuels as well as providing for a cleaner environment, particularly in big cities.

New Zealand

The territory of New Zealand is endowed with a number of high-temperature geothermal resources, which together produce in excess of 435 MWe of electric power, constituting about 5.5% of total power production from all energy sources in the country (Dunstall, 2005). With an additional 69 MWe planned for production over the next 5–10 years, it is the seventh largest geothermal power producing country.

The Wairakei geothermal field continues to be the largest geothermal power producer over the past five decades. Other producing fields are located at Reporoa (Ohaaki), Kawerau, Rotokawa, Northland (Ngawha), McLachlan-Wairakei and Mokai (Table 8.7). There are several other hot water and steam occurrences with temperatures ranging from 45 to 140 °C that are currently being used for direct heating purposes such as paper and timber drying, greenhouse heating and geothermal heat pumps. The locations of the major geothermal fields are shown in Fig. 8.15.

Iceland

Iceland is located on the mid-Atlantic ridge, an active spreading ridge dotted with volcanoes and hot springs (Fig. 8.16). Not surprisingly, the geothermal fields in the country are intimately associated with this tectonically active, volcanic zone. High-temperature geothermal resources are abundant, accounting for over 50% of the country's primary energy needs. With a small population of about 3,00,000 only, generation of electric power did not receive adequate attention during the past several decades. The total installed capacity is about 200 MWe, accounting for about 14% of the total electricity production in the country (Ragnarsson, 2005). A summary of the electric power-generating capacities of geothermal fields in Iceland is given in Table 8.8. However, the growing industrial energy requirements during the

Table 8.7. Electric power generation from geothermal resources in New Zealand (after Dunstall, 2005)

#	Name of geothermal field	Installed capacity (MWe)
1	Wairakei	165
2	Reporoa (Ohaaki)	104
3	Kawerau	14.5
4	Rotokawa	31
5	Northland (Ngawha)	11
6	McLachlan-Wairakei	55
7	Mokai	55
	Total	435.5

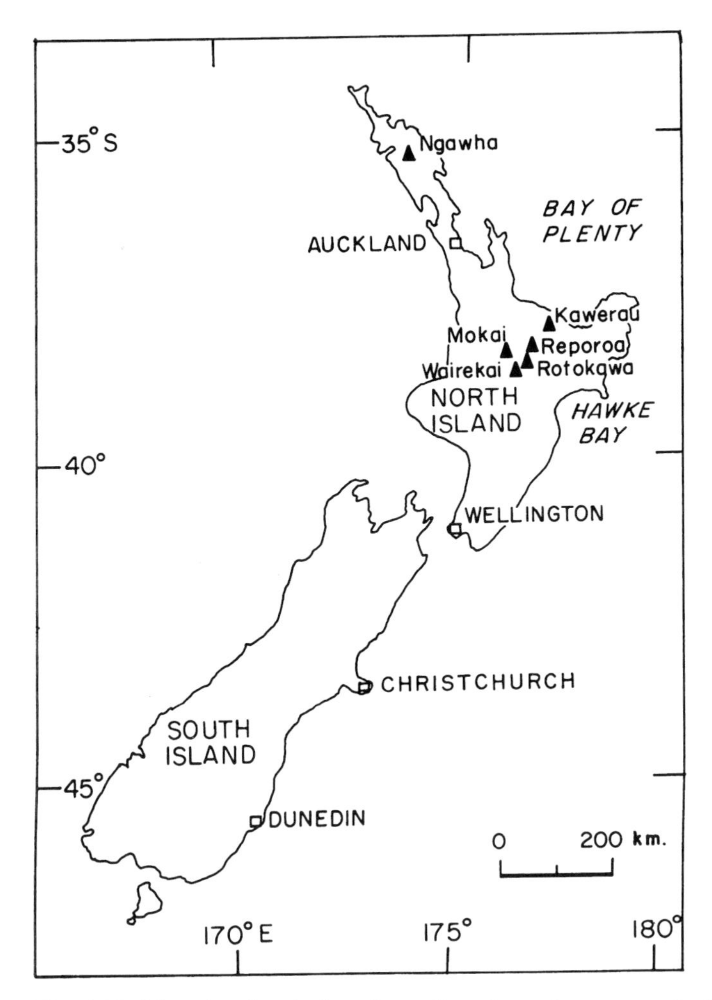

Fig. 8.15. Map showing the locations of geothermal power-producing plants (filled triangles) in New Zealand (Data source: Dunstall, 2005).

past few years have resulted in planned capacity addition of 200 MWe of power-generation capacity in the near future.

Being a very cold country, the main use of geothermal energy is for district heating. Today, about 87% of the houses are heated by geothermal energy resources, a significant jump from 43% in 1970. Practically, the whole population of Reykjavik, the capital of Iceland, is served by geothermal energy. The other important use is in snow melting on the sidewalks and parking places. The heating systems mostly utilize the heat of the spent water from the houses. Swimming pools, greenhouse heating, fish farming and several industrial applications utilize geothermal energy. In

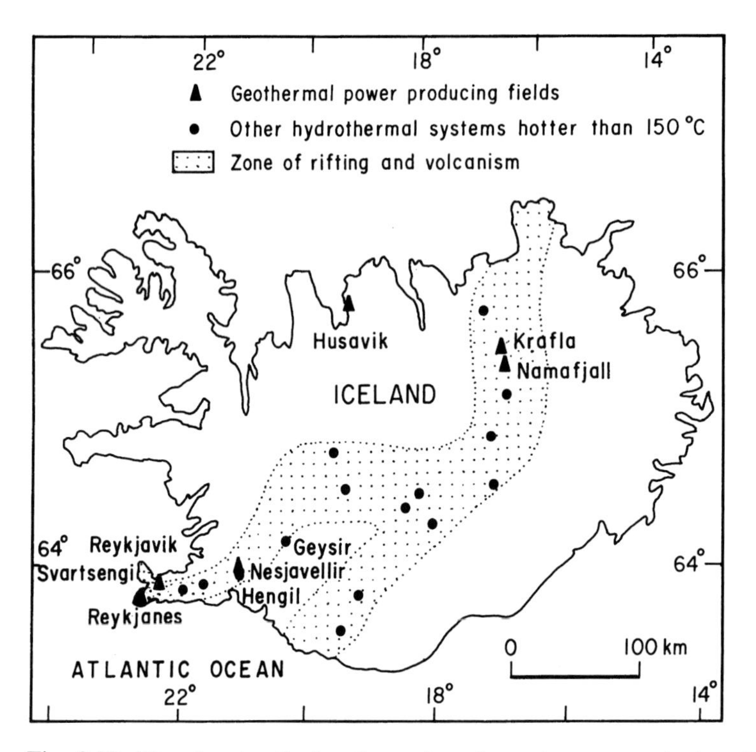

Fig. 8.16. Map showing the locations of geothermal power-producing plants (filled triangles) in Iceland (after Ragnarsson, 2005). Note that all the geothermal plants and other hydrothermal systems hotter than 150 °C are located within the zone of rifting and volcanism (shaded region).

Table 8.8. Electric power generation from geothermal resources in Iceland (after Ragnarsson, 2005)

#	Geothermal field	Installed capacity (MWe)
1	Namafjall	3.2
2	Krafla	60.0
3	Svartsengi	45.0
4	Nesjavellir	90.0
5	Reykjanes	0.5
6	Husavik	2.0
	Total	200.7

short, day-to-day life in Iceland is centered around geothermal energy resources. An up-to-date review of utilization of geothermal energy is provided in Ragnarsson (2005).

Central America

Central America is a tectonically active zone surrounded by the Cocos, Caribbean and North American plates. It is a part of the "Pacific Ring of Fire", a belt associated with spectacular geothermal manifestations—hot springs, geysers and fumaroles, which derive heat from the chain of Quaternary volcanoes dotting the region. The four Central American countries, Costa Rica, El Salvador, Guatemala and Nicaragua, together produce 424 MWe of electric power. In the individual countries, geothermal power contributes about 8.4%, 14%, 1.7% and 11.2%, respectively, to their national electricity-generating capacities, and 15%, 24%, 3% and 10%, respectively, to their national energies (Bertani, 2005). Commercial production has not yet started in Honduras. The geothermal energy potential in this region is at least a few orders of magnitude higher than what has been exploited so far. Many of the high-enthalpy geothermal fields are located within protected areas where geothermal exploitation is prohibited. Further, due to its physiographic setting, hydroelectric power is abundant and constitutes the predominant energy source in this part of the Americas. However, with growing energy needs as a result of rapid industrialization, several exploration and development programs have been initiated to tap the geothermal energy potential of the region and lend support to the hydropower-dominated energy scenario.

Locations of geothermal fields are shown in Fig. 8.17, and the installed capacities for power generation are summarized in Table 8.9. Although Costa Rica and El Salvador are the two smallest countries in central America, they together produce about three-quarters of the total geothermal power output. In Costa Rica, five operating power plants at Miravalles geothermal field have a total installed capacity of 162.5 MWe (Mainieri, 2005). The field, located on the flank of Miravalles volcano, is a water-dominated system with average temperature of 240 °C. El Salvador produces about 151 MWe of electric power through two producing fields at Ahuachapan and Berlin. Average temperatures of geothermal resources are 230 and 300 °C, respectively. Several other unexploited resources with temperatures below 200 °C are located along the volcanic chain passing through El Salvador. Of the four countries, Nicaragua is believed to possess the largest geothermal potential resulting from the occurrence of active volcanoes adjoining the Pacific coast. The Momotombo geothermal field, located along this belt, is currently the only producing field with an installed capacity of 77.5 MWe. The San Jacinto-Tizate field has been recently taken up for development. Commercial exploitation of geothermal energy in Guatemala took off quite recently in 1998. Two producing fields, Zunil and Amatitlan, together have an installed capacity of 33 MWe. Many other areas have been explored and identified for development in the coming decade.

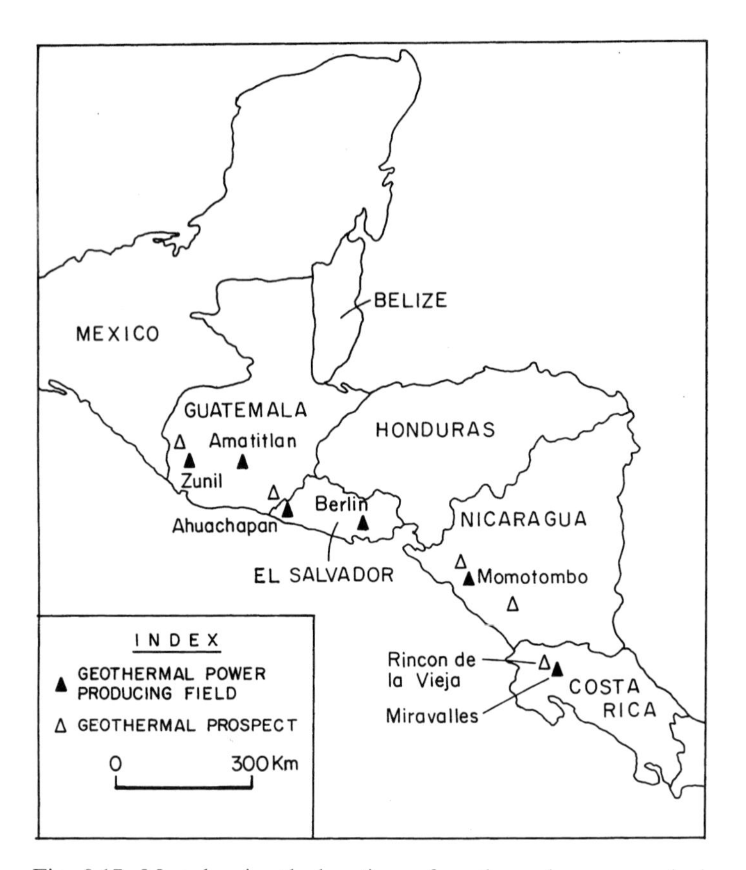

Fig. 8.17. Map showing the locations of geothermal power-producing plants (filled triangles) and major prospects (unfilled triangles) in Central America (Source of data: Mainieri, 2005; Mayorga, 2005; Rodriguez and Herrera, 2005; Roldán Manzo, 2005).

Table 8.9. Electric power generation from geothermal resources in Central America

Country	Geothermal field	Installed capacity (MWe)	Reference
Costa Rica	Miravalles	162.5	Mainieri (2005)
El Salvador	Ahuachapan	95.0	Rodriguez and
	Berlin	56.0	Herrera (2005)
Nicaragua	Momotombo	77.5	Mayorga (2005)
Guatemala	Zunil	28.0	Roldán Manzo
	Amatitlan	5.0	(2005)
	Total	424.0	

Kenya

Geothermal manifestations in Kenya are located in the intracontinental East African Rift Valley, which extends from the Triple junction in the north to Malawi in the south. Rifting occurs at a rate of 1 cm yr^{-1}. This accompanied by intense volcanism, which presumably occurred up to the Late Pliestocene, provides the source of heat for the hydrothermal systems (Mwangi, 2005). The Olkaria geothermal field, located in the Rift Valley, is the only producing field so far, with an installed capacity of 127 MWe. Geothermal power accounts for about 11% of the Kenya's total power production.

Russia

Geothermal energy is used in Russia both for power generation and for direct uses. A large part of the resources are used for space heating. Producing geothermal fields are located in the Kamchatka and Kuril islands, in the eastern part of Russia (Kononov and Povarov, 2005) (Fig. 8.18). The total installed capacity is about 79 MWe, obtained from a number of small-capacity power plants (Table 8.10). Additional power plants, with a combined installed capacity of about 150 MWe, are under construction.

Other Countries

Assessment of geothermal resource utilization has been considered in several other countries also. These are too many to discuss, however a mention may be made of

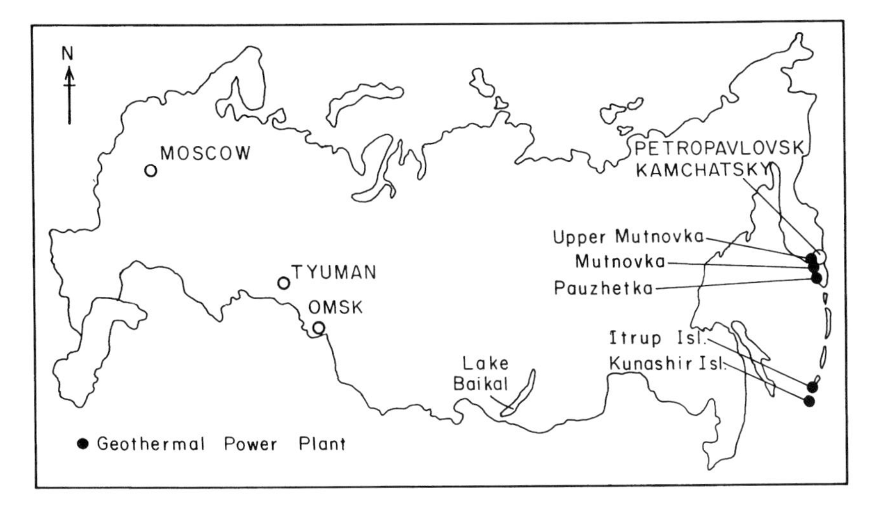

Fig. 8.18. Map showing the locations of geothermal power producing plants (filled circles) in Russia (modified after Kononov and Povarov, 2005).

Table 8.10. Electric power generation from geothermal resources in Russia (Kononov and Povarov, 2005)

Area	Geothermal field	Installed capacity (MWe)
Kamchatka	Pauzhetka	11
	Verkhne-Mutnovka	12
	Mutnovka	50
Kuril islands	Iturup islands	3.4
	Kunashir islands	2.6
	Total	79

France, Germany, Ethiopia, Papua New Guinea and Portugal, which joined the select list of geothermal power-producing countries in the late 1990s and later. China and Turkey have been producing in a very small way since the early 1970s, but are extensive users of geothermal resources for direct applications. In China, geothermal power is produced mainly from the Yangbajing fields in Tibet. A few other fields have been identified, all of which are located in Tibet. The Kizildere geothermal field is the only producing field in Turkey. Each of the previous-mentioned countries produces a few megawatts to a few tens of megawatts of electric power (Bertani, 2005). In southern Australia, deep drilling up to 4–5 km depth is underway in the Cooper Basin, where hot-dry-rock (HDR) projects have been planned in the near future. Other countries such as Sweden, Norway, Denmark, Chile and Netherlands, which do not produce electric power, have recorded the highest growth rates in direct uses of geothermal resources during the last decade (Lund et al., 2005a).

A special mention may not be out of place as far as India is concerned. In literature, there are several references to the use of geothermal springs for balneological purposes. However, India has yet to produce electric power from a geothermal field. Here, we give a brief outline of the efforts made in assessing the geothermal potential of India. Several groups of hot springs have been reported from different geological provinces of India. The locations, geological settings and temperatures of these hot springs have been compiled by previous workers (Oldham, 1882; Ghosh, 1954; Gupta, 1974; Guha, 1986; GSI, 1991 and others). A brief, chronological, account of the efforts made toward development of geothermal energy in India by various organizations is given by Rao (1997). Detailed studies regarding development of some of the geothermal fields was first examined by the Government of India in 1966 and a comprehensive report was produced in 1968, recommending preliminary prospecting in the Puga and Manikaran fields in the Himalaya (Hot Springs Committee, 1968). This was followed by systematic exploration programs including deep drilling in a number of geothermal areas in the country during the period 1970–1990 (Gupta et al., 1973, 1974; Krishnaswamy and Ravi Shanker, 1980; GSI, 1983, 1991, 2002; Thussu et al., 1987; Moon and Dharam, 1988; Gupta, 1992). Additionally, regional heat flow determinations have been made from temperature measurements in more than 200 boreholes covering a number of provinces (Roy,1997; Roy and Rao, 2000; Rao et al., 2003; Roy et al., 2003; Ray et al., 2003 and references therein). So far, no geothermal field has been identified which could clearly be associated with Quaternary or Late Tertiary magmatic heat source at shallow depths, the kind of fields that have been exploited for power generation elsewhere (Roy et al., 1996; Roy and Rao, 1996). Nevertheless, two promising areas have been considered for setting up pilot-scale, binary-cycle geothermal power plants: one, at Puga-Chumathang (1 MWe) and another, at Tattapani in central India (300 kWe). The results emerging from these experiments could guide further exploration efforts in the country. Moreover, the low-to-moderate enthalpy fluids, which constitute the discharges from the majority of hot springs, can be gainfully utilized for nonelectrical uses such as development of spas, greenhouse cultivation and extraction of rare materials like cesium.

CONCLUDING REMARKS

Utilization of geothermal energy resources for both power generation as well as direct use has increased steadily in several parts of the world since 1975. Geothermal energy has contributed immensely toward development of small nations such as Iceland, Costa Rica, El Salvador, Italy and the Phillipines. In the United States of America, geothermal energy has played a significant role in tiding over the recent energy crisis in the highly populated state of California. In countries that are not endowed with high-enthalpy resources, the available low-temperature geothermal resources have contributed to a wide spectrum of direct uses. The most phenomenal growth during the last decade has been the extensive deployment of geothermal heat pumps.

The overall growth rate of utilization of geothermal energy resources has been quite modest. In the year 2005, the installed electric-generating capacity worldwide has reached about 8,900 MWe only as against an anticipated capacity exceeding 11,000 MWe (Huttrer, 2001). It should be noted that some of the major producing fields have shown declining steam pressures and consequent capacity reductions, and a number of power plants have been shut down due to age and unanticipated technological challenges. On the other hand, several exploration and development programs are underway at a number of locations. A number of HDR pilot projects have been funded during the past five years, such as those in France, Germany and Australia, of which a few have successfully passed the demonstration stage (Klee, 2005; Rummel, 2005). Researches in enhanced geothermal systems (EGS), which involve rock fracturing, water injection and water circulation techniques to sweep heat from the unproductive areas of existing geothermal fields or new fields lacking adequate production capacity, are gaining momentum. It appears, therefore, that the vast potential of geothermal energy resources is yet to be fully exploited. The need for developing state-of-the art technologies in exploration, assessment and exploitation of geothermal resources, has become more relevant than ever before.

Chapter 9

THERMAL ENERGY OF THE OCEANS

Oceans cover two-thirds of the Earth's surface. The heat energy stored within the ocean waters is a vast energy resource. Here we discuss an interesting topic of making use of the energy stored in the ocean waters for generation of electric power and potable water. In this chapter we largely draw from Vega (1992, 1995), Avery and Wu (1994), Gupta (2003), Ravindran (2005), NIOT (2005), Kathiroli et al. (2006), and NREL (2006), and several interactions and personal communications with S. Kathiroli, P. Jalihal, and R. Singh of the National Institute of Ocean Technology, Chennai, India.

OCEAN THERMAL ENERGY CONVERSION (OTEC)

Concept

Of the total solar radiation, oceans are the largest collectors, accumulating 250 billion barrels of oil equivalent, according to an estimate. This vast amount of solar energy absorbed in the oceans can be converted into electricity by a process known as Ocean Thermal Energy Conversion, popularly known as OTEC. OTEC makes use of the difference in temperatures of warm surface water (22–27 °C) and very cold water at a depth of 1 km (4–7 °C). OTEC is particularly suitable for tropical oceans extending from 20° N to 20° S.

Globally, three concepts of OTEC have been developed and tested to varying degree of success. They are known as:

- Open-cycle
- Closed-cycle
- Hybrid design

(a) In the open-cycle plant, large quantities of warm ocean surface water are poured into a chamber where the pressures are reduced to near vacuum level. This converts water into steam. The steam is directed to a large low-pressure turbine, which generates electricity. After steam comes out of the generator, it is cooled by another set of pipes containing very cold water brought from depths. This converts steam through condensation into desalinated water.

(b) The closed-cycle plant operates more or less in the same manner as the open-cycle plant, except that it uses ammonia as an exchange fluid to run the turbine. Ammonia is repeatedly circulated throughout the cycle. This type of plant does not have the benefit of producing the desalinated water. However, closed-cycle plants are more efficient.

(c) In a hybrid plant, both closed- and open-cycle concepts are used to make it energy efficient.

Historical Background

In 1881 Jacques Arsene d' Arsonval, a French physicist, became the first man to propose tapping of the thermal energy in the ocean. His student, Georges Claude, built an experimental open-cycle OTEC Plant at Matanzas Bay, Cuba in 1930 (Claude, 1930). This produced 22 kW of electricity by using a low-pressure turbine. In 1935, Claude built another open-cycle plant aboard a 10,000 ton cargo vessel moored off the coast of Brazil. Both plants developed by Claude were destroyed by weather and waves, and the goal of producing power was never achieved. In 1956, a team of French scientists designed a 3 MW open-cycle plant for Abidjan on Africa's west coast. However, this was not completed because of competition with inexpensive hydroelectric power. Then, in 1974, the Natural Energy Laboratory of Hawaii was established which has become world's foremost laboratory and test facility for OTEC technologies. This Institute set up a 50 kW closed-cycle OTEC demonstration plant on a converted U.S. Navy Barge moored 2 km off Keahole Point. This plant produced 52 kW of gross power and 15 kW of net power.

Later in 1980, the U.S. Department of Energy installed closed-cycle OTEC heat exchangers on board a converted U.S. Navy Tanker. In 1981, Japan demonstrated—in the Republic of Nauru, in the Pacific Ocean—operation of a 100 kW closed-cycle plant, which surpassed engineering expectations. The plant produced 35 kW of net power. In 1993, at Keahole Point, Hawaii, a 50 kW plant was successfully operated.

1-MW Plant in India

India is geographically well placed as far as the OTEC potential is concerned. Around 2,000 km of coast length along the south Indian coast, a temperature difference of above 20 °C persists between the surface water and water at a depth of one km, around the year. This amounts to about 1.5×10^6 square kilometers of tropical water in the Exclusive Economic Zone around India with a power density of $0.2\,\mathrm{MW\,km^{-2}}$. Additionally, attractive OTEC plant locations are available around Lakshawdweep and Andaman & Nicobar Islands. The total OTEC potential around India is estimated to be 1,80,000 MW considering 40% of gross power for parasitic losses. This indicates the promise of OTEC for India and points out the urgent need to develop OTEC technology.

In 1993, the National Institute of Ocean Technology (NIOT), located at Chennai, was formed by the Department of Ocean Development (DOD). Ocean energy is a part of NIOT's mission-based activities. Under this mission a major thrust was given for the technology development for OTEC. In early 1997, DOD proposed to establish a 1 MW gross OTEC plant in India, which will be the first ever MW-range plant established anywhere in the world. NIOT conducted detailed surveys at the proposed OTEC site near Tuticorin, south India. Based on the temperature and bathymetric profiles, the optimization of the closed-loop systems was done.

The bathymetry of the coast around the mainland of India is such that a water depth of 1,000 m is available at a distance of 40 km from the shore. This necessitates the use of a floating platform to house the OTEC plant. There exist some locations where a shell-mounted OTEC plant can be constructed at a depth of 200 m. However, considering the future need of large plants, it was decided to design a floating OTEC plant. NIOT aims to deploy a 1 MW floating OTEC plant off the coast of Tamil Nadu near Tuticorin port on the south east coast of India. This region is known to be free of cyclones during the last four decades.

All commercial OTEC plants are expected to be in the 10–50 MW range or larger. Therefore, a 1 MW gross power output plant is selected for the present design, considering the scaling up in the future. The design of the power module is based on a closed Rankine Cycle with ammonia as the working fluid. Titanium plate heat exchangers are suitable for such an environment. A cold-water pipe made of high-density polyethylene (HDPE) of 1 m outer diameter is selected. Axial flow turbine having a higher adiabatic efficiency is chosen for power conversion and also for easy scaling up in future. Following are the baseline design conditions:

Gross power output	1 MW
Warm water temperature	29 °C
Cold-water temperature	7 °C
Depth of cold-water intake	1,000 m
Cold-water pipe (inner diameter)	0.90 m

The evaporator and condenser consist of four modules of plate heat exchangers, which will be the largest of its kind used for such an application. The ammonia side of the evaporator is coated with stainless-steel powder to enhance the overall heat transfer co-efficient. The coating is expected to improve the power output by 20–40%.

Radial inflow turbines have been previously used in the U.S.-Mini OTEC experiment. The power system flow rates and net power are very much dependent on the turbine efficiency. Parametric studies showed that a 4-stage axial turbine can improve the efficiency up to 89%.

The seawater pumps are of vertical mixed-flow type due to the low-head and high-discharge combination. Two identical pumps are to be connected in parallel on both warm- and cold-water sides, and are driven by a variable-speed drive to adjust the flow rates. As the system is highly sensitive to the flow variables, a variable-speed drive is an essential component in an OTEC system.

There are several configurations for cold-water pipe mooring. After studying several options the concept finalized consists of a floating platform connected to a cold-water pipe which itself acts as mooring for the platform. The platform assumes a great significance because it houses the entire plant, accommodates the seawater pumps, and the cold-water pipe or mooring system. NIOT has carried out studies with the help of numerical simulation as well as model tests to estimate the motions and stresses in the vessel and pipe. NIOT is expected to deploy the 1-MW OTEC plant shortly.

Economics of OTEC

The OTEC power can be cost effective only if the unit cost of power produced is comparable with the fossil-fuelled plants. OTEC system can also have other benefits like enhanced mariculture, desalination, or even air conditioning, which might reduce the cost of electricity generated.

For small plants of 1-MW range the unit power generation cost is high compared to other conventional energy sources. The production of fresh water along with power is to be considered to estimate unit cost for OTEC plants in islands. Studies predict that OTEC would be economical and production cost would be comparable to other conventional sources for higher ranges.

Future

It is envisaged that most of the future commercial OTEC plants shall be closed-cycle floating plants of 10–50-MW range. But plants of 200–400-MW range are also feasible and are economically more attractive. The commercial plants should be preceded by demonstration plants of smaller size for power cycle optimization and also for operational information. The design, development, and operation of a power system in a hostile sea environment pose a great challenge.

The capital cost of the plant depends much on the cost of heat exchangers and hence any improvement in the performance in this single component is an added advantage. New materials for the cold-water pipe is to be developed to withstand the marine conditions and also for easy fabrication and deployment. The design of the barge also requires care so as to position the seawater pumps for the required Net Positive Suction Head (NPSH). The equipment and the piping system are to be assembled on the barge in such a way that the static head and the minor losses are the least. Bio-fouling on the warm-water circuit and the release of the dissolved gases in the cold-water circuit is a problem to be attended for a considerably long period. As the floating plants are away from seashore, underwater power transmission to the land is an area that needs further study.

LOW-TEMPERATURE THERMAL DESALINATION OF SEAWATER

Availability of fresh water is becoming scarce due to increase in global population. As against an availability of some 9,000 cubic meters of fresh water per person per year in the year 1989, it dropped to 7,800 cubic meters per person per year in the year 2000. It is estimated that by the year 2025, the population would have reached 8 billion and the availability of fresh water would come down to some 5,100 cubic meters per person per year. As a matter of fact, it is often said that the future wars would be fought for water.

Although about 70% of the Earth's surface is covered by water, only 3% of the total available water is fresh. Out of this meager 3%, about 2% of water is locked up in glaciers and icecaps, and only 1% is accessible for human use. This 1% of fresh water gets continuously recycled through rain and other processes. Efforts have been made globally to desalinate seawater and make it available for human use. Reverse osmosis is the most commonly used technology to convert seawater into fresh water. However, when the salt content is in excess of $2{,}000\,\mathrm{mg\,l^{-1}}$, the reverse osmosis process becomes less effective, and the cost of frequently changing the membranes makes the whole process very expensive.

Thermal desalination provides an effective way of producing fresh water from the seawater. The seawater is heated and flash evaporated. The vapors are then condensed using the surface seawater. This is known as high temperature desalination. As a lot of energy is required in heating, this process is not efficient and economical. A more efficient and economical method of converting seawater into fresh water is to flash evaporate the surface seawater and condense the vapors using cold seawater transported from depth. This is known as the low temperature desalination technique. This approach is very effective in tropical waters where surface water temperatures are in the vicinity of 25–28 °C round the year, and very often cold water at 10–12 °C is available at depths of about 300 m.

Concept

The concept of low-temperature thermal desalination (LTTD) is described in Fig. 9.1. The warm surface water is taken to the flash chamber and is flash evaporated under appropriate low pressure. These vapors are led to the condenser where cold water brought from depth is circulated to condense the vapors and produce fresh water.

International scenario

The principle of LTTD is known for a long time. Some experiments were conducted in U.S. and Japan to test the low-temperature driven desalination technology. In Japan, a spray flash evaporation system was tested by Saga University (Nakaoka et al., 1997). In U.S., at Hawaii Islands, the National Energy Laboratory tested an open-cycle OTEC plant with fresh water and power production using a temperature

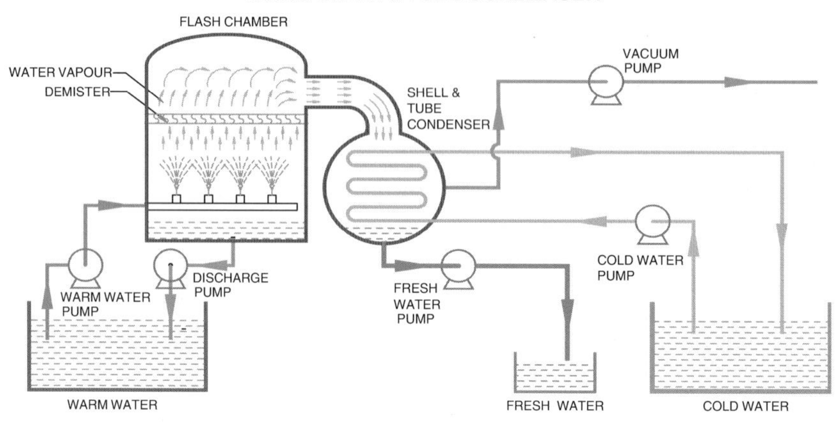

Fig. 9.1. Schematic diagram of the working principle of an LTTD plant (after Kathiroli et al., 2006).

Fig. 9.2. A 5,000 l per day capacity plant commissioned at NIOT (Kathiroli et al., 2006).

of 20 °C between surface water and water at a depth of around 500 m. The Italian government also commissioned a 25 ton h^{-1} desalination plant in 1992. As a matter of fact, most developed countries do not have shortage of fresh water. The problem is basically more severe with developing countries. LTTD is a unique technique for tropical countries in the latitude band of 20° N to 20° S. Around 35% of the global population resides in this latitude band. Experiments conducted recently in India have shown the efficacy of LTTD, which is discussed in the next section.

Initiative of NIOT in Developing LTTD in India

A pilot plant of 5000 l/day capacity was commissioned at NIOT to test the concept of LTTD (Fig. 9.2). Successful operation of this pilot plant encouraged commissioning of a larger LTTD plant.

Kavaratti LTTD Plant—100,000 l/day Capacity

To establish a 100,000 l per day (lpd) capacity LTTD plant, a number of sites suitable for commissioning such a plant were examined. As it was initially planned to set up such a plant on land, it was necessary to have a steep slope so that deep cold

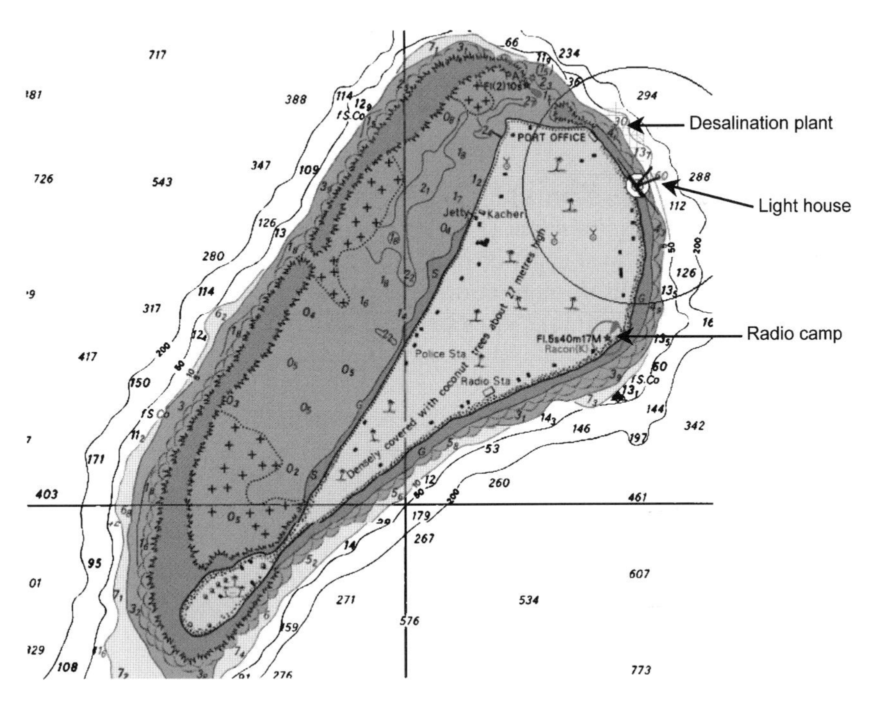

Fig. 9.3. The figure shows Kavaratti island, located 200 km west of Kochi, India, and the bathymetry around it. The island is 5.6 km in length and the maximum width is 1.2 km. Water-depth contours are shown in meters. Spot values indicate water depth in meters. Coral reefs (indicated by crosses) are located to the west of the island. The LTTD plant is located near the northeastern margin of the island.

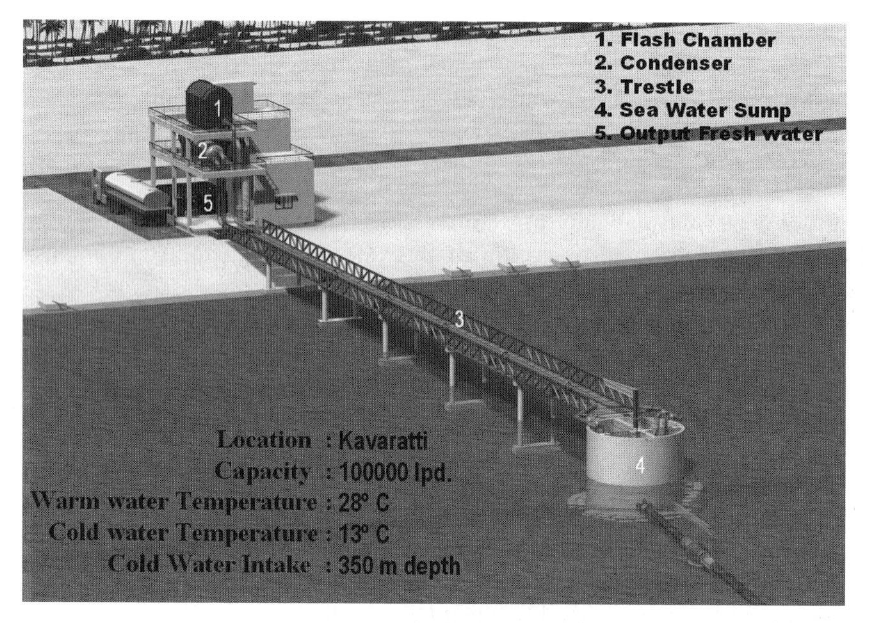

Fig. 9.4. Conceptual layout of the LTTD plant at Kavaratti island.

water could be found in the near proximity of the site. One such site was found at the Kavaratti island in the Arabian Sea (Fig. 9.3). Kavaratti island, located about 200 km west of Kochi on the west coast of India, has a population of about 10,000. It is well connected with Kochi through air and sea routes.

The Kavaratti site is very suitable for setting up a LTTD plant, as within a distance of about 400 m from the shore, the ocean deepens to some 350 m and water temperature drops to about 12 °C. Fig. 9.4 shows the conceptual layout of the LTTD plant and Fig. 9.5 shows the LTTD plant that has been operating at Kavaratti since May 2005. Fig. 9.6 shows the HDPE pipe line used to bring the cold water from depth to the plant site.

The LTTD plant consists of a flash chamber, which is rectangular (4.5 m × 2.5 m × 3 m) in cross section (Fig. 9.7). Vacuum is created in the flash chamber. Surface seawater is flashed inside the chamber. The vapors generated are sent to the condenser. The condenser is 7.5 m long and has a cylindrical shape (1.5 m diameter). The condenser consists of cupro-nickel tubes to exchange the heat (Fig. 9.8). Water vapor is condensed into fresh water with the help of deep-sea coldwater.

Using a crane barge, a 600-mm-diameter fusion welded HDPE pipe was deployed at 350-m depth to draw deep-sea cold water at 13 °C. The shore end of the HDPE pipes is connected to the cold-water sump or intake well. The sump is divided into two compartments of cold water and warm water. Coarse screen has been used to

Fig. 9.5. A picture of the 100,000 l/day capacity LTTD plant in operation at Kavaratti (Kathiroli et al., 2006).

Fig. 9.6. View of the 600 m long HDPE pipeline used to bring cold water up to the plant site.

Fig. 9.7. A view of the flash chamber at the Kavaratti LTTD plant.

filter foreign particles before the water is pumped into the main equipment. Deep cold water and warm surface seawater pumps have been positioned inside the respective compartments of the sump and are connected to the main cold water and warm water HDPE pipelines using header lines. Pipes have been routed along the trestle (Fig. 9.9) to the flash chamber and condenser. Flash chamber has been kept at an elevation of ~10 m above the high tide level to optimize energy consumption.

Specifications of the Kavaratti LTTD plant (after Kathiroli et al., 2006)

Fresh-water generation rate	100,000 l per day
Warm-water temperature	28 °C
Cold-water temperature	13 °C
Cold-water flow rate	180 kg s^{-1}
Warm-water flow rate	145 kg s^{-1}
Cold-water intake	From 350 m water depth
Vacuum maintained	Low vacuum level

Fig. 9.8. A view of the cylindrical condenser in operation at the Kavaratti LTTD plant.

Fig. 9.9. A view of sump and trestle at the Kavaratti LTTD plant.

Salient Features of the Plant

The plant does not require any pre-treatment of feed water, and it is easy to maintain because of its operational simplicity besides being environment-friendly and non-polluting. The quality of fresh water generated conforms to the Bureau of Indian Standards/World Health Organization standards (see Table 9.1). The deep seawater utilized is rich in nutrients and can be used to cultivate marine life.

Table 9.1. Comparison of quality of desalinated water obtained from the Kavaratti plant with recommended standards (Kathiroli et al., 2006)

Parameter	Desirable limit	Permissible limit	Desalinated water
Color	Unobjectionable	Unobjectionable	OK
Odor	Unobjectionable	Unobjectionable	OK
Taste	Unobjectionable	Unobjectionable	OK
pH	6.5	8.5	7–8
Total dissolved solute (ppm)	500	2,000	280
Chloride (ppm)	250	1,000	90
Total hardness	300	600	100
Total Coli form (MPN)	1	10	ND

Fig. 9.10. A view of barge-mounted desalination plant off Tuticorin coast, south India.

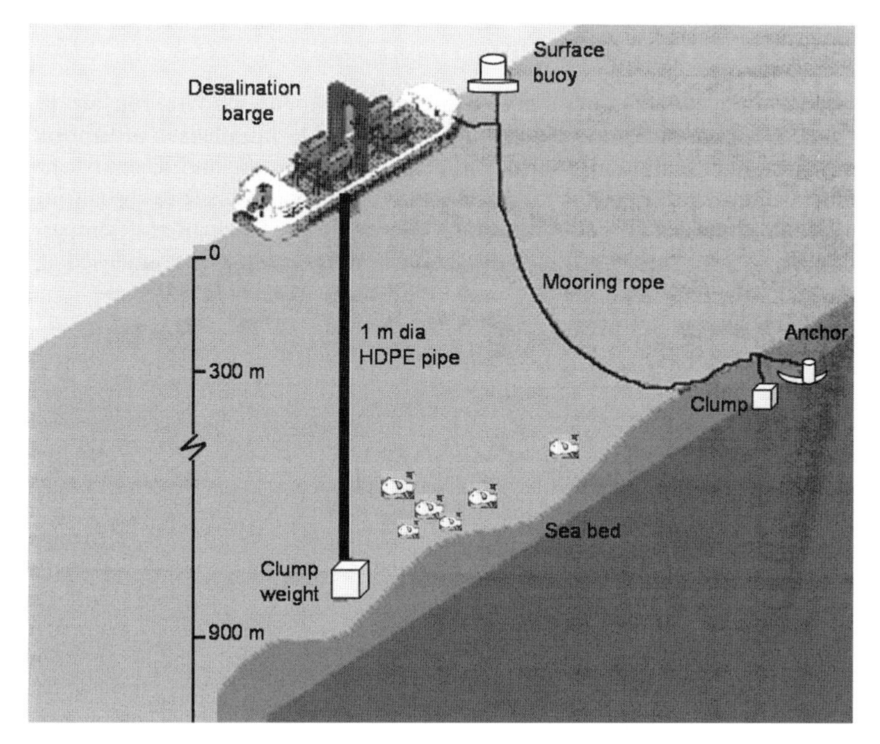

Fig. 9.11. Conceptual layout of the Tuticorin LTTD plant.

Future of Low Temperature Thermal Desalination

A 1 million l/day capacity plant is currently being designed by NIOT. This plant is barge-mounted and is being deployed offshore of Tuticorin, near Chennai, off the east coast of India (Fig. 9.10). The conceptual layout of this desalination plant is given in Fig. 9.11.

Global Scenario

The success of deploying a 100,000 l per day plant at Kavaratti, which has now successfully operated for more than a year, has given a new direction to use the thermal energy entrapped in the tropical ocean waters to solve the ever-increasing fresh-water requirement globally in the tropical regions. The NIOT is currently busy in developing a 1-million liters per day plant at Tuticorin as observed earlier, and also plans to develop a 10-million liters per day plant. The cost of producing 1,000 l water drops to one US Dollar for a 10-million liters per day plant, taking into account the initial cost of the plant and all other expenses depreciated to zero value in 10 years time.

Intangible Benefits

The cold water, which is being discharged to the sea, can be used to run an air conditioning system. Another important application of this technique is in utilization of high-temperature waste fluid in several commercial activities on land. Usually, the waste fluids, at temperatures of 50 °C or so, need to be cooled before discharge. Instead, using the same technique and making use of the water at inland temperatures of 35 °C or 36 °C, these hot fluids can be flash evaporated and condensed so that instead of commercial waste, it gets converted into useful fresh water.

REFERENCES

Allis, R.G., 2000. Review of subsidence at Wairakei field, New Zealand. Geothermics 29, 455–478.

Allis, R.G., Hunt, T.M., 1986. Analysis of exploitation-induced gravity changes at Wairakei geothermal field. Geophysics 51, 1647–1660.

Allis, R.G., Gettings, P., Chapman, D.S., 2000. Precise gravimetry and geothermal reservoir management. In: Proceedings of the 25th Workshop on Geothermal Reservoir Engineering, Stanford, 10pp.

Anderson, L.A., Johnson, G.R., 1976. Application of the self-potential method to geothermal exploration in Long Valley, California. Journal of Geophysical Research 81, 1527–1532.

Annual Energy Review, 2004. Energy Information Administration. This is the official energy statistics from the U.S. government. Website: http://www.eia.doe.gov/emeu/aer

Apuada, N.A., Olivar, R.E.R., 2005. Repeat microgravity and leveling surveys at Leyte geothermal production field, North Central Leyte, Philippines. In: Proceedings of the World Geothermal Congress, Antalya, Turkey, 24–29 April.

Armstead, H.C.H., 1976a. Environmental factors and waste disposal: summary of Section V. In: Proceedings of the 2nd U.N. Symposium on the Development and Use of Geothermal Resources, vol. 1. U.S. Government Printing Office, Washington, DC, pp. lxxxvii–xciv.

Armstead, H.C.H., 1976b. Some unusual ways of developing power from a geothermal field. In: Proceedings of the 2nd U.N. Symposium on the Development and Use of Geothermal Resources, vol. 3. U.S. Government Printing Office, Washington, DC, pp. 1897–1904.

Arnórsson, S., 1985. The use of mixing models and chemical geothermometers for estimating underground temperatures in geothermal systems. Journal of Volcanology and Geothermal Research 23, 299–235.

Arnórsson, S. (Ed.), 2000. Isotopic and chemical techniques in geothermal exploration, development and use—sampling methods, data handling, interpretation. International Atomic Energy Agency, Vienna, Austria, 351 pp.

Arnórsson, S., Gunnlaugsson, E., Svavarsson, H., 1983. The chemistry of geothermal waters in Iceland III. Chemical geothermometry in geothermal investigations. Geochimica et Cosmochimica Acta 47, 567–577.

Avery, W.H., Wu, C., 1994. Renewable Energy from the Ocean. Oxford University Press, Oxford, 446 pp.

Baksi, A.K., Watkins, N.D., 1973. Volcanic production rates: comparison of oceanic ridges, islands and the Columbia plateau basalts. Science 180, 493–496.

Banwell, C.J., 1970. Geophysical techniques in geothermal exploration. In: Proceedings of the U.N. Symposium on the Development and Utilization of Geothermal Resources. Geothermics, Special Issue l, 32–56.

Banwell, C.J., 1973. Geophysical methods in geothermal exploration. In: Armstead, H.C.H. (Ed.), Geothermal Energy, UNESCO, Paris, pp. 31–48.

Banwell, C.J., Gomez Valle, R., 1970. Geothermal exploration in Mexico 1968–1969. In: Proceedings of the U.N. Symposium on the Development and Utilization of Geothermal Resources. Geothermics, Special Issue 2 (1), 27–40.

Barbier, E., 2002. Geothermal energy technology and current status: an overview. Renewable and Sustainable Energy Reviews 6, 3–65.
Batini, F., Menut, P., 1964. Etude structurale de la zone loccastrade en vue de la recherché de vapeur pour les methodes geophysiques, gravimetriques, et electrique. In: Proceedings of the U.N. Symposium on New Sources of Energy, vol. 2. UNESCO, Paris, pp. 73–81.
Batini, F., Nicholich, R., 1984. The application of seismic reflection methods to geothermal exploration. U.N. Seminar on Utilization of geothermal energy for electric power production and space heating. Florence, Italy.
Beardsmore, G.R., 2004. The influence of basement on surface heat flow in the Cooper Basin. Exploration Geophysics 35 (4), 23941.
Beck, A.E., 1988. Thermal properties. In: Haenel, R., Rybach, L., Stegena, L. (Eds.), Handbook of Terrestrial Heat Flow Density Determination. Kluwer Academic, Dordrecht, pp. 87–124.
Benito, F., Ogena, M.S., Stimac, J.A., 2005. Geothermal energy development in the Philippines: country update. In: Proceedings World Geothermal Congress, Antalya, Turkey, April 24–29.
Bertani, R., 2005. World geothermal power generation in the period 2001–2005. Geothermics, 34, 651–690.
Biehler, S., 1971. Gravity studies in the Imperial Valley. In: Cooperative geological–geophysical–geochemical investigations of geothermal resources in the Imperial Valley area of California. Univ. California, Riverside, Education Research Service, pp. 29–41.
Birch, F., Clark, H., 1940. The thermal conductivity of rock and its dependence on temperature and composition. American Journal of Science, 238, 529–558 and 613–635.
Bjornsson, S., Einarsson, P., 1974. Seismicity of Iceland. In: Kristjansson, L. (Ed.), Geodynamics of Iceland and the North Atlantic Area. D. Reidel, Dordrecht, 225 pp.
Blackwell, D.D., 1985. A transient model of the geothermal system of the Long Valley Caldera, California. Journal of Geophysical Research 90, 11,229–11,241.
Blackwell, D.D., Bowen, R.G., Hull, D.A., Riccio, J., Steele, J.L., 1982. Heat flow, arc volcanism, and subduction in Northern Oregon. Journal of Geophysical Research 87, 8735–8754.
Blackwell, D.D., Morgan, P., 1976. Geological and geophysical exploration of the Marysville geothermal area, Montana, USA. In: Proceedings of the 2nd U.N. Symposium on the Development and Use of Geothermal Resources, San Francisco, vol. 2, pp. 895–902.
Blackwell, D.D., Spafford, R.E., 1987. Experimental methods in continental heat flow. In: Sammis, C.G., Henyey, T.L. (Eds.), Experimental Methods in Physics, vol. 24. Academic Press, Orlando, FL, Geophysics; Part B; Field Measurements, pp. 189–226.
Blackwell, D.D., Steele, J.L., 1989. Heat flow and geothermal potential of Kansas. In: Steeples, D.W. (Ed.), Geophysics in Kansas, Kansas Geological Survey Bulletin 226, 267–295.
Blackwell, D.D., Steele, J.L., 1992. Geothermal map of North America, decade of North American geology, neotectonic series. Boulder, Colorado, Geological Society of America Map CSM-007 (scale 1:5,000,000, 4 sheets).
Blakeslee, S., 1984. Seismic discrimination of a geothermal field: Cerro Prieto. Lawrence Berkley Lab., LBL-17859.
Bodvarsson, G., 1970. Evaluation of geothermal prospects and the objectives of geothermal exploration. Geoexploration 8, 7–17.
Boldizsár, T., 1963. Terrestrial heat flow in the natural steam field at Larderello. Geofisica Pura e Applicata 56, 115–122.

Boldizsár, T., 1970. Geothermal energy production from porous sediments in Hungary. In: Proceedings of the U.N. Symposium on the Development and Utilization of Geothermal Resources. Geothermics Special Issue, 2 (1), 99–109.

Brown, K.L., 1995. Impacts on the physical environment. In: Pre-Congress course on environmental aspects of geothermal development. World Geothermal Congress 1995, pp. 39–55.

Brown, K.L., 2000. Impacts on the physical environment. In: Brown, K.L. (Ed.), Environmental Safety and Health Issues in Geothermal Development, WGC 2000 Short Courses, Japan, pp. 43–56.

Browne, P.R.L., 1978. Hydrothermal alteration in active geothermal fields. Annuual Review of Earth Planetary Sciences 6, 229–250.

Cagniard, L., 1953. Basic theory of the magnetotelluric method of geophysical prospecting. Geophysics 18, 605.

Calvin, W.M., Coolbaugh, M., Kratt, C., Vaughan, R.G., 2005. Application of remote sensing technology to geothermal exploration. Geological Survey of Nevada.

Cappetti, G., Ceppatelli, L., 2005. Geothermal power generation in Italy: 2000–2004 update report. In: Proceedings of the World Geothermal Congress, Antalya, Turkey, April 24–29, 2005.

Carslaw, H.S., Jaeger, J.C., 1959. Conduction of Heat in Solids (2nd ed.). Oxford University Press, London, 510pp.

Cathles, L.M., 1977. An analysis of the cooling of intrusives by groundwater convection which includes boiling. Economic Geology 72, 804–826.

Chapman, D.S., 1986. Thermal gradients in the continental crust. In: Dawson, J.B., Carswell, D.A., Hall, J., Wedepohl, K.H. (Eds.), The Nature of the Lower Continental Crust, Geological Society, Splecial Publication 24, 63–70.

Chapman, D.S., Clement, M.D. Mase, C.W., 1981. Thermal regime of the Escalante Desert, Utah, with an analysis of the newcastle geothermal system. Journal of Geophysical Research 86, 11735–11746.

Chopra, P., Wyborn, D., 2003. Australia's first hot dry rock geothermal energy extraction project is up and running in granite beneath the Cooper Basin, NE South Australia. In: Blevin, P., Jones, M., Chappell, B. (Eds.), Magmas to Mineralisation: The Ishihara Symposium. Geoscience Australia, Record 2003/14, 43–45.

CICESE, 1977. Primer informe preliminar del estudio de microsismicidad del sistema de fallas transformadas Imperial—Cerro Prieto. CFE Archives.

Ciptomulyono, U., 2002. Small geothermal development system using slimhole-drilling techniques for remote powering. In: Geothermal Energy: The Baseload Renewable Resource. Geothermal Resources Council Transactions 26, 183–187.

Claude, G., 1930. Power from the Tropical Seas. Mechanical Engineering 52 (12), 1039–1044.

Clauser, C. (Ed.), 2003. Numerical simulation of reactive flow in hot aquifers: SHEMAT and processing SHEMAT. Springer-Verlag, Berlin, 332pp.

Clauser, C., 2006. Geothermal energy. In: Heinloth, K., (Ed.), Landolt-Börnstein, Group VIII, Advanced Materials and Technologies, vol. 3. Energy Technologies, Subvol. C Renewable Energies. Springer, Heidelberg, pp. 480–595.

Combs, J., 1976. Summary of Section IV: Geophysical techniques in exploration. In: Proceedings of the 2nd U.N. Symposium on the Development and Use of Geothermal Resources, vol. 1. U.S. Government Printing Office, Washington, DC, pp. lxxxi–lxxxvi.

Combs, J., Hadley, D., 1977. Microearthquake investigation of the Mesa geothermal anomaly, Imperial Valley, California. Geophysics 42, 17–33.
Combs, J., Gupta, H.K., Helsley, C.E., 1976a. Lateral variations of the subsurface velocity, East Mesa geothermal field, Imperial Valley, California. EOS Transaction of American Geophysical Union 57, 153.
Combs, J., Gupta, H.K., Jarzabek, D.C., 1976b. Seismic travel time delays and attenuation of anomaly associated with the East Mesa geothermal field. Imperial Valley, California. EOS Transaction of American Geophysical Union, Spring Annual Meeting Program 34 (abstract).
Combs, J., Muffler, L.J.P., 1973. Exploration for geothermal resources. In: Kruger, P., Otte, C. (Eds.), Geothermal Energy. Stanford University Press, Stanford, CA, pp. 95–128.
Combs, J., Rotstein, Y., 1976. Microearthquake studies at the Coso geothermal area, China lake, California. In: Proceedings of the 2nd U.N. Symposium on the Development and Use of Geothermal Resources, vol. 2. U.S. Government Printing Office, Washington, DC, pp. 209–216.
Compagnie Generale de Geophysique, 1963. Master curves for electrical sounding. European Association of Exploration Geophysicists.
Coolbaugh, M.F., 2003. The prediction and detection of geothermal systems at regional and local scales in Nevada using a geographic information system, spatial statistics, and thermal infrared imagery. Ph.D. thesis (unpublished), University of Nevada, Reno.
Coolbaugh, M.F., Taranik, J.V., Kruse, F.A., 2000. Mapping of surface geothermal anomalies at Steamboat Springs, Nevada using thermal infrared multispectral scanner (TIMS) and advanced visible and infrared imaging spectrometer (AVIRIS) data. In: Proceedings of the 14th Thematic Conference, Applied Geologic Remote Sensing, Environmental Research Institute of Michigan, Ann Arbor, MI, pp. 623–630.
Cooper, G.A., 2002. Proposal for the real-time measurement of drill bit tooth wear. In: Geothermal Energy: The Baseload Renewable Resource, Geothermal Resources Council Transactions 26, 189–192.
Corwin, R.F., Diaz, S., Rodriguez, J., 1978. Self-potential survey at the Cerro Prieto geothermal field, Baja California, Mexico. Geothermal Resources Council Transactions 2, 115–117.
Corwin, R.F., Hoover, D.B., 1979. The self-potential method in geothermal exploration. Geophysics 44, 226–245.
Corwin, R.F., Morrison, H.F., Diaz, S., Rodriguez, J., 1980. Self-potential studies at the Cerro Prieto geothermal field. Geothermics 9, 39–47.
Craig, H., 1963. The isotopic geochemistry of water and carbon in geothermal areas. In: Tongiorgi, E. (Ed.), Nuclear Geology in Geothermal Areas. Consiglio Nazionale delle Ricerche, Laboratorio di Geologia Nucleare, Pisa, pp. 17–53.
Craig, H., Boata, G., White, D.E., 1956. The isotopic geochemistry of thermal waters. In: Nuclear Geology in Geothermal Areas. National Research Council on Nuclear Science Series Report 19, 29–44.
Craig, H., Clarke, W.B., Beg, M.A., 1975. Excess ^{3}He in deep water on the East Pacific Rise. Earth Planetary Science Letters 26, 125–132.
Crecelius, E., Robertson, D., Fruchter, J., Ludwick, J., 1976. Chemical forms of mercury and arsenic emitted by a geothermal power plant. Abstract, 10th Annual Conference on Trace Substances in Environmental Health, University of Missouri.

Crittenden, M.D. Jr., 1981. Environmental aspects of geothermal energy development. In: Rybach, L., Muffler, L.J.P. (Eds.), Geothermal Systems: Principles and Case Histories. Wiley, New York, pp. 199–217.

D'Amore, F. (Ed.), 1992. Application of geochemistry in geothermal reservoir development. U.N. Institute for Training and Research, New York, 408pp.

Dalrymple, G.B., Silver, E.A., Jackson, E.D., 1973. Origin of the Hawaiian Islands. American Scientist 61, 294–308.

De la Funte, D.M., Summer, J.S., 1974. An aeromagnetic study of the Colorado River delta area, Baja California, Mexico. In: Conference on Geothermal Energy. California Institute of Technology, Pasadena, CA (abstract).

Deguen, J., Lebouteiller, D., Reford, M., Tiger, B., 1974. Shallow basin mapping with high-resolution aeromagnetics. 44th Meeting on Social Exploration Geophysicists, Dallas, TX (presented paper).

Del Grande, N.K., 1976. An advanced airborne infrared method for evaluating geothermal resources. In: Proceedings of the 2nd U.N. Symposium on the Development and Use of Geothermal Resources, vol. 2. U.S. Government Printing Office, Washington, DC, pp. 947–953.

Dench, N.D., 1973. Well measurements. In: Armstead, H.C.H. (Ed.), Geothermal Energy. UNESCO, Paris, pp. 85–96.

Denlinger, R.P., Kovach, R.L., 1981. Seismic reflection investigations at Castle Rock Springs in The Geysers geothermal area. In: McLaughlin, R.J., Donnelly-Nolan, J.M. (Eds.), Research in the Geyser—Clear Lake Geothermal Area, Northern California. U.S.G.S. Prof. Paper 1141, pp. 117–128.

Díaz, E., 1978. Pozos de Gradiente Cerro Prieto Oeste. Internal Report Comisión Federal de Electricidad, Mexicali.

Dickinson, D.J., 1973. Aerial infrared survey of Kawerau, Rotorua and Taupo urban areas—1972. Geophysics Division, New Zealand Department of Science and Industrial Research, Report 89, 53pp.

Dickinson, D.J., 1976. An airborne infrared survey of the Tauhara geothermal field, New Zealand. In: Proceedings of the 2nd U.N. Symposium on the Development and Use of Geothermal Resources, vol. 2. U.S. Government Printing Office, Washington, DC, pp. 955–961.

Dickson, M.H., Fanelli, M., 1995. Geothermal energy, Wiley, Chichester, 214pp.

DiPippo, R., 2004. Second Law assessment of binary plants generating power from low-temperature geothermal fluids. Geothermics 33, 565–586.

Dobrin, M.B., Lawrence, P.L., Sengbush, R.L., 1954. Surface and near surface waves in the Delaware basin. Geophysics 19, 695–715.

Dobrin, M.B., Savit, C.H., 1988. Introduction to geophysical prospecting (4th ed.). McGraw Hill, New York.

Dominguez, A.B., Vital, B.F., 1976. Repair and control of geothermal wells at Cerro Prieto, Baja California, Mexico. In: Proceedings of the 2nd U.N. Symposium on the Development and Use of Geothermal Resources, vol. 2, U.S. Government Printing Office, Washington, DC, pp. 1495–1499.

Donaldson, I.G., 1968. The flow of steam water mixtures through permeable beds: a simple simulation of a natural undisturbed hydrothermal region. New Zealand Journal of Science 11, 3–23.

Duffield, W.A., 1975. Late Cenozoic ring faulting and volcanism in the Coso range area of California. Geology 3, 335–338.
Duffield, W.A., Sass, J.H., 2003. Geothermal energy—clean power from the Earth's heat. U.S. Geological Survey Circular 1249, 36pp.
Duffy, T.S., Hemley, R.J., 1995. Some like it hot: the temperature structure of the Earth. Reviews of Geophysics 33 (Suppl.) (U.S. National report to IUGG, 1991–1994).
Dunstall, M.G., 2005. 2000–2005 New Zealand country update, In: Proceedings of the World Geothermal Congress, Antalya, Turkey, April 24–29.
Dupart, A., Omnes, G., 1976. The cost of geophysical programs in geothermal explorat-ion. In: Proceedings of the 2nd U.N. Symposium on the Development and Use of Geo-thermal Resources, vol. 2. U.S. Government Printing Office, Washington, DC, pp. 963–970.
Dziewonski, A., Landisman, M., Bloch, S., Sato, Y., Asano, S., 1968. Progress report on recent improvements in the analysis of surface wave observation. Journal of Physics of the Earth (Japan) 16, 1–26.
Elders, W.A., 1979. The geological background of the geothermal fields of the Salton Trough. In: Elders, W.A. (Ed.), Guidebook: Geology and Geothermics of the Salton Trough (Field Trip No. 7, Geological Society of American Annual Meeting, San Diego); University of California, Riverside Campus Museum Contributions, No. 5, pp. 1–19.
Elders, W.A., Bird, D.K., Williams, A.E., Schiffman, P., 1984. Hydrothermal flow regime and magmatic heat source of the Cerro Prieto geothermal stystem, Baja California, Mexico. Geothermics 13, 24–47.
Ellis, A.J., Mahon, W.A.J., 1964. Natural hydrothermal systems and experimental hot water/ rock interactions. Geochimica et Cosmochimica Acta 28, 1323–1357.
Ellis, A.J., Mahon, W.A.J., 1967. Natural hydrothermal systems and experimental hot water/ rock interactions (Part 2). Geochimica et Cosmochimica Acta 31, 519–539.
Ellis, A.J., Mahon, W.A.J., 1977. Chemistry and geothermal systems, Academic Press, 392pp.
Energy Alternatives:A Comparative Analysis, 1975. The science and public policy program, University of Oklahoma, Norman, OK.
Erkan, K., Blackwell, D.D., Leidig, M., 2005. Crustal thermal regime at The Geysers/Clear Lake area, California. In: Proceedings of the World Geothermal Congress, Antalya, Tur-key, April 24–29, 9pp.
Eysteinsson, H., 2000. Elevation and gravity changes at geothermal fields on the Reykjanes peninsula, SW Iceland. In: Proceedings of the World Geothermal Congress 2000, Japan, pp. 559–564.
Farmer, V.C. (Ed.), 1974. The infrared spectra of minerals. Mineral Society Monograph, vol. 4, London.
Faust, C.R., Mercer, J.W., 1976. Mathematical modeling of geothermal systems. In: Pro-ceedings of the 2nd U.N. Symposium on the Development and Use of Geothermal Re-sources, vol. 3. U.S. Government Printing Office, Washington, DC, pp. 1635–1641.
Fausto, J.J., Jiménez, M.E., Esquer, I., 1981. Current state of the hydrothermal geochemistry studies at Cerro Prieto. In: Proceedings of the 3rd Symposium on the Cerro Prieto geo-thermal field, March 24–26, San Francisco, California; Lawrence Berkeley Laboratory Report LBL-11967, pp. 188–220.
Fausto, J.J., Sánchez, A., Jiménez, M.E., Esquer, I., Ulloa, F., 1979. Hydrothermal geo-chemistry of the Cerro Prieto geothermal field. In: Proceedings of the 2nd Symposium on

the Cerro Prieto geothermal field, October 17–19, Mexicali, Baja California, Mexico, Comisión Federal de Electricidad, pp. 199–223.

Finger, J., Jacobson, R., 2000. Slimhole drilling, logging, and completion technology—an update. In: Proceedings of the World Geothermal Congress 2000, Kyushu—Tohoku, Japan, May 28 – June 10, pp. 2335–2339.

Finn, C.B.P., 1993. Thermal physics. Chapman and Hall, (2nd ed.) 249pp.

Fischer, W.A., Davis, B.A., Souza, T., 1966. Fresh water springs of Hawaii from infrared images. U.S. Geological Survey on Hydrologic Atlas, HA-218.

Flovenz, O.G., Georgeson, L.S., 1982. Prospecting for near-vertical aquifers in low-temperature geothermal areas in Iceland. Transaction, Geothermal Resource Council, 6, 19–22.

Fournier, R.O., 1979. A revised equation for the Na/K geothermometer. Geothermal Resources Council Transactions, 3, 221–224.

Fournier, R.O., 1981. Application of water geochemistry to geothermal exploration and reservoir engineering. In: Rybach, L., Muffler, L.J.P. (Eds.), Geothermal Systems: Principles and Case Histories. Wiley, pp. 109–143.

Fournier, R.O., Potter, R.W., 1979. Magnesium correction to the Na–K–Ca geothermometer. Geochimica et Cosmochimica Acta, 43, 1543–1550.

Fournier, R.O., Potter, R.W., 1982. An equation correlating the solubility of quartz in water from 25 to 900 °C at pressures up to 10,000 bars. Geochimica et Cosmochimica Acta, 46, 1969–1973.

Fournier, R.O., Truesdell, A.H., 1973. An empirical Na–K–Ca geothermometer for natural waters. Geochimica et Cosmochimica Acta, 37, 1255–1275.

Fournier, R.O., Truesdell, A.H., 1974. Geochemical indicators of subsurface temperatures, 2. Estimation of temperature and fraction of hot water mixed with cold water. U.S. Geological Survey, Journal Research, 2 (3), 263–269.

Fridleifsson, I.B., 2001. Geothermal energy for the benefit of the people. Renewable and Sustainable Energy Reviews, 5, 299–312.

Fridleifsson, I.B., 2003. Status of geothermal energy amongst the world's energy sources. Geothermics, 32, 379–388.

Frischknecht, F.G., 1967. Fields about an oscillating magnetic dipole over a two-layer earth and application to ground and airborne electromagnetic surveys. Quaterly, Colorado School of Mines, 62, 326.

Furlong, K.P., Chapman, D.S., 1987. Thermal state of the lithosphere. Reviews of Geophysics, 25, 1255–1264.

Garcia, S.D., 1976. Geoelectric study of the Cerro Prieto geothermal area, Baja California, Mexico. In: Proceedings of the 2nd U.N. Symposium on the Development and Use of Geothermal Resources, vol. 2, U.S. Government Printing Office, Washington, DC, pp. 1009–1011.

Gertson, R.C., Smith, R.B., 1979. Interpretation of a seismic refraction profile across the Roosevelt Hot Springs, Utah and vicinity. University of Utah, Department of Geology and Geophysics, Report. IDO/78–1701.a.3, March, 120pp.

Gettings, P., Harris, R.N., Allis, R.G., Chapman, D.S., 2002. Gravity signals at the Geysers geothermal system. In: Proceedings of the Geothermal Research Council Transactions, 26, 425–429.

Ghosh, P.K., 1954. Mineral springs of India. Records, Geological Survey of India, 80, 541–558.

Giggenbach, W.F., 1988. Geothermal solute equilibria: derivation of Na–K–Mg–Ca geoindicators. Geochimica et Cosmochimica Acta, 52, 2749–2765.

Giggenbach, W.F., 1991. Chemical techniques in geothermal exploration. In: D'Amore, F. (coordinator), Application of geochemistry in geothermal reservoir development. UNITAR/UNDP publication, Rome, pp. 119–142.

Giggenbach, W.F., 1992. Isotopic shifts in waters from geothermal and volcanic systems along convergent plate boundaries and their origin. Earth and Planetary Science Letters, 113, 495–510.

Giggenbach, W.F., Poreda, R.J., 1993. Helium isotopic and chemical composition of gases from volcanic-hydrothermal systems in the Philippines. Geothermics, 22, 369–380.

Glowacka, E., González, J., Fabriol, H., 1999. Recent vertical deformation in Mexicali Valley and its relationship with tectonics, seismicity, and the exploitation of the Cerro Prieto geothermal field, Mexico. Pure and Applied Geophysics, 156, 591–614.

Glowacka, E., González, J., Nava, F.A., 2000. Subsidence in Cerro Prieto geothermal field, Baja California, Mexico. In: Proceedings of the World Geothermal Congress, Tohoku-Kyushu, Japan, pp. 591–596.

Glowacka, E., Sarychikhina, O., Nava, F.A., 2005. Subsidence and stress change in the Cerro Prieto geothermal field, Baja California, Mexico. Pure and Applied Geophysics, 162, 2095–2110.

Goldstein, N.E., Halfman. S., Corwin, R. F., Alvarez, J.R., 1989. Self-potential anomaly changes at the East Mesa and Cerro Prieto geothermal fields. In: Proceedings of the 14th Workshop on Geothermal Reservoir Engineering, Stanford University, pp. 145–153.

Goldstein, N.E., Mozley, E., Wilt, M., 1982. Interpretation of shallow electrical features from electromagnetic and magnetotelluric surveys at Mount Hood, Oregon. Journal of Geophysical Research, 87, 2815–2828.

Grannell, R.B., 1980. The use of surface gravity methods in monitoring subsurface reservoir changes, with case studies at Cerro Prieto, Mexico, and Heber, California. Geothermal Research Council Transactions, 4, 49–52.

Grant, M.A., Truesdell, A.H., Mañón, A., 1984. Production induced boiling and cold water entry in the Cerro Prieto geothermal reservoir indicated by chemical and physical measurements. Geothermics, 13, 117–140.

Grauch, V.J.S., 2001. High-resolution aeromagnetic data, a new tool for mapping intrabasinal faults: an example from the Albuquerque basin, New Mexico, Geology, 29, 367–370.

Grauch, V.J.S., 2002. High-resolution aeromagnetic survey to image shallow faults, Dixie Valley geothermal field, Nevada. U.S. Geological Survey Open File Report, 02-384, 13pp.

Grauch, V.J.S., Hudson, M.R., Minor, S.A., 2001. Aeromagnetic expression of faults that offset basin fill, Albuquerque basin, New Mexico. Geophysics, 66, 707–720.

GSI, 1983. Status note on geothermal exploration in India by Geological Survey of India and a profile of the exploration strategy. Geological Survey of India, Kolkata.

GSI, 1991. Geothermal Atlas of India. Geological Survey of India, Special Publication, 19, 143pp.

GSI, 2002. Geothermal energy resources of India. Geological Survey of India, Special Publication, 69, 210pp.

Guha, S.K., 1986. Status of geothermal exploration for geothermal resources in India. Geothermics, 15, 665–675.

Gupta, H.K., 1980. Geothermal resources: an energy alternative. Elsevier, Amsterdam, 227pp.

Gupta, H.K., 1992. Reservoir induced earthquakes. Elsevier, Amsterdam, 355pp.
Gupta, H.K., 2003. Energy from the oceans. Prof. C. Karunakaran Endowment Lecture Series, No. 4, Centre for Earth Science Studies, Thiruvananthapuram, India, May 28, 15pp.
Gupta, H.K., Nyman, D.C., 1977a. Short period surface wave dispersion studies of geothermal areas. EOS Transactions of American Geophysical Union, 58, 541 (abstract).
Gupta, H.K., Nyman, D.C., 1977b. Short period surface wave dispersion studies in the East Mesa geothermal field, California. Geothermal Resources Council Transactions, 1, 123–126.
Gupta, H.K., Nyman, D.C., Landisman, M., 1977. Shield like upper mantle structure inferred from long period Rayleigh and Love wave dispersion investigations in the Middle East and South East Asia. Bulletin of the Seismological Society of America, 67, 103–119.
Gupta, H.K., Rastogi, B.K., 1976. Dams and earthquakes. Elsevier Scientific Publishing Company, Amsterdam, 229pp.
Gupta, H.K., Ward, R.W., Lin T.-L., 1982. Seismic wave velocity investigations at The Geysers–Clear Lake geothermal field, California. Geophysics, 47, 819–824.
Gupta, M.L., 1974. Geothermal resources of some Himalayan hot spring areas. Himalayan Geology, 4, 492–515.
Gupta, M.L., 1992. Geothermal resources of India: geotectonic associations, present status and perspectives. In: Gupta, M.L., Yamano, M. (Eds.), Terrestrial Heat Flow and Geothermal Energy in Asia, Oxford and IBH, Delhi, pp. 311–329.
Gupta, M.L., Rao, G.V., Hari Narain, 1974. Geothermal investigations in the Puga Valley hot spring region, Ladakh, India. Geophysical Research Bulletin, 12, 119–136.
Gupta, M.L., Rao, G.V., Singh, S.B., Rao, K.M.L., 1973. Report on thermal studies and resistivity surveys in the Puga Valley geothermal field, Ladakh district, J&K. NGRI Technical Report 73–79, National Geophysical Research Institute, Hyderabad, India.
Gutiérrez-Negrín L.C.A., Quijano-León J.L., 2005. Update of geothermics in Mexico. In: Proceedings of the World Geothermal Congress, Antalya, Turkey, April 24–29.
Haenel, R., Rybach, L., Stegena, L. (Eds.), 1988. Handbook of terrestrial heat flow density determination. Kluwer Academic, Dordrecht, 486pp.
Hagiwara, Y., 1966. Studies of the thermal state of the Earth. The 20th paper: mountain formation and thermal conduction. Bulletin of the Earthquake Research Institute, 44, 1537–1551.
Halfman, S.E., Lippmann, M.J., Zelwer, R., Howard, J.H., 1984. A geologic interpretation of geothermal fluid movement in Cerro Prieto field, Baja California, Mexico. American Association Of Petroleum Geologists Bulletin, 68, 18–30.
Halfman, S.E., Maiibn, A., Lippmann, M.J., 1986. Update of the hydrogeologic model of the Cerro Prieto field based on recent well data. Geothermal Resources Council Transactions, 10, 369–375.
Hamilton, R.M., Muffler, L.J.P., 1972. Microearthquakes at The Geysers geothermal area, California. Journal of Geophysical Research, 77, 2081–2086.
Hayakawa, M., 1966. Geophysical study of Matsukawa geothermal area, Iwata Prefecture, Japan. Bulletin of Volcanology, 27, 499–515.
Hayakawa, M., 1970. The study of underground structure and geophysical state in geothermal areas by seismic exploration. Geothermics, Special Issue, 2, 347–357.
Healy, J., 1970. Pre-investigation geological appraisal of geothermal fields. In: Proceedings of the U.N. Symposium on the Development and Utilization of Geothermal Resources. Geothermics, Special issue, 2 (1), 571–577.

Hellman, M.J., Ramsey, M.S., 2004. Analysis of hot springs and associated deposits in Yellowstone National Park using ASTER and AVIRIS remote sensing. Journal of Volcanology and Geothermal Research, 135, 195–219.

Henley, R.W., Truesdell, A.H., Barton, P.B. Jr., Whitney, J.A., 1984. Fluid-mineral equilibria in hydrothermal systems. Reviews in Economic Geology, 1, 267pp.

Hill, D.P., 1976. Structure of Long Valley caldera, California from a seismic refraction experiment. Journal of Geophysical Research, 81, 145–153.

Hill, D.P., Mooney, W.D., Fuis, G.S., Healy, J.H., 1981. Evidence on the structure and tectonic environment of the volcanoes in the Cascade Range, Oregon and Washington, from seismic refraction/reflection experiments. Presented at the 51st Annual International Meeting and Expositions, Society of Exploration Geophysicists, Los Angeles.

Hill, D.P., Mowinckel, P., Pake, L.G., 1975. Earthquakes, active faults and geothermal areas in the Imperial Valley, California. Science, 188,1306–1308.

Hill, R., 2004. Geopowering the West. Geothermal Resources Council Bulletin, 33, 63–66.

Hiramatsu, Y., Kokado, J., 1958. Eine Untersuchung über die Kühlung von Gruben durch den Wetterstrom. Bergbau Archives, 19, 16–40.

Hiriart, G., 2001. A bright future—geothermal energy development in Mexico and other Latin American countries has great potential. Geothermeral Research Council Bulletin, 30, 145–149.

Hiriart, G., Andaluz, J.I., 2000. Strategies and economics of geothermal power development in Mexico. In: Proceedings of the World Geothermal Congress 2000, Kyushu-Tohoku, Japan, May 28 – June 10, pp. 799–802.

Hochstein, M.P., Dickinson, D.J., 1970. Infrared remote sensing of thermal ground in the Taupo region, New Zealand. In: Proceedings of the U.N. Symposium on the Development and Utilization of Geothermal Resources. Geothermics, Special Issue, 2 (1), 420–423.

Hochstein, M.P., Hunt, T.M., 1970. Seismic, gravity and magnetic studies, Broadlands geothermal field, New Zealand. In: Proceedings of the 2nd U.N. Symposium on the Development and Utilization of Geothermal Resources. Geothermics, Special Issue, 2 (1), 333–346.

Hodder, D.T., 1970. Application of remote sensing to geothermal prospecting. In: Proceedings of the 2nd U.N. Symposium on the Development and Utilization of Geothermal Resources. Geothermics, Special Issue, 2 (1), 368–380.

Hodder, D.T., Martin, R.C., Calamai, A., Cataldi, R., 1973. Remote sensing of Italian geothermal steam field by infrared scanning. In: Proceedings of the 1st Pan-American Symposium on Remote Sensing.

Hofmeister, A.M., Criss, R.E., 2005. Earth's heat flux revised and linked to chemistry. Tectonophysics, 395, 159–177.

Hoover, D.B., Long, C.L., 1976. Audio-magnetotelluric methods in reconnaissance geothermal exploration. In: Proceedings of the 2nd U.N. Symposium on the Development and Use of Geothermal resources, U.S. Government Printing Office, Washington, DC, vol. 2, pp. 1059–1064.

Hot Springs Committee, 1968. Report of Hot Springs Committee. Government of India, unpublished.

Howard, J.H. (Ed.), 1975. Present status and future prospects for non-electrical uses of geothermal resources. NATO CCMS Report, No. 40. Lawrence Livermore Laboratory, University of California, Livermore, Calif., 161pp.

Hudson, R.B., 1995. Electricity generation. In: Dickson, M.H., Fanelli, M. (Eds.), Geothermal Energy,Wiley, Chichester, pp. 39–72.
Human Development Report, 2004. United Nations Development Program (UNDP). Website: http://hdr.undp.org/2004
Hunsbedt, A., London, A.L., Kruger, P., 1976. Laboratory studies of stimulated geothermal reservoirs. In: Proceedings of the 2nd U.N. Symposium on the Development and Use of Geothermal Resources, U.S. Government Printing Office, Washington, DC, vol. 3, pp. 1663–1671.
Hunt, G., 1977. Spectral signatures of particulate minerals in the visible and near infrared. Geophysics, 42 (3), 501–513.
Hunt, T., 2001. Five lectures on environmental effects on geothermal energy utilization. United Nations University Geothermal Training Programme 2000, Report 1, Reykjavík, Iceland, 109pp.
Hunt, T.M., Lattan, J.H., 1982. A survey of seismic activity near Wairakei geothermal field, New Zealand. Journal of Volcanology and Geothermal Research, 14, 319–334.
Hurtig, E., Grobwig, S., Jobmann, M., Kühn, K., Marschall, P., 1994. Fibre-optic temperature measurements in shallow boreholes: experimental application for fluid logging. Geothermics, 23, 355–364.
Hutton, V.R.S., Galanopoulos, D., Dawes, G.J.K., Pickup, G.E., 1989. A high-resolution magnetotelluric survey of the Milos geothermal prospect. Geothermics, 18, 521–532.
Huttrer, G.W., 2001. The status of world geothermal power generation 1995–2000. Geothermics, 30, 1–27.
Ibrahim R., Fauzi A., Suryadarman, 2005. The progress of geothermal energy resources activities in Indonesia. In: Proceedings of the World Geothermal Congress, Antalya, Turkey, April 24–29.
International Energy Annual, 2003. Energy Information Administration. Website: www.eia.-doe.gov
International Energy Outlook, 2005. Energy Information Administration. Website: www.eia.-doe.org
Isherwood, W.F., 1976. Complete Bouger gravity map of the Geysers area, California. U.S. Geologycal Survey Open File Report 76–357.
Isherwood, W.F., Mabey, D.R., 1978. Evaluation of Baltazor known geothermal resource area, Nevada. Geothermics, 7, 221–229.
Ishido, T., Kikuchi, T., Sugihara, M., 1987. The electrokinetic mechanism of hydrothermal-circulation related and production-induced self-potentials. In: Proceedings of the 12th Workshop Geothermal Reservoir Engineering, Stanford University, SGP-TR-109, p. 285–290.
Ishido T., Pritchett, J.W., 2000. Using numerical simulation of electrokinetic potentials in geothermal reservoir management. In: Proceedings of the World Geothermal Congress 2000,vol. 4, 2629–2634.
Iwata, S., Nakano, Y., Granados, E., Butler, S., Robertson-Tait, A., 2002. Mitigation of cyclic production behavior in a geothermal well at the Uenotai geothermal field, Japan. In: Geothermal Energy: The Baseload Renewable Resource, Geothermal Resources Council Transactions, 26, 193–196.
Iyer, H.M., 1978. Status of seismic methods in geothermal application. In: Memories from an APSMAGS Workshop. FT Burgwin Research Center, Taos, NM, pp. 11–26.

Iyer, H.M., Hitchcock, T., 1976. Seismic noise survey in Long Valley, California. Journal of Geophysical Research, 81, 821–840.
Iyer, H.M., Oppenheimer, D.H., Hitchcock, T., 1979. Abnormal *P*-wave delays in The Geysers–Clear Lake geothermal area, California. Science, 204, 495–497.
Iyer, H.M., Stewart, R.M., 1977. Teleseismic technique to locate magma in the crust and upper mantle. In: Dick, H.J.B. (Ed.), Magma Genesis, Oregon Deptartment, of Geological Mining Industry Bulletin, 96, 281–299.
Jackson, D.D., Hill, J.H., 1976. Possibilities for controlling heavy metal sulfides in scale from geothermal brines. UCRL-52977, Lawrence Livermore Laboratory, University of California, Livermore, Calif., 14pp.
Jacob, M., 1964. Heat transfer. Wiley, New York, NY, 1, 758 pp.
Jaeger, J.C., 1961. The effect of the drilling fluid on temperatures measured in boreholes. Journal of Geophysical Research, 66, 563–569.
James, R., 1976. Possible serious effect of steam on water flow measurements. In: Proceedings of the 2nd U.N. Symposium on the Development and Use of Geothermal Resources, U.S. Government Printing Office, Washington, DC, vol. 3, pp. 1703–1706.
James, R., McDowell, G.D., Allen, M.D., 1970. Flow of steam water mixtures through a 12-inch diameter pipeline. In: Proceedings of the U.N. Symposium on the Development and Utilization of Geothermal Resources. Geothermics, Special Issue, vol. 2,1581–1597.
Jeanloz, R., Morris, S., 1986. Temperature distribution in the crust and mantle. Annual Review of Earth and Planetery Sciences, 14, 377–415.
Jessop, A.M., 1990. Thermal geophysics. Elsevier, Amsterdam, 306pp.
Jones, P.H., 1970. Geothermal resources of the Northern Gulf of Mexico basin. In: Proceedings of the U.N. Symposium on the Development and Utilization of Geothermal Resources. Geothermics, Special Issue, 2 (1), 14–26.
Kalina, A.L., 1984. Combined cycle system with novel bottoming cycle. ASME Journal of Engineering for Gas Turbines and Power, 106, 737–742.
Kappelmeyer, O., Haenel, R., 1974. Geothermics with special reference to application. Gebrüder Borntraeger, Berlin, 238pp.
Kathiroli, S., Jalihal, P., Singh, R., 2006. Low temperature thermal desalination plant at Kavaratti, Lakshadweep. Geological Society of India, 67, 820–822.
Kauahikaua, J., 1981. Interpretation of time-domain electromagnetic soundings in the East Rift geothermal area of Kilauea volcano, Hawaii. U.S. Geological Survey Open File Report, 81–979.
Kawazoe, S., Shirakura, N., 2005. Geothermal power generation and direct use in Japan. In: Proceedings of the World Geothermal Congress, Antalya, Turkey, April 24–29.
Keller, G.V., 1970. Induction methods in prospecting for hot water. In: Proceedings of the Symposium on the Development and Utilization of Geothermal Resources. Geothermics, Special Issue, 2 (1), 318–332.
Keller, G.V., Frischknecht, F.C., 1966. Electrical methods in geophysical prospecting. Pergamon, New York, NY, 519pp.
Keller, G.V., Rapolla, A., 1974. Electric prospecting methods in volcanic areas. In: Civetta, L., Gasparini, P., Luongo, G., Rapolla, A. (Eds.), Physical Volcanology. Elsevier, Amsterdam, pp. 133–166.
Keller, G.V., Taylor, K., Santo, J.M., 1981. Megasource EM method for detecting buried conductive zones in geothermal exploration. In: Proceedings of the 51st Annual International Meeting and Exposition, Society Exploration Geophysicisits, Los Angeles.

Klee, G., 2005. The European hot-dry-rock project in the tectonic regime of the upper Rhine graben. In: Rummel, F. (Ed.), Rock Mechanics with Emphasis on Stress, Oxford & IBH, New Delhi, pp. 169–183.

Kononov, V., Povarov, O., 2005. Geothermal development in Russia: country update report 2000–2004. In: Proceedings of the World Geothermal Congress, Antalya, Turkey, April 24–29.

Kononov, V.I., Polak, B.G., 1976. Indicators of abyssal heat recharge of recent hydrothermal phenomenon. In: Proceedings of the 2nd U.N. Symposium on the Development and Use of Geothermal Resources, vol. 1, U.S. Government Printing Office, Washington, DC, pp. 767–773.

Kratt, C., 2005. Geothermal exploration with remote sensing from 0.45 to 2.5 μm over Brady–Desert Peak, Churchill County, Nevada. M.S. thesis, Univesity of Nevada, Reno.

Krishnaswamy, V.S., Ravi Shanker, 1980. Scope of development, exploration and preliminary assessment of the geothermal resource potentials of India. Records, Geological Survey of India, 111, Pt. II, pp. 17–40.

Kristmannsdóttir, H., Ármannsson, H., 2003. Environmental aspects of geothermal energy utilization. Geothermics. 32, 451–461.

Kruse, F., 1999. Mapping hot spring deposits with AVIRIS at Steamboat Springs, Nevada. In: Proceedings of the AVIRIS Airborne Geoscience Workshop, JPL Publishers, Pasadena, pp. 1–7.

Lachenbruch, A.H., 1978. Heat flow in the Basin and Range Province and thermal effects of tectonic extension. Pure and Applied Geophysics, 117, 34–50.

Lachenbruch, A.H., Sass, J.H., 1977. Heat flow in the United States and thermal properties of the crust. In: Heacock, J.G. (Ed.), The Earth's Crust, Geophysics Monograph, vol. 20, American Geophysical Union, Washington, DC, pp. 626–675.

Lachenbruch, A.H., Sass, J.H., Galasnis, S.P. Jr., 1985. Heat flow in southernmost California and the origin of the Salton Trough. Journal of Geophysical Research, 90, 6709–6736.

Lachenbruch, A.H., Sass, J.H., Munroe, R.J., Moses, T.H., Jr., 1976b. Geothermal setting and simple heat conduction models for the Long Valley Caldera. Journal of Geophysical Research, 81, 769–784.

Lachenbruch, A.H., Sorey, M.L., Lewis, R.E., Sass, J.H., 1976a. The near-surface hydrothermal regime of Long Valley Caldera. Journal of Geophysical Research, 81, 763–768.

Lange, A.L., Westphal, W.H., 1969. Microearthquakes near the Geysers, Sonoma County, California. Journal of Geophysical Research, 74, 4377–4382.

Layton, D.W., Anspaugh, L.R., O'Banion, K.D., 1981. Health and environmental effects. Document on geothermal energy – 1981. Lawrence Livermore Laboratory, Report UCRL-53232, Livermore, California, 61 pp.

Lewis, C.R., Rose, S.C., 1970. A theory relating high temperatures and over-pressures. Journal of Petroleum Technology, 22, 11–16.

Lienau, P., 1995. Industrial applications. In: Dickson, M.H., Fanelli, M. (Eds.), Geothermal Energy, Wiley, Chichester, pp. 169–206.

Lienau, P.J., 1992. Data acquisition for low-temperature geothermal well tests and long-term monitoring. U.S. Department of Energy Report, 42pp.

Lindal, B., 1973. Industrial and other applications of geothermal energy. In: Armstead, H.C.H. (Ed.), Geothermal Energy, UNESCO, Paris, pp. 135–148.

Lippmann, M.J., 1983. Overview of Cerro Prieto studies. Geothermics, 12, 265–289.

Lippmann, M.J., Golstein, N.E., Halfman, S.E., Witherspoon, P.A., 1983. Exploration and development of the Cerro Prieto geothermal field. Lawrence Berkeley Laboratory Report. LBL-15594, 16pp.

Lippmann, M.J., Truesdell, A.H., Halfman-Dooley, S.E., Mañón, M.A., 1991. A review of the hydrogeologic-geochemical model for Cerro Prieto. Geothermics, 20, 39–52.

Lippmann, M.J., Truesdell, A.H., Mañón, M.A., Halfman, S.E., 1989. The hydrogeologic-geochemical model of Cerro Prieto revisited. Lawrence Berkeley Laboratory, University of California, Report LBL-26819, 10pp.

Lippmann, M.J., Truesdell, A.H., Rodríguez, M.H., Pérez, A., 2004. Response of Cerro Prieto II and III (Mexico) to exploitation. Geothermics, 33, 229–256.

Lorensen, L.E., Walkup, C.M., Mones, E.T., 1976. Polymeric and composite materials for use in systems utilizing hot, flowing geothermal brine. In: Proceedings of the 2nd U.N. Symposium on the Development and Use of Geothermal Resources, vol. 3, U.S. Government Printing Office, Washington, DC, pp. 1725–1731.

Lubimova, E.A., 1969. Thermal history of the Earth. In: Hart, P.J. (Ed.), The Earth's Crust and Upper Mantle. AGU Monograph, 13, 63–77.

Lumb, J.T., 1981. Prospecting for geothermal resources. In: Rybach, L., Muffler, L.J.P. (Eds.), Geothermal Systems: Principles and Case Histories. Wiley, pp. 77–108.

Lund, J.W., 2004. 100 years of geothermal power production. Geo-Heat Center Bulletin, 25, 3, 11–19.

Lund, J.W., Freeston, D.H., Boyd, T.L., 2005a. Direct application of geothermal energy: 2005 worldwide review. Geothermics, 34, 691–727.

Lund, J.W., Bloomquist, R.G., Boyd, T.L., Renner, J., 2005b. The United States of America country update. In: Proceedings of the World Geothermal Congress, Antalya, Turkey, April 24–29.

Lund, J.W., Freeston, D.H., 2001. Worldwide direct uses of geothermal energy 2000. Geothermics, 30,29–68.

Lund, J.W., Lienau, P.J., Lunis, B.C., 1998. Geothermal direct-use engineering and design guidebook, Geo-Heat Center, Klamath Falls, Oregon, 470pp.

Lyons, D.J., van de Kamp, P.C., 1980. Subsurface geological and geophysical study of the Cerro Prieto geothermal field, Baja California, Mexico. Lawrence Berkeley Laboratory Report, LBL-10540, CP-11, University of California, 95pp.

Macdonald, W.J.P., Muffler, L.J.P., 1972. Recent geophysical exploration of the Kawerau geothermal field, North Island, New Zealand. New Zealand Journal of Geology and Geophysics, 15 (3), 303.

Mainieri, P.A., 2005. Costa Rica country update. In: Proceedings of the World Geothermal Congress, Antalya, Turkey, April 24–29.

Majer, E.L., 1978. Seismological investigations in geothermal areas. Lawrence Berkeley Laboratory Report, LBL-7054, 225pp.

Majer, E.L., McEvilly, T.V., 1979. Seismological investigations at The Geysers geothermal field. Geophysics, 44, 246–269.

Mañón, A., de la Pena, A., Puente, I., 1978. El Campo Geotermico de Cerro Prieto—Estudios Realizados y Programas Futuros. Paper presented at the OLADE Symposium on Geothermal Exploration in Ecuador.

Mañón, A., Sánchez, A., Fausto, J.J., Jiménez, M.E., Jacobo, A., Esquer, I., 1979. Preliminary geochemical model of the Cerro Prieto geothermal field. Geothermics, 8, 211–222.

Marinelli, G., 1964. Les anomalies thermiques et les champs géothermiques dans le cadre des intrusions récentes en Toscane. In: Proceedings of the U.N. Conference on New Resources of Energy, vol. 2, pp. 288–291.
Marsh, S.E., Lyon, R.J.P., Honey, F., 1976. Evaluation of NOAA satellite data for geothermal reconnaissance studies. In: Proceedings of the 2nd U.N. Symposium on the Development and Use of Geothermal Resources, U.S. Government Printing Office, Washington, DC, vol. 2, pp. 1135–1141.
Marshall, T., Braithwaite, W.R., 1973. Corrosion control in geothermal systems. In: Armstead, H.C.H. (Ed.), Geothermal Energy. UNESCO, Paris, pp. 151–160.
Marshall, T., Tombs, A., 1969. Delayed fracture of geothermal bore casing steels. Australian Corrossion Engineering, 13, 2–8.
Martini, B.A., Silver, E.A., Pickles, W.L., Cocks, P.A., 2003. Hyperspectral mineral mapping in support of geothermal exploration: examples from Long Valley Caldera, California and Dixie Valley, Nevada. Geothermal Resources Council Transaction, 27, 657–662.
Matsuo, K., 1973. Drilling for geothermal steam and hot water. In: Armstead, H.C.H. (Ed.), Geothermal Energy. UNESCO, Paris, pp. 73–83.
Matsushima, N., Kikuchi, T., Tosha, T., Nakao, S., Yano, Y., Ishido, T., Hatakeyama, K., Ariki, K., 2000. Repeat SP measurements at the Sumikawa geothermal field, Japan. In: Proceedings of the World Geothermal Congress, vol. 4, pp. 2725–2730.
Mayorga, A.Z., 2005. Nicaragua country update. In: Proceedings of the World Geothermal Congress 2005, Antalya, Turkey, April 24–29, 5pp.
Mazor, E., Truesdell, A.H., 1984. Dynamics of a geothermal field traced by noble gases: Cerro Prieto, Mexico. Geothermics, 13, 91–102.
McDougall, I., 1971. Volcanic island chains and sea floor spreading. Nature, 231, 141–144.
McEuen, R.B., 1970. Delineation of geothermal deposits by means of long-spacing resistivity and airborne magnetics. In: Proceedings of the U.N. Symposium on the Development and Utitization of Geothermal Resources. Geothermics, Special Issue, 2 (1), 295–302.
McKenzie, W.F., Truesdell, A.H., 1977. Geothermal reservoir temperatures estimated from the oxygen isotope compositions of dissolved sulfate and water from hot springs and shallow drillholes. Geothermics, 5, 51–61.
McNitt, J.R., 1973. The role of geology and hydrology in geothermal exploration. In: Armstead, H.C.H. (Ed.), Geothermal Energy. UNESCO, Paris, pp. 33–40.
McNitt, J.R., 1976. Summary of United Nations geothermal exploration experiences. In: Proceedings of the 2nd U.N. Symposium on the Development and Use of Geothermal Resources, vol. 2, U.S. Government Printing Office, Washington, DC, pp. 1127–1134.
Meidav, T., 1970. Application of electrical resistivity and gravimetry in deep geothermal exploration. In: Proceedings of the U.N. Symposium on the Development and Utilization of Geothermal Resources. Geothermics, Special Issue, 2 (1), 303–310.
Meidav, T., 1972. Electrical resistivity in geothermal exploration. Annual Meeting on Society, of Exploration Geophysicists, Anaheim, California (presented paper).
Meidav, T., Furgerson, R., 1972. Resistivity studies of the Imperial Valley, California. Geothermics, 1 (2), 47–62.
Meidav, T., Rex, R.W., 1970. Investigation of geothermal resources in the Imperial Valley and their potential value for desalination for water and electricity production. Institute of Geophysics and Planetary Physics, University of California, Riverside, California, 54pp.

Meidav, T., Tonani, F., 1976. A critique of geothermal exploration techniques. In: Proceedings of the 2nd U.N. Symposium on the Development and Use of Geothermal Resources, vol. 2, U.S. Government Printing Office, Washington, DC, pp. 1143–1154.

Mercado, G.S., 1968. Localization of zones of maximum hydrothermal activity by means of chemical ratios. Cerro Prieto Geothermal Field, Baja California, Mexico. CFE Archives.

Mercado, G.S., 1976. Movement of geothermal fluids and temperature distribution in the Cerro Prieto geothermal field, Baja California, Mexico. In: Proceedings of the 2nd U.N. Symposium on the Development and Use of Geothermal Resources, vol. 1, U.S. Government Printing Office, Washington, DC, pp. 492–498.

Mercado, G.S., 1977. Geothermal chemistry of some geothermal fields of Mexico. Paper presented at the Geological Society of America, 1977 Annual Meeting, Washington.

Mercado, G.S., Samaneigo, F., 1978. An example of fluid migration between the layers of the Cerro Prieto geothermal field. Lawrence Berkeley Laboratory Report, LBL-8883, 95–107.

Mizutani, Y., Rafter, T.A., 1969. Oxygen isotopic composition of sulfates—Part 3: Oxygen isotope fractionationi the bisulfate ion–water system. New Zealand Journal of Science, 12, 54–59.

Molina, B.R., Banwell, C.J., 1970. Chemical studies in Mexican geothermal fields. In: Proceedings of the U.N. Symposium on the Development and Utilization of Geothermal Resources. Geothermics, Special; Issue, 2 (2), 1377–1391.

Moon, B.R., Dharam, P., 1988. Geothermal energy in India, present status and future prospects. Geothermics, 17, 439–449.

Morgan, W.J., 1972. Plate motions and deep mantle convection. In: Shagham, R. (Ed.), Studies in Earth and Space Sciences. Geological Society of America Memoirs 132, 7–22.

Moskowitz, B., Norton, D., 1977. A preliminary analysis of intrinsic fluid and rock resistivity in active hydrothermal systems. Journal of Geophysical Research, 82, 5787–5795.

Muffler, L.J.P., 1976. Tectonic and hydrologic control of the nature and distribution of geothermal resources. In: Proceedings of the 2nd U.N. Symposium on the Development and Use of Geothermal Resources, vol. 1, U.S. Government Printing Office, Washington, DC, pp. 499–507.

Muffler, L.J.P., 1979. Assessment of geothermal resources of the United States—1978. U.S. Geological Survey Circular, 790, 163pp.

Muffler, L.J.P., White, D.E., 1969. Active metamorphism of Upper Cenozoic sediments in the Salton Sea geothermal field and the Salton Trough, southeastern California. Bulletin of Geological Society of America, 80, 157–182.

Mundry, E., 1966. Berechnung des gestörten geothermischen Feldes mit Hilfe des Relaxations—Verfahrens. Z. Geophys., 3 (32), 157–162.

Mwangi, M., 2005. Country update report for Kenya 2000–2005. In: Proceedings of the World Geothermal Congress, Antalya, Turkey, April 24–29.

Nakaoka, T., Ikegami, Y., Tuda, M., Uehara, H., 1997. Bulletin of the Society of Sea Water Science, Japan, 51 (6), 375 [in Japanese].

Narasimhan, T.N., Goyal, K.P., 1984. Subsidence due to geothermal fluid withdrawal in man-induced land subsidence. In: Holzer, T.L. (Ed.), Review on Engineering Geology, Geological Society of America, Boulder, Colorado, 6, pp. 35–66.

Nash, G.D., Johnson, G.W., Johnson, S., 2004. Hyperspectral detection of geothermal system-related soil mineralogy anomalies in Dixie Valley, Nevada: a tool for exploration. Geothermics, 33, 695–711.

Nehring, N.L., D'Amore, F., 1984. Gas chemistry and thermometry of the Cerro Prieto, Mexico, geothermal field. Geothermics, 13, 75–89.

Nettleton, L.L., 1971. Elementary gravity and magnetics for geologists and seismologists. Society of Exploration Geophysicists, Tulsa, Okla, 121pp.

Newman, G.A., Hoversten, M., Gasperikova, E., Wannamaker, P.E., 2005. 3D magnetotelluric characterization of the Coso geothermal field. In: Proceedings of the 13th Workshop on Geothermal Reservoir Engineering, Stanford University, Stanford, California, Jan 31–Feb 2.

Nicholson, K., 1993. Geothermal fluids: chemistry and exploration techniques. Springer-Verlag, Berlin, 263pp.

NIOT, 2005. Annual report 2004–2005. National Institute of Ocean Technology, Chennai, India, pp. 3–5.

Nishino, T., Kainuma, N., Matsuyama, K., Muraoka, A., 2000. Self-potential survey in Hachijojima. In: Proceedings of the World Geothermal Congress, vol. 4, pp. 2761–2766.

Noble, J.W., Ojiambo, S.B., 1976. Geothermal exploration in Kenya. In: Proceedings of the 2nd U.N. Symposium on the Development and Use of Geothermal Resources, vol. 1, U.S. Government Printing Office, Washington, DC, pp. 189–204.

Norton, D.L., 1984. Theory of hydrothermal systems. Annual Review of Earth and Planetary Sciences, 12, 155–177.

NREL, 2006. Ocean thermal energy conversion. National Renewable Energy Laboratoy. Website: http://www.nrel.gov/otec (as on April 28).

Nusselt, W., Jurges, W., 1922. Die Kühlung einer ebenen Wand durch einen Luftstrom. Gesund.-Ing., 45, 641–642.

Nyman, D.C., Landisman, M., 1977. The display equalized filter for frequency-time analysis. Bulletin of the Seismology Society of America, 67, 393–404.

Oldham, T., 1882. Thermal springs of India. Memoirs, Geological Survey of India, 10, part 2.

Palmason, G., 1971. Crustal structure of Iceland from explosion seismology. Society Science Islandica Publication, 40, 187.

Palmason, G., 1976. Geophysical methods in geothermal exploration. In: Proceedings of the 2nd U.N. Symposium on the Development and Use of Geothermal Resources, vol. 2, U.S. Government Printing Office, Washington, DC, pp. 1175–1184.

Palmason, G., Friedman, J.D., Williams, R.S. Jr., Jonsson, J., Saemundsson, K., 1970. Aerial infrared surveys of Reykjanes and Torfajökull thermal areas, Iceland, with a section on cost of exploration surveys. In: Proceedings of the U.N. Symposium on the Development and Utilization of Geothermal Resources. Geothermics, Special Issue, 2 (1), 399–412.

Papadopulos, S.S., Wallace, R.H., Jr., Wesselman, J.B., Taylor, R.E., 1975. Assessment of onshore geopressured-geothermal resources in the Northern Gulf of Mexico basin. In: White, D.E., Williams, D.L. (Eds.), Assessment of Geothermal Resources of the United States—1975. U.S. Geological Survey Circular, 726, 125–146.

Pasternak, A.D., 2000. Global energy futures and human development: a framework for analysis. Lawrence Livermore National Laboratory, U.S. Deptartment of Energy, Report UCRL-ID-140773, 25pp.

Pollack, H.N., Hurter, S.J., Johnson, J.R., 1993. Heat flow from the Earth's interior: analysis of the global data set. Reviews of Geophysics, 31, 267–280.

Popov, Y.A., Berezin, V.V., Semionov, V.G., Korosteliov, V.M., 1985. Complex detailed investigations of the thermal properties of rocks on the basis of a moving point source. Izvestiya, Physics of the Solid Earth, 1, 64–70.

Popov, Y.A., Pimenov, V.P., Tertychnyi, V., 2001. Developments of geothermal investigations of oil and gas fields. Oil-Gas Review, 2, 4–11.

Popov, Y. A., Pribnow, D.F.C., Sass, J.H., Williams, C.F., Burkhardt, H., 1999. Characterization of rock thermal conductivity by high-resolution optical scanning. Geothermics, 28, 253–276.

Press, F., Siever, R., 1998. Understanding Earth. W.H. Freeman and Co., 2nd ed., 682pp.

Pribnow, D.F.C., Williams, C.F., Sass, J.H., Keating, R., 1996. Thermal conductivity of water saturated rocks from the KTB pilot hole at temperatures of 25–300 °C. Geophysical Research Letters, 23, 391–394.

Quijano-León, J.L., Gutiérrez-Negrín, L.C.A., 2003. An unfinished journey: 30 years of geothermal-electric generation in Mexico. Geothermal Research Council Bulletin, 30, 198–203.

Ragnarsson, A., 2005. Geothermal development in Iceland 2000–2004. In: Proceedings of the World Geothermal Congress, Antalya, Turkey, April 24–29.

Raleigh, C.B., Healy, J.H., Bredehoeft, H.D., 1972. Faulting and crustal stress at Rangely, Colorado. In: Heard, H.C., Bord, I.Y., Carter, N.L., Raleigh, C.B. (Eds.), Flow and Fracture of Rocks. American Geophysical Union, Geophysical Monograph, 16, 275–284.

Ramey, H.J., Jr., 1976. Pressure transient analysis for geothermal wells. In: Proceedings of the 2nd U.N. Symposium on the Development and Use Geothermal Resources, vol. 3, U.S. Government Printing Office, Washington, DC, , pp. 1749–1757.

Ramey, H.J., Kruger Jr., P., London, A.L., Brigham W.E., 1976. Geothermal reservoir engineering research at Stanford University. In: Proceedings of the 2nd U.N. Symposium on the Development and Use of Geothermal Resources, U.S. Government Printing Office, Washington, DC, vol. 3, pp. 1763–1771.

Rao, R.U.M., 1997. Book review: Geothermal energy in India (Geological Survey of India Special Publication, 45, 1996). Journal of Geological Society of India, 49, 746–748.

Rao, R.U.M., Rao, G.V., Reddy, G.K., 1982. Age dependence of continental heat flow—fantasy and facts. Earth and Planetary Science Letters, 59, 288–302.

Rao, R.U.M., Roy, S., Srinivasan, R., 2003. Heat flow researches in India: results and perspectives. Memoirs, Geological Society of India, 53, 347–391.

Rapolla, A., Keller, G.V. (Eds.), 1984. Geophysics of geothermal areas: state–of- the art and future development. In: Proceedings of the 3rd Course, School of Geophysics, "Ettore Majorana" International Centre for Science, Culture, Erice, Italy, 1980. Colorado School of Mines Press.

Ravindran, M., 2005. Harnessing of the ocean thermal energy resource. In: Gupta, H.K. (Ed.), Oceanology, Universities Press, pp. 26–38.

Ray, L., Kumar, P.S., Reddy, G.K., Roy, S., Rao, G.V., Srinivasan, R., Rao, R.U.M., 2003. High mantle heat flow in a precambrian granulite province: Evidence from southern India. Journal of Geophysical Research, 108, B2, 10.1029/2001JB000688, pp. ETG 6-1–6-13.

Razo, A., Arellano, F., Fouseca, H., 1980. CFE resistivity studies at Cerro Prieto. Geothermics, 9, 7–14.

Razo, M., Arellano, F., 1978. Prospección Electrica de la Porcion Norte del valle de Mexicali y Campo Geotermico de Cerro Prieto, Baja California. Informe, pp. 6–78, CFE Archives.

Razo, M.A., Fonseca, L.H., 1978. Prospección gravimetrica y magnetometrica en el valle de Mexicali, B.C., Mexicali. Comisión Federal de Electricidad Report, 34pp.

Reed, M.J., Renner, J.L., 1995. Environmental compatibility of geothermal energy. In: Sterret, F.S. (Ed.), Alternative Fuels and the Environment, CRC Press.

Richter, C.F., 1958. Elementary seismology. W.H. Freeman and Co., San Francisco, California, 768pp.
Risk, G.F., Macdonald, W.J.P., Dawson, G.B., 1970. DC resistivity surveys of the Broadlands geothermal region, New Zealand. In: Proceedings of the U.N. Symposium on the Development and Utilization of Geothermal Resources. Geothermics, Special Issue, 2 (1), 287–294.
Rivera, R., Bermejo, F., Castillo, F., Pérez, H., Abraján, A., 1982. Update of temperature behavior and distribution in Cerro Prieto II and III. In: Proceedings of the 4th Symposium Cerro Prieto, Comisión Federal de Electricidad, Mexicali.
Robinson, P.T., Elders, W.A., Muffler, L.J.P., 1976. Quaternary volcanism in Salton Sea geothermal field. Geological Society of American Bulletin, 87, 347–360.
Robinson, R.J., Morse, R.A., 1976. A study of the effects of various reservoir parameters on the performance of geothermal reservoirs. In: Proceedings of the 2nd U.N. Symposium on the Development and Use of Geothermal Resources, vol. 3, U.S. Government Printing Office, Washington, DC, pp. 1773–1779.
Rodriguez, J.A., Herrera, A., 2005. El Salvador country update. In: Proceedings of the World Geothermal Congress, Antalya, Turkey, April 24–29.
Roldán Manzo A.R., 2005. Geothermal power development in Guatemala 2000–2005. In: Proceedings of the World Geothermal Congress, Antalya, Turkey, April 24–29.
Ross, H.P., Green, D.J., Mackelprang, 1996. Electrical resistivity surveys, Ascension Island, South Atlantic Ocean. Geothermics, 25, 489–506.
Ross, H.P., Moore, J.N., 1985. Geophysical investigations of the Cove Fort—Sulfurdale geothermal system, Utah. Geophysics, 50, 1732–1745.
Ross, H.P., Nielson, D.L., Moore, J.N., 1982. Roosevelt hot springs geothermal system, Utah—Case study. Bulletin of American Association of Petroleum Geology, 66, 879–902.
Roy, R.F., Beck, A.E., Touloukian, Y.S., 1981. Thermophysical properties of rocks. In: Touloukian, Y.S., Judd, W.R., Roy, R.F. (Eds.), Physical Properties of Rocks and Minerals, McGraw-Hill/CINDAS Data Series on Material Properties Vol. II-2, Mc-Graw Hill Book Co., New York, pp. 409–502.
Roy, S., 1997. Geothermal Studies in the India Shield: Ground Temperature History and Crustal Thermal Structure. Ph.D. thesis (Unpublished), Banaras Hindu University, Varanasi, India, 218pp.
Roy, S., Rao, R.U.M., 1996. Regional heat flow and the perspective for the origin of hot springs in the Indian shield. In: Pitale, U.L., Padhi, R.N. (Eds.), Geothermal Energy in India, Geological Survey of India, Spl Publication, 45, 39–40.
Roy, S., Rao, R.U.M., 2000. Heat flow in the Indian shield. Journal of Geophysical Research, 105, 25,587–25,604.
Roy, S., Sundar, A., Rao, R.U.M., 1996. Heat flow and hot springs over the Deccan Volcanic Province, India. Abstract, IASPEI Regional Assembly in Asia, Tangshan, China, August 1–4.
Roy, S., Ray, L., Kumar, P.S., Reddy, G.K., Srinivasan, R., 2003. Heat flow and heat production in the Precambrian gneiss–granulite province of southern India. Memories, Geological Society of India, 50, 177–191.
Rummel, F., 2005. Heat mining by hot-dry-rock technology. In: Proceedings of International Conference of ECOMINING, Bukharest, October 9pp.
Rummel, F., Kappelmeyer, O. (Eds.), 1993. Geothermal energy—future energy source? Verlag C.F. Müller Karlsruhe, 98pp.

Ryan, G.P., 1981. Equipment used in direct heat projects. In: Geothermal Resources Council Transactions, 5, 483–485.
Rybach, L., 1976. Radioactive heat production in rocks and its relation to other petrophysical parameters. Pure and Applied Geophysics, 114, 309–318.
Rybach, L., 1989. Heat flow techniques in geothermal exploration. First Break, 7, 8–16.
Salem, A., Furuya, S., Aboud, E., Elawadi, E., Jotaki, H., Ushijima, K., 2005. Subsurface structural mapping using gravity data of Hohi geothermal area, Central Kyushu, Japan. In: Proceedings of the World Geothermal Congress, Antalya, Turkey, April 24–29.
Sass, J.H., Blackwell, D.D., Chapman, D.S., Costain, J.K., Decker, E.R., Lawver, L.A., Swanberg, C.A., 1981. Heat flow from the crust of the United States. In: Touloukian, Y.S., Judd, W.R., Roy, R.F. (Eds.), Physical Properties of Rocks and Minerals, McGraw Hill Book Co., pp. 503–548.
Sass, J.H., Lachenbruch, A.H., Munroe, R.J., 1971. Thermal conductivity of rocks from measurements on fragments and its application to heat flow determinations. Journal of Geophysical Research, 76, 3391–3401.
Sass, J.H., Priest, S.S., Lachenbruch, A.H., Galasnis, S.P., Moses, T.H., Kennelly, J.P., Munroe, R.J., Smith, E.P., Grubb, F.V., Husk, R.H., Mase, C.W., Norton, G., Groat, C.G., 2005. Summary of supporting data for USGS regional heat-flow studies of the Great Basin, 1970–1990. USGS Open-file Report 2005–1207 online version 1.0.
Sass, J.H., Sammel, E.A., 1976. Heat flow data and their relation to observed geothermal phenomena near Klamath Falls, Oregon. Journal of Geophysical Research, 81, 4863–4868.
Sass, J.H., Walters, M.A., 1999. Thermal regime of the Great Basin and its implications for enhanced geothermal systems and off-grid power. Geothermal Resources Council Transactions, 23, 211–218.
Sawkins, F.J., Chase, C.G., Darby, D.G., Rapp, G. Jr., 1974. The evolving earth: a text in physical geology. MacMillan Publishing Company, New York.
Schlanger, S.O., Combs, J., 1975. Hydrocarbon potential of marginal basins bounded by an island arc. Geology, 3, 397–400.
Schock, R.N., Duba, A., 1975. The effect of electrical potential on scale formation in Salton Sea brine. UCRL-51944, Lawrence Livermore Laboratory, University of California, Livermore, California, 14pp.
Seipold, U., 2001. Der Wärmetransport in Kristallinen Gesteinen under den Bedingungen der Kontinentalen Kruste. Scientific Technical Report STR01/13, Geoforschungszentrum (GFZ), Potsdam.
Sestini, G., 1970. Heat flow measurements in non-homogeneous terrains: its application to geothermal areas. In: Proceedings of the U.N. Symposium on the Development and Utilization of Geothermal Resources. Geothermics, Special Issue, 2 (1), 424–436.
Shaw, H.R., Jackson, E.D., 1973. Linear island chains in the Pacific: results of thermal plumes or gravitational anchors. Journal of Geophysical Research, 78, 8634–8652.
Sigvaldason, G.E., 1973. Geochemical methods in geothermal exploration. In: Armstead, H.C.H. (Ed.), Geothermal Energy. UNESCO, Paris, pp. 49–59.
Sill, W.R., 1983. Self-potential modeling from primary flows. Geophysics, 48, 76–86.
Simmons, G., 1967. Interpretation of heat flow anomalies, 1. Contrast in heat production. Reviews of Geophysics, 5 (1), 42–52.
Smith, J.H., 1973. Collection and transmission of geothermal fluids. In: Armstead, H.C.H. (Ed.), Geothermal Energy, UNESCO, Paris, pp. 97–106.

Smith, R.L., Shaw, H.R., 1973. Volcanic rocks as geological guides to geothermal exploration and evaluation. EOS Transactions of American Geophysical Union, 54, 1213.
Smith, R.L., Shaw, H.R., 1975. Igneous related geothermal systems. In: White, D.E., Williams, D.L. (Eds.), Assessment of Geothermal Resources of the United States—1975. U.S. Geological Survey Circular, 726, 58–83.
Soda, M., Takahashi, Y., Aikawa, K., Kubota, K., Eijima, Y., 1976. Experimental study on transmission phenomenon in steam water mixtures flowing through a large pipeline for geothermal power station. In: Proceedings of the 2nd U.N. Symposium on the Development and Use of Geothermal Resources, vol. 3, U.S. Government Printing Office Washington, DC, pp. 1789–1795.
Stacey, F.D., 1992. Physics of the Earth. Brookfield Press, Brisbane, 2nd ed.
Ståhl, G., Pátzay, G., Weiser, L., Kálmán, E., 2000. Study of calcite scaling and corrosion processes in geothermal systems. Geothermics, 29, 105–119.
Stanley, W.D., Jackson, D.B., Hearn, B.C., Jr., 1973. Preliminary results of geoelectrical investigations near Clear Lake, California. U.S. Geological Survey Open-File Report, 20pp.
Stanley, W.D., Jackson, D.B., Zohdy, A.A.R., 1976. Deep electrical investigations in the Long Valley geothermal area, California. Journal of Geophysical Research, 81, 810–820.
Steeples, D.W., Iyer, H.M., 1976. Teleseismic *P*-wave delays in geothermal exploration. In: Proceedings of the 2nd U.N. Symposium on the Development and Use of Geothermal Resources, vol. 2, U.S. Government Printing Office, Washington, DC, pp. 1199–1206.
Stein, C.A., Stein, S., 1992. A model for the global variation in oceanic depth and heat flow with lithospheric age. Nature, 359, 123–129.
Stevens, L., 2000. Pressure, temperature and flow logging in geothermal wells. In: Proceedings of the World Geothermal Congress, Kyushu-Tohoku, Japan, May 28–June 10, pp. 2435–2437.
Stoces, B., Cernik, A., 1931. Bekampfung hoher Grubentemperaturen. Springer, Berlin.
Strangway, D.C., Swift, C.M., Jr., Holmer, R.C., 1973. The application of audio-frequency magnetotellurics (AMT) to mineral exploration. Geophysics, 38 (6), 1159–1175.
Studt, F.E., 1964. Geophysical prospecting in New Zealand's hydrothermal fields. In: Proceedings of the U.N. Conference on New Sources of Energy, UNESCO, Paris, vol. 2 (1), 380pp.
Sudarman, S., Suroto, Pudyastuti, K., Aspiyo, S., 2000. Geothermal development progress in Indonesia: country update 1995–2000. In: Proceedings of the World Geothermal Congress 2000, Kyushu-Tohoku, Japan, May 28–June 10, pp. 455–460.
Sussman, D., Javellana, S.P., Benavidez, P.J., 1993. Geothermal energy development in the Philippines: an overview. Geothermics, 22, 353–367.
Swanson, D.A., 1972. Magma supply rate at Kilauea volcano, 1952–1971. Science, 175, 169–170.
Takahashi, Y., Hayashida, T., Soezima, S., Aramaki, S., Soda, M., 1970. An experiment on pipeline transportation of steam–water mixtures at Otake geothermal field. In: Proceedings of the U.N. Symposium on the Development and Utilization of Geothermal Resources. Geothermics, Special Issue, 2 (1), 882–891.
Thompson, G.A., Burke, D.B., 1973. Rate and direction of spreading in Dixie Valley, Basin and Range province, Nevada. Geological Society of America Bulletin, 84, 627–633.
Thorhallsson, S., 2003. Geothermal well operation and maintenance. IGC2003 Short Course, Geothermal Training Programme, The United Nations University, Iceland, September, pp. 195–217.

Thussu, J.L., Prasad, J.M., Saxena, R.K., Gyan Prakash, Muthuraman, K., 1987. Geothermal energy resource potential of Tattapani hot spring belt, Surguja, Madhya Pradesh. Records, Geological Survey of India, 115, part 6, 55–65.

Tipler, P.A., (1999). Physics for scientists and engineers, 4th ed., W.H. Freeman.

Tolivia, M.E., 1976. Evaluation of the geothermal potential of Cerro Prieto, Baja California (Mexico). In: Proceedings of the 2nd U.N. Symposium on the Development and Use of Geothermal Resources, vol. 1, U.S. Government Printing Office, Washington, DC, pp. 279–287.

Tolivia, M.E., Hoashi, J., Miyazaki, M., 1976. Corrosion of turbine materials in geothermal steam environment in Cerro Prieto, Mexico. In: Proceedings of the 2nd U.N. Symposium on the Development and Use of Geothermal Resources, vol. 3, U.S. Government Printing Office, Washington, DC, pp. 1815–1820.

Tonani F., 1980. Some remarks on the application of geochemical techniques in geothermal exploration. In: Proceedings of the Adv. Eur. Geothermal Research, 2nd Symposium, Strasbourg, pp. 428–443.

Tosha, T., Ishido, T., Matsushima, Y., Nishi, U., 2000. Self-potential variation at the Yanaizu-Nishiyama geothermal field and its interpretation by the numerical simulation. In: Proceedings of the World Geothermal Congress, vol. 3, pp. 1871–1876.

Tripp, A.C., Ward, S.H., Sill, W.R., Swift, C.M., Jr., Petrick, W.R., 1978. Electromagnetic and Schlumberger resistivity sounding in the Roosevelt hot springs KGRA. Geophysics, 43, 1450–1469.

Truesdell, A.H., 1976. Summary of Section III: geochemical techniques in exploration. In: Proceedings of the 2nd U.N. Symposium on the Development and Use of Geothermal Resources, vol. 1, U.S. Government Printing Office, Washington, DC, pp. liii–lxxx.

Truesdell, A.H., Hulston, J.R., 1980. Isotopic evidence on environments of geothermal systems. In: Fritz, P., Fontes, J.Ch. (Eds.), Handbook of Environmental Isotope Geochemistry, Vol. 1: The Terrestrial Environment, Elsevier, Amsterdam, pp. 179–226.

Truesdell, A.H., Lippmann, M.J., 1998. Effects of pressure drawdown and recovery on the Cerro Prieto Beta reservoir in the CP-III area. In: Proceedings, of the 23rd Workshop on Geothermal Reservoir Engineering, Stanford, CA, Report, SGP-TR-158, pp. 90–98.

Truesdell, A.H., Lippmann, M.J., Gutiérrez Puente, H., 1997. Evolution of the Cerro Prieto reservoirs under exploitation. Geothermal Resources Council Transactions, 21, 263–269.

Truesdell, A.H., Nehring, N.L., Thompson, J.M., Janik, C.J., Coplen, T.B., 1984. A review of progress in understanding the fluid geochemistry of the Cerro Prieto geothermal system. Geothermics, 13, 65–74.

Truesdell, A.H., Thompson, J.M., Coplen, T.B., Nehring, N.L. Janik, C.J., 1981. The origin of the Cerro Prieto geothermal brine. Geothermics, 10, 225–238.

Uchiyama, M., Matsuura, S., 1970. Measurement and transmission of steam in Matsukawa geothermal power plant. In: Proceedings of the Symposium on the Development and Utilization of Geothermal Resources. Geothermics, Special Issue, 2 (2), 1572–1580.

United Nations, Statistical Office. Demographic Year Book. This annual compilation is the source for world data on population.

United Nations, Statistical Office. World Energy Supplies Series, 1970–1973.

Vakin, E.A., Polak, B.G., Sugrobov, V.M., Erlikh, E.N., Belousov, V.I., Pilipenko, G.F., 1970. Recent hydrothermal systems of Kamchatka. In: Proceedings of the Symposium the

Development and Utilization of Geothermal Resources. Geothermics, Special Issue, 2 (2), 1116–1133.
Valette, J.N., Esquer-Patiño, I., 1979. Geochemistry of the surface emissions in the Cerro Prieto geothermal field. In: Proceedings, of the 2nd Symposium on the Cerro Prieto Geothermal Field, Mexicali, Baja California, Mexico, Comisión Federal de Electricidad, October 17–19, pp. 241–250.
Valette-Silver, J.N., Esquer-Patino, I., Elders, W.A., Collier, P.C., Hoagland, J.R., 1981. Hydrothermal alteration of sediments associated with surface emissions from the Cerro Prieto geothermal field. In: Proceedings of the 3rd Symposium on Cerro Prieto Geothermal Field, San Francisco, Lawrence Berkeley Laboratory Report, LBL-11967, pp. 140–145.
Vanyan, L.L., 1967. Electromagnetic depth soundings. Consultants Bureau, New York, NY, 312pp.
Vaughan, R.G., Calvin, W.M., Taranik, J.V., 2003. SEBASS hyperspectral thermal infrared data: surface emissivity measurement and mineral mapping. Remote Sensing of Environment, 85, 48–63.
Vega, L.A., 1992. Economics of ocean thermal energy conversion (OTEC). In: Seymour, R.J. (Ed.), Ocean Energy Recovery: The State-of-the Art, American Society of Civil Engineers, New York, pp. 152–181.
Vega, L.A., 1995. Ocean thermal energy conversion. In: Encyclopedia of Energy Technology and the Environment,Wiley, New York, pp. 2104–2119.
Verhoogen, J., 1980. Temperatures within the Earth. In: Energetics of the Earth, National Academy of Sciences, Washington, DC, USA, pp. 29–66.
Volpi, G., Manzella, A., Fiordelisi, A., 2003. Investigations of geothermal structures by magnetotellurics (MT): an example from the Mt. Amiata area, Italy. Geothermics, 32, 131–145.
Vozoff, K., 1991. The magnetotelluric method. In: Nabhighian, M.N. (Ed.), Electromagnetic Methods in Applied Geophysics, vol. 2B, Society of Exploration Geophysicists, Tulsa, pp. 641–711.
Wahl, E., Yen, I., 1976. Scale deposition and control research for geothermal utilization. In: Proceedings of the 2nd U.N. Symposium on the Development and Use of Geothermal Resources.,vol. 3, U.S. Government Printing Office, Washington, DC, pp. 1855–1864.
Wannamaker, P.E., Rose, P.E., Doerner, W.M., McCulloch, J., Nurse, K., 2005. Magnetotelluric surveying and monitoring at the Coso geothermal area, California, in support of the enhanced geothermal systems concept: survey parameters, initial results. In: Proceedings, of the World Geothermal Congress, Antalya, Turkey, April 24–29.
Ward, P.L., 1972. Microearthquakes: prospecting tool and possible hazard in the development of geothermal resources. Geothermics, 1, 3–12.
Ward, P.L., Bjornsson, S., 1971. Microearthquake swarms and the geothermal areas of Iceland. Journal of Geophysical Research, 76, 3953–3982.
Ward, P.L., Palmason, G., Drake, C., 1969. Microearthquake surveys and the Mid-Atlantic Ridge of Iceland. Journal of Geophysical Research, 74, 665–684.
Ward. R.W., Butler, D., Iyer, H.M., Laster, S., Lattanner, A., Majer, E., Mass, J., 1979. Seismic methods. In: Nielson, D.L. (Ed.), Program Review, Geothermal Exploration and Assessment Technical Program, Report, DOE/ET/27002-6, University of Utah Res. Inst., Earth Science Lab.
Ward, S.H., 1983. Controlled source electromagnetic methods in geothermal exploration. U.N. University Geothermal Traning Program, Iceland, Report, 1983–1984.

Ward, S.H., Wannamaker, P.E., 1983. The MT/AMT electromagnetic method in geothermal exploration. U.N. University of Geothermal Training Program, Iceland, Report, 1983–1985.
Ward, S.H., Parry, W.T., Nash, W.P., Sill, W.R., Cook, K.L., Smith, R.B., Chapman, D.S., Brown, F.H., Whelan, J.A., Bowman, J.R., 1978. A summary of the geology, geochemistry and geophysics of the Roosevelt hot springs thermal area, Utah. Geophysics, 43, 1515–1542.
Welhan, J., Poreda, R., Lupton, J., Craig, H., 1979. Gas chemistry and helium isotopes at Cerro Prieto. Geothermics, 8, 241–244.
Welhan, J., Poreda, R., Rison, W., Craig, H., 1988. Helium isotopes in geothermal and volcanic gases of the western United States. I. Regional variability and magmatic origin. Journal of Volcanology and Geothermal Research, 34, 185–199.
White, D.E., 1965. Geothermal energy. U.S. Geological Survey of Circular, 519, 17 pp.
White, D.E., 1969. Rapid heat flow surveying of geothermal area, utilizing individual snowfalls as calorimeters. Journal of Geophysical Research, 74, 5191–5201.
White, D.E., 1970. Geochemistry applied to the discoveries, evaluation, and exploitation of geothermal energy resources. Rapporteurs Report. In: Proceedings of the U.N. Symposium the Development and Utilization of Geothermal Resources. Geothermics. Special Issue, 1, 58–80.
White, D.E., 1973. Characteristics of geothermal resources. In: Kruger, P., Otte, C. (Eds.), Geothermal Energy, Stanford University Press, Stanford, California, pp. 69–94.
White, D.E, 1974. Diverse origins of hydrothermal fluids. Economic Geology, 69, 954–973.
White, D.E., Muffler, L.P.J., Truesdell, A.H., 1971. Vapor-dominated hydrothermal systems compared with hot-water systems. Economic Geology, 66, 75.
Whiteford, P.C., 1976. Assessment of the audio-magnetotelluric method for geothermal resistivity surveying. In: Proceedings of the 2nd U.N. Symposium on the Development and Use of Geothermal Resources, vol. 2, U.S. Government Printing Office, Washington, DC, pp. 1255–1261.
Williams, C.F., Grubb, F.V., 1998. Thermal constraints on the lateral extent of The Geysers vapor-dominated reservoir. In: Proceedings of the 23rd Workshop on Geothermal Reservoir Engineering, Stanford University, Stanford, California, Jan 26–28, 7pp.
Williams, C.F., Sass, J.H., 1996. The thermal conductivity of rock under hydrothermal conditions: measurements and applications. In: Proceedings of the 21st Workshop on Geothermal Reservoir Engineering, Stanford University, Stanford, California, Jan 22–24, 7pp.
Williams, C.F., Sass, J.H., Grubb, F.V., 1997. Thermal signature of subsurface fluid flow near the Dixie Valley geothermal field, Nevada. In: Proceedings of the 22nd Workshop on Geothermal Reservoir Engineering, Stanford University, Stanford, California, Jan 27–29, 8pp.
Wilt, M.J., Goldstein, N.E., 1981. Resistivity monitoring at Cerro Prieto. Geothermics, 10, 183–193.
Wilt, M.J., Goldstein, N.E., 1984. Interpretation of dipole–dipole resistivity monitoring data at Cerro Prieto. Geothermics, 13, 13–25.
Wilt, M.J., Goldstein, N.E., Razo, A., 1980. LBL resistivity studies at Cerro Prieto. Geothermics, 9, 15–26.
Wisian, K.W., 2000. Insights into extensional geothermal systems from numerical modeling. In: Proceedings of the World Geothermal Congress, Kyushu-Tohoku, Japan, May 28–June 10, pp. 1947–1952.

Wisian, K.W., Blackwell, D.D., Richards, M., 1999. Heat flow in the Western United States and extensional geothermal systems. In: Proceedings of the 24th Workshop on Geothermal Reservoir Engineering, Stanford University, Stanford, California, Jan 25–27, 8pp.

Wisian, K.W., Blackwell, D.D., Bellani, S., Henfling, J.A., Normann, R.A., Lysne, P., Förster, A., Schrötter, J., 1998. Field comparison of conventional and new technology temperature logging systems. Geothermics, 27, 131–141.

Wood, B., 1973. Geothermal power. In: Armstead, H.C.H. (Ed.), Geothermal Energy, UNESCO, Paris, pp. 109–121.

World Population Prospects: The 2004 Revision, 2005. Population division of the Dept. of Economic and Social Affairs of the United Nations Secretariat, United Nations, New York. Website: http://www.un.org

Wright, P.M., Ward, S.H., Ross, P., West, R.C., 1985. State-of-the-art geophysical exploration for geothermal resources. Geophysics, 50, 2666–2699.

Yasukawa K., Suzuki I., Ishido T., 2001. Reservoir monitoring by relative self-potential observation at the Nigorikawa Basin, Hokkaido, Japan. Geothermal Resources Council Transactions, 25, 705–710.

Young, C.Y., Ward, R.W., 1980. 3D Q^{-1} model of Coso hot springs, KGRA. Journal of Geophysical Research, 85, 2459–2470.

Young, C.Y., Ward, R.W., 1981. Attenuation of teleseismic *P*-waves in The Geysers—Clear Lake region, California. USGS Prof. Paper 1141, U.S. Geological Survey, Reston, Va., pp. 149–160.

Young, H.D., 1992. University Physics. Addison Wesley, 7th ed.

Zablocki, C.J., 1976. Mapping thermal anomalies on an active volcano by the self-potential method, Kilauea, Hawaii. In: Proceedings of the 2nd U.N. Symposium on Development and Use of Geothermal Resources, vol. 2, U.S. Government Printing Office, Washington, DC, pp. 1299–1309.

Zohdy, A.A., Anderson, L.A., Muffler, L.J.P., 1973. Resistivity, self-potential and induced polarization surveys over a vapor-dominated geothermal system. Geophysics, 38, 1130–1144.

Zoth, G., Haenel, R., 1988. Thermal conductivity. In: Haenel, R., Rybach, L., Stegena, L. (Eds.), Handbook of Terrestrial Heat Flow Density Determination, Kluwer Academic, Dordrecht, pp. 449–466.

SUBJECT INDEX